Ancient Greek Cosmogony

Ancient Greek Cosmogony

Andrew Gregory

Duckworth

First published in 2007 by
Gerald Duckworth & Co. Ltd.
90-93 Cowcross Street, London EC1M 6BF
Tel: 020 7490 7300
Fax: 020 7490 0080
inquiries@duckworth-publishers.co.uk
www.ducknet.co.uk

A catalogue record for this book is available
from the British Library

ISBN 978 0 7156 3477 6

Typeset by Ray Davies
Printed and bound in Great Britain by
MPG Books Ltd, Bodmin, Cornwall

Contents

Abbreviations

Publications and texts

DK = Diels & Kranz, *Die Fragmente der Vorsokratiker*. Zurich: Weidmann, 1966.

KRS = Kirk, Raven & Schofield, *The Presocratic Philosophers*, 2nd edn. Cambridge: Cambridge University Press, 1983.

LSJ = Liddell, Scott & Jones, *A Greek-English Lexicon*. Oxford: Clarendon Press, 1996.

LS = Long & Sedley, *The Hellenistic Philosophers*. Cambridge: Cambridge University Press, 1987.

Simplicius, *Physics* = Simplicius' commentary on Aristotle's *Physics*.

Simplicius, *On the Heavens* = Simplicius' commentary on Aristotle's *On the Heavens*.

SVF = *Stoicorum Veterum Fragmenta*, ed. H. von Arnim. Lipsiae: Teubner, 1903-24.

Some scientific terms

BB = Big Bang
QM = Quantum Mechanics
SS = Steady State

Preface

This book is about ancient theories of how the cosmos began. It is mainly, but not exclusively about the Greeks, as they formulated and discussed problems in cosmogony to a far greater extent than other ancient cultures. It covers a period of roughly 600 BC to 600 AD. This is not a general history of Greek philosophy, or even of Greek science. I mention this because while this book follows the general historical outline of Greek philosophy, it places the emphases in different places and has rather different relative chapter lengths to a standard history of Greek philosophy or science. Nor is this a book primarily about Greek cosmology, about the nature and organisation of the cosmos for the Greeks. Some mention has to be made of what the Greeks thought their cosmogonies led to, but this book concentrates on the origins of the cosmos, and to a lesser extent on the ultimate fate of the cosmos.

This book is accessible to those with little or no knowledge of Greek. All Greek words are translated and/ or transliterated, and technical discussions of issues of text and translation have, wherever possible, been relegated to the footnotes. This book is also accessible to those with little or no knowledge of modern ideas in cosmogony. These are explained as they are introduced and there is no mathematical formulation of these ideas beyond a very basic level.

I would like to thank Jan Opsomer, who read the entire draft and supplied a great number of helpful comments, corrections and suggestions. I would also like to thank various audiences, both in Europe and Australia for their comments on earlier versions of some of the material here. Deborah Blake at Duckworth has as ever been unfailingly helpful, supportive and efficient, and the cover design is her suggestion

This book was written with the assistance of a Research Leave Grant from the Arts and Humanities Research Council, and I would also like to thank the AHRC's referees for their comments on the initial proposal.

For Sheelagh, with love.

Introduction

I want to deal with two central issues under the banner of ancient Greek cosmogony in this book. Firstly, there is the question of accounts of the absolute origins of matter, space and time. Secondly, there is the question of the origins of the *kosmos*, where there is a pre-existing physical state which is not sufficiently well ordered to be considered a *kosmos*. This contrast, between chaos and *kosmos*, is not present in modern cosmogony and cosmology, where cosmos is taken to be synonymous with universe, but is critical to many ancient accounts. A *kosmos* for the ancient Greeks was not only a well ordered place, but had connotations of moral and aesthetic beauty as well.[1] I will use the term *kosmos* (plural, *kosmoi*) for these ordered systems, which typically would have a central earth, a sun, moon and five planets and a sphere of stars.[2] The Greek term *to pan* ('the all') I will translate as universe. In some accounts there is one *kosmos* in the universe, in others the *kosmos* is the universe or there may be an unlimited number of *kosmoi* in the universe.

I. Central aims

The central aim of this book is to try to explain why the ancient Greeks held their theories of cosmogony. Much interesting work has been done by ancient philosophers to show how Greek cosmogonies differ from creation myths and theogonies, and how they relate to philosophical issues such as change, persistence through change and matter theory. The direction of explanation here has tended to be that because someone held certain core philosophical beliefs about change/ matter/ space, then we can understand, at least in outline, why they held their views on cosmogony. As we see views change on these philosophical issues, so we see views change in cosmogony. This approach has been very valuable, but it is also somewhat limited. It has tended to treat cosmogony as a single issue, where Greek thinkers may have addressed a range of problems in cosmogony. It has also tended to see the traditional philosophical issues as the main generators of debate and change, where there may have been significant discussion of the cosmogonical issues themselves leading to change.

In this book then I want to effect a slight change of emphasis, and perhaps not surprisingly, treat the issues in cosmogony more centrally. This is not to say that Greek thinkers were cosmogonists rather than philosophers. In some cases though, particularly with the Presocratics, cosmogony is clearly a major issue and we may have under-emphasised the roles that a range of problems in cosmogony may have played in the formation of their thought. In particular, I will argue that we can only properly understand the early atomists and Empedocles if we consider them in relation to the cosmogonical challenge of the Milesians and Heraclitus as well as the philosophical challenge of Parmenides. I aim to show that at least some Greek thinkers were aware of a wide range of issues in cosmogony, and that in some cases this can open up new perspectives on their thought in general. I also hope to show that there is an interesting progression of ideas in cosmogony, which may throw some new light on relations between ancient thinkers.

II. Perennial problems

A key argument of this book will be that there are perennial philosophical and scientific problems relating to cosmogony. This is not to say that these problems are insoluble, or that no progress has been made in relation to them.[3] On the contrary, I believe we now have a much more sophisticated understanding of these problems and a greater and more sophisticated array of possible solutions, even if we often lack definitive solutions. Once these problems have been recognised and formulated, they need to be dealt with by each subsequent thinker. An example here is Parmenides on whether something (indeed, anything) can come from nothing.[4] This question will then affect all subsequent Greek thinking on cosmogony. Although we may have a more sophisticated understanding of the problem (nature of necessity, of time, etc.) and more sophisticated possible answers, this remains a fundamental problem for Big Bang cosmogony. As ever with Eleatic problems, the more literally we take nothing (no matter, energy, time, space or laws of physics) the more interesting and difficult the problems become. While there may be perennial core problems in cosmogony, they manifest themselves in varying forms at different times and in different circumstances. The problems as perceived by the ancient Greeks are related to but not identical to the problems addressed by modern cosmogony.

There are also perennial types of solutions to these philosophical and scientific problems. The following problem, phrased in modern terminology, has troubled both ancients and moderns. Can we adequately explain the genesis of our universe as a unique entity using only our knowledge of physics? The answer of many ancients, and many moderns is no, even if their reasons for that answer are somewhat different. The options taken by both ancients and moderns to solve this problem have striking structural similarities. The ancient atomists, and the modern multiple universe theorists both hold that our *kosmos*/ universe is not unique but is one of an infinite array of *kosmoi*/ universes. Thus we can explain the order of our *kosmos*/ universe as one accident among many accidents. A variation on this is that there is only one *kosmos*/ universe in existence at one time, but that there are an unlimited number of *kosmoi*/ universes through time, a view held by Empedocles and some modern theorists. Plato, many modern theologians and some versions of the modern anthropic principle all hold that there is a unique *kosmos*/ universe, but we need further design principles to explain the order of this unique entity. Aristotle, and some modern philosophers and theorists hold that there is no such thing as cosmogony and that the *kosmos*/ universe has always existed.

In arguing that the Greeks engaged with perennial problems in cosmogony, I do not wish to argue that what is interesting about the Greeks is that they 'anticipated' modern views. I am not at all sure what is meant by 'anticipated' in this sort of historiographical context, nor why it should be interesting or important that an 'anticipation' took place. Rather, I believe they began debates which are still continuing. I also wish to distinguish between two sorts of similarities. The Stoics believed that the sun would eventually expand and burn up the entire solar system in a process called *ekpurôsis*. The modern view on the fate of the solar system is that the sun will eventually expand into a red giant star which indeed will engulf the earth though not the entire solar system.[5] The reasons why the Stoics believed in *ekpurôsis*, as far as we can determine, have nothing in common with why we believe the sun will turn into a red giant star. Could they be said to have anticipated the modern theory? Perhaps, but I do not see that we can give any merit to the Stoics for having done so nor do I see that it is a particularly interesting thing to say.

There are, however, issues in cosmogony where the Greeks asked structurally similar questions to some that are asked in modern cosmogony, and gave structurally similar answers, even if the content of questions and answers with the Greeks was of considerably less scientific sophistication. The issue of whether there is one *kosmos* or there are multiple *kosmoi* is one such issue, and there are several more, such as the question of whether the physical processes of cosmogony are unique, or form part of the ongoing nature of the *kosmos*.[6]

This book will seek to establish that the Greeks did indeed address these problems, that as far as we are aware they were the first group to do so, and that they were the first to formulate some of the problems explicitly. I will also try to show that the Greeks give versions of the perennial solutions to these problems, and that they give coherent solutions. By coherent solutions here I mean that in answer to any one problem in cosmogony, each of the Greeks follow one type of solution, rather than mixing incompatible solutions or giving more than one solution, and in answer to a set of problems in cosmogony, they have a coherent strategy for giving types of solutions. That is important in itself, but is also significant in indicating that the Greeks have grasped the nature of the perennial cosmogonic problems.

A related question I wish to pursue in this book is a more general investigation of how ancient philosophers treated origins questions. Cosmogony is clearly the key issue, but there are also questions of zoogony, anthropogony and in some cases stoichogony (the origin of the elements). The philosophical principles used in questions of cosmogony are not specific to those issues, but can be seen being consistently applied to related issues. Perhaps the clearest example of this is Plato, who is adamant that there is one well designed *kosmos*, that species are well designed and not the result of a multiplicity of accidents, and that the basic constituents of the physical world have a small number (two!) of well chosen shapes, rather than the unlimited multiplicity of shapes and sizes supposed by Leucippus and Democritus. A further issue to cosmogony is the ultimate fate of the *kosmos*. I will attempt to show that views on the fate of the *kosmos* are related to the same sorts of principles that drive views on its origins. In places the fate of the *kosmos* will be a controversial topic which directly affects views on cosmogony, especially where there is a debate as to whether a thinker believes in one *kosmos* or successive single *kosmoi*.

III. Classifying ancient cosmogonies

One contrast that can be drawn in ancient cosmogony is the age-old one between teleologists and anti-teleologists. There are other factors to take into consideration though, as we will find that there are reasons outside the teleology/ anti-teleology divide for why some thinkers advocate multiple *kosmoi*. It is a significant fact about the ancients though that no one believed there to be a single *kosmos* which came about by chance. A single *kosmos* was always ordered in some way, though there were many ideas on how this ordering might come about. It may be misleading to assume that prior to Plato these *kosmoi* were ordered teleologically, at least in the sense that has become common since Plato. Where many co-existent *kosmoi* were supposed, chance rather than teleology was the dominant factor in cosmogony, although as we will see, there were considerable differences between how Leucippus and Democritus on the one hand and Epicurus and Lucretius on the other treat the justification of multiple *kosmoi*. Where many successive *kosmoi* were supposed, the situation is more complex. Some thinkers, most notably Empedocles, do so because of their views on chance and explanation. Others, such as the Stoics, do so in order to make a belief in a degenerating *kosmos* compatible with belief in a providential god, while the Christian Origen does so in an attempt to avoid some theological problems with infinities.

Furley's distinction between the closed world and the open universe is a useful one for the classification and analysis of *kosmoi*.[7] It is perhaps less useful for cosmogonies, though in fairness it is intended for cosmologies rather than cosmogonies. A closed world is one in which there is a finite amount of space which is taken up by a single *kosmos* (Plato, Aristotle), while an infinite universe has an infinite amount of space and may have an unlimited number of *kosmoi*. Certainly it is the case that closed *kosmoi* in the ancient world are always accompanied by some form of design principle or teleological explanation. This though, I shall argue, is due to the fact that these are unique *kosmoi* rather than their being closed *kosmoi*. There is a greater diversity of cosmogony with open universes. The two groups of atomists, Leucippus and Democritus, and Epicurus and Lucretius, who advocate unlimited *kosmoi* in an unlimited space, have cosmogony occurring by chance rather than by any design. However, there are others, arguably the Milesians and the Stoics, who have a single *kosmos* within an unlimited space, who employ design principles and/ or teleology. Both groups of atomists have *kosmoi* which evolve and then decay, which can broadly be contrasted with the closed worlds which are more permanent in their nature.[8] However, the open universes of the Milesians all have cosmogonies where the *kosmos* evolves into something permanent.

One of the main issues I will pursue in this book, of whether the *kosmos* is the unique result of design or is one of a series of accidents, also crosses the closed world/ open universe distinction. There are open universes with design principles, and closed worlds with a multiplicity of accidents, most notably with Empedocles. Whether the processes of cosmogony are unique or part of the ongoing physical framework of the *kosmos* also crosses the distinction between open universes and closed *kosmoi*.

There are some related issues to the origins of the *kosmos*, these being the origins of matter, space, time and motion. While there was a considerable diversity of views on the nature of matter now, the nature of matter originally and how matter has changed in between, prior to the Christian era the Greeks have no doubt that matter does not have an ultimate origin. Nor, again prior to the Christian era, is there any question of the generation of space.[9] Plato is the first to suggest an origin for time, at least in the sense that time as we perceive and measure it may have had an origin along with the *kosmos*. Aristotle is keen on the question of the origins of motion, not least because he believes that this creates serious problems for many of his predecessors.

IV. The ancient evidence

I have suggested that the Greeks began debates in cosmogony which continue to the present day, and are part of the intrinsic nature of cosmogony. In particular there is a theme of one *kosmos* by design or multiple *kosmoi* by accident. How well should we expect the Greeks to fit into these patterns, and how well do they fit? The argument is not that as a matter of necessity every Greek thinker accords to these schemes. Rather, it is that an intelligent and well informed person who is thinking seriously about philosophical cosmogony will address these perennial problems and will follow one of the standard solution types. Others who are not quite so bright or well informed, or have a different agenda may fall outside these patterns. In terms of the major thinkers, I believe the fit to be remarkably good but not perfect. There are some candidates for a mis-fit, most notably Diogenes of Apollonia, where the evidence could go either way. A more severe blow to this historiography would be a front rank ancient thinker clearly interested in cosmogony who in some way fails to fit.

This book presents the case for a good fit. Often this means taking sides in scholarly controversies over the interpretation of individuals or schools of thinkers. Sometimes this is relatively straightforward, as with Heraclitus where I argue in line with the modern consensus that he believed in one *kosmos* rather than successive *kosmoi*. Sometimes this will be rather more controversial, as with Anaximander where I refurbish the old Kirk/Kahn view that Anaximander believed in one *kosmos* rather than many co-existent *kosmoi* or successive *kosmoi*. For the Presocratics, the individual pieces of evidence, whether they be putative fragments or testimonia, are often keenly contested, both in terms of interpretation and whether they genuinely represent the writing or thought of the subject. We have no completely reliable sources for the Presocratics. The modern line on Heraclitus involves rejecting the evidence of Aristotle, Theophrastus and several later writers. Aristotle, who is probably our best general source on the Presocratics, is by no means entirely reliable. Sometimes he is careless or vague about whom he attributes a theory to. Sometimes his evidence is ambiguous between competing interpretations. His evidence is often seriously biased by attempting to impose his own categories of thought on his predecessors. Sometimes, to the best of our knowledge, he is simply wrong.[10] But we cannot discount Aristotle's evidence, as we have so little evidence from elsewhere. One objection to some of the arguments I put forward in this book, given that I reject some of the ancient evidence, might be that some of the evidence I adduce in favour of the positions I take comes from sources which I have already shown to be suspect. Clearly, I cannot simply accept the positive evidence for my views and reject the negative from these sources. As I believe all the sources to be suspect on at least some issues, we need to treat the individual pieces of evidence on each issue independently and on their merits. That does not always work in favour of the thesis I propose in this book, as at times there is reason to be suspicious of the positive evidence as well.

Where it is unclear when we should accept a piece of evidence or not, we need to strike a balance between the following two considerations. To avoid circularity, I cannot argue both that we should adopt a specific interpretation because it fits my general thesis, and that we should adopt my general thesis because it is supported by this interpretation. On the other hand, we might hypothetically accept a piece of evidence on the grounds that it will allow us a coherent account, adopting the principle: If we accept this piece of evidence, then we get a coherent picture both of the cosmogony of this thinker and how they fit into the development of Greek cosmogony.

This book then offers one arrangement of the evidence, aimed at producing a good but not perfect fit between Greek cosmogonical concerns and solutions and perennial cosmogonical problems and solutions. The merits of this arrangement are that it gives coherent and interesting positions to all those who took problems in cosmogony seriously, and that it opens up some interesting new lines of investigation concerning the relation of some thinkers.

V. Philosophical cosmogony

What sort of cosmogony were the Greeks doing? I believe the best term for their activities is philosophical cosmogony. I would distinguish this from modern scientific cosmogony because modern cosmogony is predominantly empirical. There is a debate as to when cosmogony became a science in the modern sense, and even if we should currently call cosmogony a science. Generally it is held that cosmogony became a science around 1920-30, when the first calculations were made on the age of the universe using Hubble's data on the recession rates of stars.[11] Two problems big bang cosmogony faces is that it proposes a singular event which as such cannot be repeatedly studied, and as space and time are created with the big bang, the usual causal type of explanations are ruled out. Steady state cosmology, which denied any single instant of creation and was popular in the middle of the twentieth century, exploited these and other factors to brand big bang cosmogony unscientific.

I have described modern cosmogony as predominantly empirical here because while modern cosmogony is now an empirical science, there still remain considerable philosophical issues to be resolved. In particular, as I discuss a little later, there is a debate on what should constitute an explanation in cosmogony. It is also the case that while Greek cosmogony was predominantly a philosophical activity, observations were used to support cosmogonical theories. That may seem a little odd, but there is a perennial question in cosmogony about whether the processes which bring the *kosmos* into being then cease, and so are specific to cosmogony, or are part of the ongoing physical processes of the *kosmos*. Some of the Greeks opted for the latter, and so were able to point to how a physical process produces the result which they believe to be important for cosmogony.

A second distinction which needs to be drawn in relation to Greek philosophical cosmogony is to separate it from creation myths and theogonies. Clearly, many societies prior to the Greeks had myths concerning the origins of the world. I am not concerned with using the term cosmogony for tales which predate the term *kosmos*, as, for instance, I am quite happy that there is science before the term science is used. Rather my concern is whether the world that is envisaged as coming into being can properly be called a *kosmos*. Some creation tales will begin with primeval water, or primeval chaos, out of which will be formed the earth and the heavens.

However, if that new world is subject to interference in its processes on the whims of capricious gods then I do not think we can call such a world a *kosmos* in the sense meant by the Presocratics. In Chapter 1 I argue that the key issues which separate myths from philosophical accounts are parsimony, invariance, consistency and a rejection of the supernatural. Arguably, Thales produces the first philosophical cosmogony (arguable because the evidence is relatively small) and there is certainly philosophical cosmogony in Anaximander.

A further argument here that we find the first philosophical cosmogony with the Greeks relates to the perennial problems in cosmogony. I do not see that the creation tales address, either explicitly or implicitly, or formulate any of the philosophical problems of cosmogony. The Greeks formulate and address philosophical problems in cosmogony in a way that simply had not occurred before.

VI. The history of cosmogony

I am aware that the approach I take in his book may be controversial, and some may even consider it to be anachronistic. A critical question here is whether there are these perennial question and solution types in cosmogony or not. By perennial here I mean something quite strong. Are there problem and solution types which are inherent in, or intrinsic to or part of the natural structure of cosmogony once the project of philosophical cosmogony has been launched? My view is that there are. Those underlying problems existed as much for the Greeks as they do for us, although the Greeks formulated them in a different manner. We can then attempt to explain why the Greeks adopted certain positions in cosmogony as perennial solution types to perennial problems.

So to be clear here, I am not arguing that what is important or interesting about the ancient Greeks is that they anticipate or move toward modern answers. Nor am I arguing that the Greeks can only be understood in relation to modern answers. Nor am I arguing that Greek and modern issues are identical. The Greeks had their own formulations of these issues specific to their own intellectual context, which this book seeks to investigate. I do believe though that a comparison of Greek formulations of problem and solution types throughout history, and not just with the modern period, is justified and interesting. It helps to make the perennial thesis plausible, throws light on how these problem and solution types can be formulated differently in different conditions and is interesting in itself. I do think it is useful to have a knowledge of the problems and possible solutions relevant to modern cosmogony and to other periods in order to be able better to spot when the Greeks are addressing these issues. I hope that the results of this book, in finding treatment of a greater number of issues of cosmogony than was previously thought, as well as greater debate between thinkers, are a vindication of this method.

Establishing that there are similarities between some ancient and modern problem and solution types does not establish that there are perennial problems in the strong sense. One might argue that there is an accidental concordance between ancient and modern problem and solution types. One might also argue that I have 'read in' modern concerns where no such concerns existed for the ancients. One aim of this book is to show that the concordances between ancient and modern problem and solution types are too many and too tight, and the ancients' treatment of the problems too systematic for this to be a matter of accident, or to have been 'read in'.

One might also argue that the concordance of ancient and modern ideas is a matter of tradition, or even cultural prejudice. We deal with these problems in this way not because they are inherent in the subject but because we have always dealt with these questions, and in fact the choice is quite arbitrary. Or more strongly, Western culture, wishing to see its roots in an idealised version of classical antiquity, deals with the problems bequeathed by antiquity. Seeking to privilege both itself and classical antiquity, it then attempts to treat those problems as atemporal and in some way more significant than problems addressed by other cultures.

This cannot be entirely a matter of tradition though, for modern science has rejected some problems and many solution types from antiquity. Indeed, the scientific revolution of the seventeenth century saw a conscious and wide spread rejection of ancient ideas in both philosophy and science. The origins of cosmogony I take as an open question for historical enquiry. I argue they are with the Greeks, but if asked where the origins of predictive mathematical astronomy lay, I would argue with Babylonians, or if asked where the origins of medical practices lay I would argue with the Babylonians and the Egyptians. On the development of science, I would argue that an extremely significant role was played by the Arabic/ Islamic culture of the seventh to the sixteenth centuries, and would also want to recognise significant contributions from the Chinese and Indian cultures. So I do not see my work or this book as part of a scheme to aggrandise Western culture or classical antiquity.

The history I have constructed credits the Greeks with an appreciation of perennial problems and solutions. It also attempts to see debate and increasing sophistication between successive contributors. In some places one can establish this reasonably firmly. The ancients seem to understand the unique universe/ multiple universes debate very well, opting either for multiple accidents or for design, with no one mixing the two and no one taking the view that the *kosmos* could come about in the absence of design or multiple accidents. In other places, this is rather more conjectural and underdetermined. To what extent then are we justified in arranging the evidence to produce a dialectical and progressive history? At least to the same extent that the history of Greek mainstream philosophy is constructed, as it faces similar problems and uses similar criteria. If we can show that the Greeks were discussing a recognisable philosophical issue intelligently, and even better, that we can see a tradition of discussion of that problem with overt or implied criticism of previous thinkers and an increase in sophistication over time, we believe we have written good history of philosophy. I would argue that there are perennial problem and solution types once the project of philosophy is launched. Greek and modern formulations of these problem and answer types are of course different. To give a simple example, Descartes is credited with shifting attention in epistemology from 'what is the extent of what we know' to 'can I know anything for certain'. Yet we can still recognise an investigation into the problems of epistemology took place in antiquity, and that the problem and solution types are related though not identical to those proposed by early modern and modern philosophers. It helps to be aware of modern epistemological questions, and epistemological questions throughout history, in order to spot when and how the Greeks are addressing epistemological issues. I do not then believe my approach to be anachronistic. There are perennial problem and solution types in cosmogony, but the Greeks have their own versions of these, which this book seeks to detail and explain.

VII. Cosmogony and Meno's paradox

Ancient cosmogony is of course crude when compared to modern theories. There is a difficult problem in cosmogony though, which can be expressed deceptively simply: What is cosmogony *like*? As we have no direct experience of cosmogony, we are forced into analogies. If we think the Greek analogies crude, it is well to remember that the analogy used by the current dominant theory of cosmogony is that the initial moment was a huge explosion, and we live in the aftermath of that explosion.

There is a sense in which much of modern science, at least on a practical level, has managed to answer Meno's (the learner's) paradox. Plato has Meno ask:

> How will you search for this thing, Socrates, not knowing at all what sort of thing it is? For what sort of thing that you do not know will be proposed in your search? Or even supposing that you should meet with this thing, how will you know that it is this thing which you do not know?[12]

In much of modern science, we at least believe we know, at a practical level, the form of the answer we are looking for, and can then fill in the specific content. So in a chemical analysis, if we are able to say 'The crystals in this test tube are sodium chloride' we at least know we have given the right sort of answer.[13] In cosmogony, we do not yet know what sort of answer is appropriate. This is not just a matter of competing empirical theories, but of fundamentally different types of explanation. Ought we to suppose there are an infinity of universes, and that we can explain our universe as one amid a multiplicity of accidental universes? If there is only one universe, is the anthropic principle an acceptable form of explanation? If space and time are created with the big bang, and causal explanations dependent on prior spatio-temporal relations are thereby ruled out, what sort of explanation can we seek for a big bang? If we do not find an explanation in cosmogony sufficiently satisfying, is the explanation inadequate or is our wish for something more satisfying misplaced? Modern cosmogony still struggles with all of these problems. If we still struggle to be sure of the form of explanation appropriate to cosmogony, then we should have considerable sympathy for the ancients struggling with a similar problem.

VIII. Resources in ancient cosmogony

A further important factor in understanding ancient cosmogony is being aware of the resources available to the ancients. I have argued elsewhere that the lack of a conception of gravity seriously affects the nature and development of Greek astronomy and cosmology.[14] Here I will argue that that is true for cosmogony as well, especially where a *kosmos* is supposed to be ordered from a pre-existing chaos.

The ancients sometimes considered the *kosmos* to be alive in some respects, and at times modelled cosmogonical processes on biological ones. It is an easy criticism of the ancients that they employed hylozoism or anthropomorphism too readily. It is rare to find the ancients saying, for instance, that the *kosmos is* an animal. Much more often they say it is *like* an animal. The interesting question then is in what respects do they think it to be like or unlike an animal, and why. Plato, though he believes the *kosmos* to be intelligent, ensouled and alive, takes pains to point out the respects in which his *kosmos* is not like an animal as well in the *Timaeus*.[15]

Much the same is true of ancient cosmogony. The ancients may use biological analogues, but rarely do they say that cosmogony is simply a biological process, rather it is like a biological process. We need to investigate then what aspects of biological processes they employ and for what ends. I shall also argue that in some cases we have been to quick to assume that a biological analogy is being used, where something more subtle may be going on. Of course, biological analogues are used, and I offer two arguments why we should not be overly critical of this. Firstly, biological analogues are sometimes used by the ancient Greeks to enforce the idea of regularity rather than to oppose it.[16] Secondly, we are looking here at a time before science set itself against the use of biological analogies in cosmogony. It was perfectly reasonable of the Greeks to explore this possibility at this stage of the development of science.

This book then seeks to map the development of philosophical cosmogony among the ancient Greeks, from Thales up to the debates about Christian creation in late antiquity, around the time of Simplicius and Philoponus. This cut off point is somewhat arbitrary, and I hope to examine subsequent cosmogony in Islam and the Christian West in a further volume.

1

Mythological Accounts of Creation

This is a brief introductory chapter on mythological accounts of creation prior to the establishment of philosophical cosmogony. Its purpose is to serve as a background for the theories of the Milesians in the next chapter, and to clarify the differences between creation tales and philosophical cosmogony.

I. Philosophical cosmogony and creation tales

As I believe that philosophical cosmogony is something quite different from mythological creation tales, I will begin by giving the reasons why I believe this to be so. To a certain extent I agree with Kirk, that while there are many myths, there is no such thing as myth or mythical thinking.[1] Myths have such a diversity of functions in such a diversity of societies and situations that it is extremely difficult, if not impossible, to pick out factors which are common to all myths or to give any sensible definition of what is involved in mythical thinking. I would disagree with Kirk to the extent that we must have some general notion of what constitutes myth in order to be able to identify something as a myth, and to separate myths from non-myths. My approach is to try to say what it is that philosophy and philosophical cosmogony do that myths do not, without seeking any positive characterisation of myth.

The key issue which separates myths from philosophy in my view is justification. Philosophical views always require intellectual justification, and the nature of that justification also needs to be justified. A consequence of justification is that theories need to be parsimonious. This is not to say that reductionism or nominalism are always good philosophy. Rather, when we make an ontological commitment in philosophy, we need to be able to justify that commitment. With myths generally, and with creation myths in particular, I see no recognition of or commitment to the principle of parsimony. Indeed for some myths it may be preferable to be ontologically profligate in order to better fulfil their function. Arguably, Thales is the first to give us a parsimonious philosophical cosmogony. Hesiod, while he may attempt a systematisation of theogony, is not parsimonious in his theogony.

It has been argued that an important factor in the move from mythology to philosophy was a depersonification of the factors controlling the universe, a move from 'thou to it', to impersonal forces.[2] The issue is one of invariance rather than personification. That is, do the factors controlling the behaviour of the universe act in a regular and predictable manner? One might have a depersonified force which is not invariant. One might have a personal God which sees to it that the universe behaves in a perfectly regular manner. The latter is of considerable importance for the philosophical and scientific ideas of theists from Plato onwards. It is quite acceptable for the behaviour of the universe to change, as long as that change occurs in a regular manner, and not on the whim of some capricious god.

It is fundamental to philosophy that theories in themselves are non-contradictory, are as general as possible and have no exceptions, and that a group of theories do not contradict each other. Holding one theory will exclude holding other contradictory theories. Myths are not subject to these restrictions, and while many myths are not contradictory in themselves, as Wilson puts it: 'The Egyptians accepted various myths and discarded none of them.'[3] One problem with rejecting parsimony, invariance or non-contradiction is that one is left with a situation where virtually anything goes. Ontologically one can suppose there to be any number of things, and those things could act in an arbitrary and indeed a contradictory manner.

It has also been argued that a rejection of the supernatural is important in distinguishing myths from philosophy. I agree that it is, though I suspect this is a consequence of the acceptance of parsimony and invariance rather than an independent factor. As a definition and analysis of what it is to be supernatural would take us too far astray here, I will simply treat this as a further independent factor which separates myth from philosophy.

II. The origins of philosophical cosmogony

We have then some criteria to separate myth from philosophical cosmogony. I prefer not to use the terms rational or irrational in relation to this sort of discussion. The term irrational is too often taken as pejorative, when I have no wish to denigrate the achievements of cultures prior to the Greeks. They achieved many marvellous things, especially in terms of healing practices and the observation and prediction of the motions of the heavens. The Greeks had the considerable advantage of knowledge of those achievements. Irrational is also often taken as 'completely irrational', and that would be most odd for cultures with such achievements, who were clearly capable of reason. I would say much the same about the terms philosophical and pre-philosophical. Even if the myths produced by societies prior to the Greeks do not employ parsimony, invariance, consistency or a distinction between the natural and the supernatural those societies clearly contained people who were articulate, capable of systematic and abstract reasoning and able to resolve intellectual problems.

There is, however, the following consideration. Philosophical cosmogony has not always existed, nor indeed have mythical creation tales, if only for the simple reason that humankind has not always existed. So there must have been humans, and human societies, who did not have creation tales, and similarly did not have philosophical cosmogony. At some stage we have to say that these things came into being.[4] I do not want to suggest that the Greeks conjured up philosophy and philosophical cosmogony as it were *ex nihilo*, as a 'Traveller without luggage' as Vernant puts it.[5] They had creation tales to build on, and there was the systematic theogony of Hesiod. Nor do I wish to suggest that once the Greeks had philosophy and philosophical cosmogony that every Greek then followed these ideas, or that even a majority of them did. Belief in myth, interventionist gods, magic, mysticism and the supernatural was widespread and commonplace. While I might disagree with Dodds on some instances of 'irrationality',[6] he clearly showed that philosophy did not have a complete dominance over Greek thought. The history of Greek philosophy and science is the history of a relatively small group of people in a society dominated by a belief in myth and interventionst theology, and is all the more remarkable for being so.

We also need to consider the role of the 'irrational' in our own society. We consider ourselves to be in possession of philosophy and science in general, and to have not only philosophical cosmogony but properly scientific cosmogony. Yet many in our society, even if we restrict ourselves here to the West, believe in mysticism, astrology, the supernatural and gods who interfere in the processes of the world. We consider ourselves to have scientific cosmogony, but what proportion of people believe in big bang cosmogony as opposed to biblical creation, and what proportion of those who believe in a big bang believe the cause of that bang was god? It cannot then be a decisive objection to a society being said to have philosophy, science or philosophical cosmogony that many in that society hold contrary views. The key issue is whether we can identify a group of people who do hold the requisite views. With philosophical cosmogony, we can do that for the first time with the ancient Greeks. It may be a small group, far smaller than today's group, but that is only to be expected for the origins of a new and radical idea.

III. Biological analogues and invariance

One issue that is commonly thought to separate myth from philosophy and science is the use of animism, anthropomorphism and hylozoism. This might then be argued to lessen the gap between the Greek philosophers and their predecessors, as clearly at least until Plato these ideas are very much in play. As I commented above on gods and depersonification though, the central issue on these matters is invariance. Paradoxically, at least to the modern eye, the Greeks used animistic ideas to explain why something behaved invariantly. A good example here is Plato. Why is it, he asks, that the heavenly bodies can move in such a precise and regular manner? Plato considers the heavenly bodies to be gods, to be animate, to be intelligent and to have souls, and so too the *kosmos* as a whole. Why does Plato adopt this position? Plato has a low opinion of what matter can do alone and unaided. Even if matter is properly formed, then physical principles are not sufficient to explain how the *kosmos* acquires or sustains order.[7] For Plato, regular and orderly behaviour requires further explanation, and so the motions of the heavens require explanation. In the *Laws* he puts it like this:

> Those who engaged in these matters accurately would not have been able to use such wonderfully accurate calculations if these entities did not have souls.[8]

It is worth looking closely at what sort of entities Plato postulates here, what they do and what they are supposed to explain. The *Timaeus* describes the *kosmos* in these terms:

> Thus in accordance with the likely account it is necessary to say that this living, ensouled intelligent *kosmos* has in truth come about through the foresight of god.[9]

The *Timaeus* states that the *kosmos* does not have many of the things we would usually associate with an animal (or even, within either a Pagan or a Christian tradition, with a god). From *Timaeus* 33b onwards we are told that this 'animal' is perfectly spherical, has neither eyes nor ears as there is nothing external to see or hear, there is no air for it to breathe, no need of any organ to receive food or to excrete the remains, as it is entirely self-sufficient and nothing comes in or goes out. As it needs neither hands to defend itself nor feet to stand on, and has no need of legs or feet to propel itself, it has no limbs, and at 34b we are told it is a god. To say the least, this is a somewhat strange animal, and certainly could not be considered to be an anthropomorphic god, even if it does have intelligence and soul.[10] If the *kosmos* is like an animal, then we have an analogue for the integrity and the internal organisation of the *kosmos*, and the ongoing order of the *kosmos*. A typical ancient contrast here is between the body with life and mind and what happens to the body after death.[11] What does this *kosmos* do though? All it does is revolve uniformly in one place. In the *Timaeus* we are told that the *kosmos* was given:

> A motion proper to its body, that of the seven motions which is best suited to reason and intelligence. Therefore he made it move in a circle, revolving of itself uniformly and in the same place, and he took from it all trace of the other six motions and kept it free from their wanderings.[12]

Here I would emphasise that this is entirely regular circular motion with no deviations. This is so for the fixed stars and the planets as well.[13] The stars are spherical and intelligent like the *kosmos*, are divine and are living creatures.[14] As with the *kosmos*, they have motions befitting their intelligence, and so we are told in the *Timaeus* that:

> To each of these he gave two motions, one being uniform and in the same place, always thinking the same thoughts concerning the same things, the other being a forward motion obeying the revolution of the same and similar. With regard to the other five motions, they were motionless and still, in order that each might attain the greatest possible perfection.[15]

The essential point here then is that the celestial bodies have no freedom of action (or no desire to deviate from regular circular motion). They have the intelligence to carry out their assigned duties, and not to do anything else. It is their intelligence which explains their regular and orderly behaviour. That they move, and do so without any external compulsion, is explained by their having souls. For Plato, the soul is a principle of motion.[16] So in the *Laws* we find that:

> If the whole course and motion of the heavens and everything that are in them are similar in nature to the motion and revolution and calculations of intelligence (*nous*), and work in a related manner, then clearly we must say that the best soul has charge of the whole *kosmos* and leads it along these selfsame courses.[17]

Although the *kosmos* and the heavenly bodies have life, soul, divinity and intelligence, they have these in a highly circumscribed and attenuated manner, from a modern point of view.[18] If this can be described as vitalism, it is a highly depersonalised one where the attributes of animate beings that are required for the description of the *kosmos* have been carefully sorted from those that are not. This allows us to distinguish quite sharply between Plato and the mythological and magical traditions. The key matter here is that these souls/ gods are not capricious.[19] They always act for the best so will always act in the same manner. There is in principle nothing unpredictable or irregular about their behaviour.[20]

A further important consideration here is the nature of physical law. Physical law can be modelled on mathematical law, and indeed since the work of Galileo, Descartes and others in the seventeenth century, we express our laws of physics in terms of inviolable equations. There was another possibility though, which was to model physical law on civil law. Such laws could in principle be broken, and then the law-breaker would be subject to punishment.[21] However, if the heavenly bodies have the best sort of soul in charge of their motions, as that sort of soul will clearly always do what is right, we can rely on the heavenly bodies moving in an entirely uniform and invariant manner.[22]

We cannot then simply identify animism, anthropomorphism or hylozoism with variable or unpredictable behaviour for the Greeks, or indeed for the cultures which preceded them. That is important for the Milesians, as they are often considered to be hylozoists, and this is sometimes taken to be an indication that they have not yet entirely escaped from mythology. More generally, we will also see many of the Greeks using biological analogies for cosmogony. The *kosmos* may be thought to emerge from a seed, or its early stages likened to the development of an embryo. Again, we need to look at each case on its merits as to whether capricious behaviour is involved. I would also reiterate something I said in the introduction. It is not easy to say what cosmogony is *like*. In the early stages of philosophical cosmogony, biological analogues may well have looked at least as plausible as physical analogues. Indeed, it is only really since the seventeenth century that science has rejected biological analogues in cosmology and cosmogony.

IV. Two questions for myths

I do not propose to give an exhaustive catalogue of creation tales prior to the Greeks. I merely want to indicate the general style and tenor of these tales, to give a background to early Greek cosmogony. The Hebrew tradition I will leave for the moment, as the outlines of Genesis are well enough known and can be easily accessed. I will of course examine the Hebrew tradition in a later chapter on the debate between Christianity and Greek philosophy. Down to Hesiod, there is no debate that these are creation tales rather than philosophical cosmogony. The two central considerations are:

(1) Is the result of the creation processes a *kosmos*? If we mean by *kosmos* a world order that is parsimonious, invariant, consistent and natural then the answer is no. The main consideration here would be the arbitrary interference of gods in the world.
(2) Are the processes of creation philosophical cosmogony? If by that we mean that these processes are parsimonious, invariant, consistent and natural, again the answer is no. Again the main issue will be arbitrary actions of the gods.

V. Egyptian creation myths

We have several creation tales from ancient Egypt. Given that the Nile flooded and receded each year, it is no great surprise that a central feature in Egyptian creation tales is the emergence of dry land from water. One might also add that the Nile mud was exceptionally fertile, and given the profundity of life-forms which emerged from it, might be thought to have life-giving properties. The first passage is from dedication ritual in a royal pyramid, dating from around the twenty-fourth century BC.

O Atum-Kheprer, you were on high on the primeval hill. You arose as did the *ben*-bird of the *ben*-stone in the *Ben*-House in Heliopolis. You spat out what was Shu, and sputtered out what was Tefnut. You put your arms around them as the arms of *ka*, for your *ka* was in them. So also, O Atum, put your arms around King Nefer-ka-Re, as the arms of *ka*. For the *ka* of King Nefer-ka-Re is in it, enduring for the course of eternity ... O Great Ennead which is in Heliopolis, Atum, Shu, Tefnut, Geb, Nut, Osiris, Isis, Seth and Nepthys whom Atum begot, spreading wide his heart in joy at his begetting you in your name of the Nine Bows.[23]

The following passage was inscribed in the coffins of nobles, and originally dates from at least the twentieth century BC. This is Wilson's translation and glosses:

'I am Atum when he was alone in Nun. I am Re in his first appearances, when he began to rule that which he had made.'

Who is he? This 'Re, when he began to rule that which he had made' means that Re began to appear as a king, as one who was before the liftings of Shu had taken place, when he was on a hill which is in Hermopolis ...

'I am the great god who came into being by himself.'

Who is he? 'The great god who came into being by himself' is water; he is Nun, the father of the gods. Another version: he is Re.

'He who created his names, the Lord of the Ennead.'

Who is he? He is Re, who created the names of the parts of his body. That is how these gods who follow him come into being.

'I am he among the gods who cannot be repulsed.'

Who is he? He is Atum, who is in the sun disc. Another version: He is Re, when he rises on the Eastern horizon of heaven.

'I am yesterday, while I know tomorrow.'

Who is he? As for 'yesterday', that is Osiris. As for 'tomorrow', that is Re on that day which the enemies of the All-Lord are annihilated and his son Horus is made ruler ....

The following passage is from The Theology of Memphis, a document dating from around the seventh century BC, but the tale itself is likely to be much older. This passage is significant as a precursor to some Christian accounts of creation, as it gives a more intellectual account, from the mind (heart) of the creator god and enacted by his speech (tongue).

There came into being as the heart and there came into being as the tongue in the form of Atum. The mighty Great One is Ptah, who transmitted life to all gods, as well as to their *ka*s, through this heart, by which Horus became Ptah, and through his tongue, by which Thoth became Ptah.

Thus it happened that the heart and tongue gained control over every other member of the body, by teaching that he is in every body and in every mouth of all gods, all men, all cattle, all creeping things, and everything that lives, by thinking and commanding everything that he wishes.

His Ennead is before him in the form of teeth and lips. That is the equivalent of the semen and hands of Atum. Whereas the Ennead of Atum came into being by his semen and his fingers, the Ennead of Ptah, however, is the teeth and lips of his mouth, which pronounced the name of everything, from which Shu and Tefnut came forth, and which was the fashioner of the Ennead.[24]

VI. Babylonian creation myths

The main Babylonian account of creation is to be found in the epic Enuma Elish (When Above), which takes its name from its opening words. It is likely that this was composed during the first Babylonian dynasty (2057-1758 BC).[25] Like the Egyptians with the Nile, the Babylonians had the flooding and receding of the Tigris and Euphrates, so again we have the first dry land emerging from water. These are the opening few lines of an epic that stretches over seven tablets, most of which are around 130 lines long.[26]

> When above the heavens had not yet been named
> And below the earth had not been called by a name
> When only Apsu primeval, their begetter existed
> And mother Ti'amat, who gave birth to them all
> When their water still mixed together
> And no dry land had been formed and not even a marsh could be seen
> When none of the gods had yet been brought into being
> When they had not yet been called by their names, and their destinies had not yet been fixed
> Then were the Gods created in the midst of them.[27]

There are many other Babylonian creation tales, the most significant of which is the tale known as the Eridu Genesis. The tablet this is taken from has been dated to the sixth century BC, though of course the tale itself may be much older. This is an excerpt from the forty lines of Eridu:

> All the lands were sea
> The spring which is in the sea was a water pipe
> Then Eridu was made, Esagila built
> Esaglia, whose foundations Lugaldukuga laid within the Apsu
> Babylon was made, Esaglia was completed
> The Anunnaki goda he created at one time
> The holy city, the dwelling of their heart's delight, they called it solemnly.
> Marduk constructed a reed frame on the face of the waters
> He created dirt and poured it out by the reed frame
> In order to settle the gods in the dwelling of their heart's delight
> He created mankind

There are many similar tales of the creation of the world, the creation of man and the roles played by the gods and water.

VII. Early Greek creation myths

Let us now turn to some of the pre-Milesian Greek accounts of creation. Again, this is not intended to be a comprehensive catalogue, rather a representative selection. In Homer, we can find the idea of Okeanos, a river which surrounds the earth, the earth being considered to be flat and the sky hemispherical. Homer describes Okeanos:

> Okeanos, generator of the gods and mother Tethys.[28]
> Okeanos, who is the generator of all.[29]

Once Okeanos has generated, we then have a theogony following on. There are many variations on Orphic creation myths. Here I follow the fourfold grouping suggested by KRS. The first of these has night as the ultimate origin of everything. So we are told that:

> The theology of Orpheus ... made Night the origin.[30]

> Night again bore Gaia and broad Ouranos.[31]

Further gods are then generated. Another two Orphic options we can represent schematically as follows:[32]

Chronos	Aither Chaos	Egg	Phanes		
Matter	Water Earth	Unaging Chronos/ Heracles Necessity Adrasteia	Aether Chaos Erebos	Egg	God without body

Finally, there is a version where water is the origin of all things:

> Water was the origin of all things, according to Orpheus, and from water mud was deposited, and from these a life form was generated, a snake with the head of a lion growing on it, and between these grew the face of a God, Heracles and Kronos by name. This Heracles generated an enormous egg, completely filled with the force of its creator, broke into two due to rubbing. The upper part became Ouranos, and the lower part became Ge, and a double bodied God came forth. Ouranos and Ge had intercourse and generated the females Cloetho, Lachesis and Atropos.[33]

This may all seem a little messy, but that is the state of the evidence. That may be due to our Neoplatonic sources, who are often more interested in producing an intelligible world/ physical world interpretation of their sources than in strict historical accuracy. Alternatively, this may be due in part to the nature of myth, where there is no great necessity to have one definitive account, and the rather loose intellectual nature of Orphism. I suspect all three contribute to a rather fragmented Orphic corpus. Aristophanes gives us the following account:

> First was Chaos, Night, black Erebos and wide Tartarus, and neither Ge nor Air nor Ouranos existed. In the unlimited bosom of Erebos, black-winged Night bears first a wind-egg.[34]

Once again a theogony follows on from this. There is also the notion of cosmogony by the separation of the earth and the sky. A typical example is Euripides:

> The earth and the heavens were one form. When they had been separated from one another they bore all things and gave them up into the light, trees, birds, animals, those nourished by the salty sea and mankind.[35]

The Derveni papyrus, while giving us some interesting insights into early Greek mythology, offers nothing fundamentally different in structure.[36] We still have theogony with a proliferation of gods.[37]

VIII. Hesiod

I come to Hesiod (*c.* 700 BC) last, not because he is chronologically later than any of the mythologies we have looked at, but because Hesiod is often seen as either the link between myth and philosophy, or even sometimes as the first to produce a philosophical cosmogony. It should be noted though that there are considerable similarities between Hesiod's account and the Hittite myth found on the Kumarbi stone.[38] Pherecydes (*c.* 550 BC) is an interesting character,[39] but Aristotle puts him in a different tradition to the philosophical thought of the Milesians, and here I would agree with Aristotle.[40] Alkman (*c.* 600 BC) is another possible bridge between myth and philosophy, but it is clear he has multiple divinities, and what little we know of him is badly corrupted by later commentators.[41] The most significant passage in Hesiod is from the *Theogony*:

> First a chasm was generated, then broad-breasted Gaia (earth), a safe seat for all forever,[42] and misty Tartarus in a recess of the broad earth, and Eros, most beautiful of the immortal gods, relaxer of limbs, who subdues in their breasts the sense and intelligent counsel of all gods and all men. From the chasm, Erebos and black night were generated. From night, aether and day were generated, whom she bore after sex with Erebos. Earth first generated something equal to herself, starry Ouranos, in order that she should be entirely covered, to be a safe seat for the blessed gods forever. Then she generated large mountains, happy haunts of the divine Nymphs who live in the wooded mountains. She also bore the unharvested sea, with its rolling

> swell, Pontos, but not in union of love. Then having had sex with Ouranos she bore deep-eddying Okeanos.[43]

The first issue here is how to translate the Greek word *chaos*, which I have rendered here as chasm. *Chaos* does not mean chaos in the modern sense, nor does it mean void in any sense where the void would be unlimited. It can mean the space between earth and sky, and it can mean a gaping or yawning chasm, where the chasm is a limited space. So the first event in cosmogony here could be the separation of the earth and sky generating the space between,[44] or the generation of a limited chasm.[45]

Later in antiquity, Philo considered this passage to show that for Hesiod the *kosmos* is generated but indestructible. Generated because first there is chaos and only then is there earth.[46] He gives Aristotle's opinion that chaos must be space because a body must have something to contain it. What Aristotle says is that:

> This view would mean that Hesiod was correct in saying that firstly there was chaos. He says 'First a chasm was generated, then broad-breasted Gaia', as necessarily first there must be space for something to be in. Which is to say, as with the many, that everything must be somewhere and in some place.[47]

According to Philo, the *kosmos* is indestructible for Hesiod because 'he did not disclose that it would be dissolved or destroyed'.[48] We might take it from the statement that the earth is 'a safe seat for all forever' that at least the earth will never be destroyed.

Hesiod doubtless brings order and systematisation to theogony. It is also the case that the end product is more ordered than the starting material, even if we dispose of the idea that *chaos* means chaos. Is that enough though, for the beginnings of philosophical cosmogony? There are two important considerations here.

Can we consider the world of Hesiod to be a *kosmos*? The gods still interfere with processes on earth. The gods have sex with mortals not only to start the race of heroes, but to keep them going. In *Works and Days* it is clear that Zeus will punish behaviour he disapproves of.[49] As a more general point we might compare justice as the whim and subsequent interference of Zeus against justice occurring of necessity in Anaximander's account of the opposites.[50] I agree with Xenophanes who says that:

> Homer and Hesiod have ascribed to the gods all those things which are shameful and reproachful among men: theft, adultery and deceiving each other.[51]

Hesiod gives us the tale of Prometheus, where deceit and jealousy among the gods leads to their interfering with the world by bringing mankind the gift of fire.[52] Hesiod also gives us the tale of Pandora's Jar, a gift from the gods containing countless troubles for men.[53] I do not see how we can call this a *kosmos* in the same way that we call the worlds of the Presocratic philosophers *kosmoi*. I do not see a recognition of the importance of parsimony in the account of the world, there certainly is not invariance, and the natural world is suffused with supernatural happenings.

Secondly, can we consider the processes which bring about Hesiod's world to be philosophical cosmogony? If we take Hesiod in a suitable manner, we can see the Theogony generating earth, sky and sea. If Hesiod stopped there, we could perhaps see some parsimony, in that he invokes only enough gods to match the essential parts of the world.[54] However, this is very far from the case and there is a huge overuse of gods, titans, heroes, monsters, etc. who have no aetiological function in cosmogony at all. One might argue that Hesiod's work is multi-functional, giving a philosophical cosmogony and some myths within the one tale. Even then, though, we have many gods reproducing in many ways, and there is no real sense of any parsimony here. The theogonical process is dependent on the whims of the gods. Not content with sexual reproduction, there is also both male and female parthenogenesis without there being any cosmogonical reason for three processes.[55] There is no sense of a distinction between or a separation of the natural and the supernatural, especially given the gods' habit of having sex with mortals. I do not see that we can call Hesiod's cosmogony a philosophical cosmogony in the same sense that we call the work of the Presocratics philosophical cosmogony. Again, I see no recognition of the importance of parsimony, the processes are not invariant and there is not a satisfactory distinction or separation of natural and supernatural processes. It remains to be seen whether the origins of philosophical cosmogony are with Thales and the other Milesians, but they are certainly not with Hesiod.

IX. Oriental influences

West has argued that there was a considerable Oriental influence in early Greek cosmogony.[56] I would draw a distinction between the content of early Greek cosmogony, and the way in which that content is treated. It may well be the case that Thales, for instance, when he says that everything comes from water, is drawing upon Egyptian or Babylonian or other possible sources. The way that the early Greeks treat their cosmogony, however, producing a *kosmos* from processes acceptable to philosophical cosmogony, is without precedent in other cultures, both in terms of the end result and the nature of the processes involved.

I am not concerned that there are no precedents here. If there is innovation and progress, then there must be ideas without precedents. If we believe the theory of evolution, there was a time when there were no humans, no creation tales and no cosmogony. The first creation tale must have been without precedent, and creation tales different from the first one must also have been in a sense without precedent. This is not to say that progress is always made by radical new ideas, but hopefully Kuhn's work in the history and philosophy of science has shown that science at least sometimes progresses in a revolutionary manner, swapping an old paradigm for an incommensurable new one.[57]

There is a great temptation, when confronted with the proposal that a society originated something, to argue that societies preceding them surely had this too. At some point we have to resist this regression. None of this is to denigrate the societies who immediately preceded the Greeks. They produced much marvellous technology, their healing practices were effective, they made accurate observations and predictions of the heavens and were capable of solving equations. It is highly likely that they did these things better than societies preceding them, who had their own achievements. It is quite possible that they too did some things without precedent. They did not though begin philosophical cosmogony.

It is sometimes said that a subject evolved, or was the work of many hands, and so its origins are diffuse and difficult, if not impossible to pin down.[58] I take it that at one time there were neither humans, technology, science or cosmogony, while all those things exist now. If biologists can find the origins of *Homo sapiens* within meaningful temporal and geographical windows, then it should not be beyond philosophers and historians of science to do the same for the origins of cosmogony.

There may then have been many influences on the content of Greek cosmogony. Certainly one would want to look to the Egyptians and Babylonians, but as West has argued there are many other possibilities as well.[59] I would raise one final historiographical question though. Do we allow multiple independent sources of ideas? What grounds do we have to deny this? It may be the case that the conditions for the origination of an idea are similar in many places. Both the Egyptians and the Babylonians had flood-plain agricultures, and both had creation tales where the first earth appeared from water. One may have influenced the other, there may be a common source, or there is the possibility of independent origination. The notion that some form of sexual reproduction among the gods generated the world is so widespread as to defy any attempt to find a single origin. If we do allow multiple independent sources, however, on what grounds do we deny the possibility of independent innovation to the Greeks?

2

The Milesians

The Milesian thinkers give us the first cosmogonies as opposed to mythical creation stories or theogonies. The nature of these cosmogonies is contested. Here I argue a common line for Thales (*c.* 625 – *c.* 546 BC), Anaximander (610-547 BC) and Anaximenes (*c.* 585 – *c.* 525 BC), that they believed that their principal substance, be it water, the unlimited or air, had an inherent ability to steer the processes leading to the formation of the *kosmos*.

I. Thales: general principles

We do not have any detail of Thales' cosmogony, but we do have some indications of the relevant general philosophical principles. Aristotle tells us that:

> Most of the first philosophers thought of matter as the only principle of all things. That from which all things are, that from which a thing first comes to be, and into which it is ultimately destroyed, the substance persisting but changing in its qualities. This they say is the element and *archê* of the things that are and because of this they say there is no absolute coming to be or destruction, but nature is always preserved.[1]

I do not propose to translate *archê*, which can mean 'beginning', 'origin', 'first cause' or even 'first place' in a political sense. It is wise to keep all these in mind rather than be tied to one. After a brief excursus, Aristotle continues:

> There must be some natural substance, either one or many, which is preserved while other things come to be. On the number and form of this *archê* there is no agreement, but Thales, the originator of this sort of philosophy, said that it is water (and so declared the earth to be upon water).
>
> Taking his hypothesis from observing the nurture of all things to be moist, the warm itself coming to be from this and living by this (that from which they come to be being the *archê* of all things). He took the hypothesis from this and the fact that the seeds of all things have a moist nature, water being the natural *archê* of all things moist.[2]

Aristotle also says:

> Others say the earth rests on water. This is the oldest account we have, given they say by Thales of Miletus, that it remains in place by floating like a piece

> of wood or something similar (of these things none rest on air, but on water), as if the account concerning the earth did not apply to the water supporting the earth.[3]

The only further information we have on Thales' cosmology/ cosmogony comes from Heraclitus Homericus:

> The moist nature, since it is easily remodelled into each thing, is accustomed to undergoing many varied changes. That which is exhaled becomes air, and the finest part of this is kindled into aether, and when water collapses it changes into mud and land. Therefore of the four elements Thales declared water to be basic, and, as it were, the cause.[4]

This is a later source and the Greek has a good deal of Stoic phraseology, so it is difficult to tell how reliable this is in reporting Thales' actual opinions. The general scheme of transformation from water is plausible enough though.

II. Thales and soul

Thales also seems to believe that there is soul pervading the universe. Aristotle tells us that:

> Thales supposed the soul to be capable of generating motion, as he said that the magnet has soul because it moves iron.[5]

> Some believe soul pervades the whole *kosmos*, and perhaps this is the source of Thales' view that everything is full of gods.[6]

Aetius tells us that for Thales:

> The mind of the *kosmos* is God, the all is ensouled and is full of demons, extending to the elemental moisture, and there is a divine power which moves it.[7]

Cicero comments:

> Thales of Miletus, who was the first to investigate these questions, said that water was the principal substance, god being the mind which fashioned everything out of water.[8]

What should we make of this evidence? Clearly Thales requires some sort of explanation of self-motion, and so supposes soul to be the source of this. It may well be that Thales' view was that soul pervaded the universe, and this was later corrupted, even by Aristotle's time, to the idea that everything is full of gods, or perhaps Thales meant that there was a unitary cosmic intelligence but expressed this slightly loosely as 'all things are full of gods'.[9] Soul here would not be something separate or independent from water, but would be an aspect or quality of water. The later doxography is suspect because of the tendency of some Stoic, Neoplatonist and Christian commentators to attribute a belief in god, and a god who organises the *kosmos*, wherever possible. Nevertheless, the passages here from Aetius and Cicero still have some value. Thales may well have believed in some form of cosmic soul or intelligence, which Aetius calls the mind of God, capable of giving things motion, even if it is highly likely that the demons are a latter accretion. The Cicero passage is particularly interesting, because one thing we lack with Thales is any idea of how a *kosmos* came about, other than it came from water. One possibility is that a cosmic intelligence inherent in water guided the processes which brought a *kosmos* into being out of primordial water. Cicero then interpreted that cosmic intelligence as a god. Here I disagree with KRS, who consider this passage to be 'recognizably fictitious'.[10] The attribution of a god who is the cosmic intelligence is perhaps so, but with so little information on Thales' cosmogony we cannot rule out the idea that a cosmic intelligence inherent in matter guided the formation of the *kosmos*. Later in this chapter I will argue that Anaximander and Anaximenes had similar ideas, with the unlimited and air respectively steering the formation of the *kosmos*.

I am not overly concerned here whether Thales or the other Milesians considered their principal substance to be divine or not. They clearly have in mind some form of pantheism/ panpsychism, and one where the principal substance behaves in an invariant and predictable manner.[11] This is enough to separate off the Milesians from mythical accounts of creation and theogony. Nor do I see it as problematic that Thales chooses water when there were many water-based creation tales before him. Thales produces a *kosmos* by means which are parsimonious, invariant, non-contradictory and natural.[12]

While we have little information on the processes of cosmogony in Thales, if the other Milesians are anything to go by here (and differences in this would certainly be commented on by Aristotle and later commentators),[13] then the processes of cosmogony are ongoing processes that we can see happening in the world around us. Unlike theogony and myth where there are singular incidents based on the arbitrary wills of superbeings, Thales' cosmogony needs nothing more than the ordinary cycle of the elements beginning from water.[14] Aristotle does not seem to believe that Thales borrowed from myth, and gives what he believes are Thales' reasons for choosing water.[15] O'Grady points out that quite apart from any mythological influences, water is a good choice for a theory of matter/ change in that it can readily be seen to evaporate to form air, to silt to form earth, is critical to all life-forms and might be thought to be critical to spontaneous generation as well.[16] If Thales did rationalise prior mythological belief, that in itself is no small achievement. Should we accept that Thales was interested in cosmogony, as opposed to an analysis of the world around him? The later commentators seem happy to attribute some form of cosmogony to him, though it might be argued that is due to misconstruing the 'out of water, into water' view as being to do with cosmogony rather than an analysis of what the world around us is. I would add a historical dimension to this, however. The Hippocratics had to fight to establish their view that all disease has a natural cause against the view that disease was a punishment from the gods, contesting this issue in the most difficult case, that of the 'sacred disease', epilepsy. So too the Milesians may have struggled to establish their views, and one key issue would be whether the *kosmos* had a natural origin, against the Egyptian, Babylonian and early Greek creation tales. Is it coincidence that Thales chooses water, or does that pick up earlier Babylonian-based creation tales and give a statement that cosmogony out of water can now be construed naturally?

III. Thales, vortices and multiple *kosmoi*

Does Thales believe in any sort of multiple *kosmoi*? The evidence of the doxographers is against this. Pseudo-Plutarch tells us bluntly that 'Thales and his followers believed in one *kosmos*.'[17] Stobaeus gives a long list of those who believed in one *kosmos* only, and Thales is at the head of this list.[18] Simplicius does not mention Thales in his classification of those believing in one *kosmos* or multiple *kosmoi*.[19] This is significant, not only for Thales but for Anaximander and Anaximenes as well. All three believe, in schematic form, that 'everything is generated out of x and is destroyed into x', whether x be water, the unlimited or air, something the doxographers were well aware of. This belief, on its own, was not sufficient for the doxographers to attribute a belief in successive *kosmoi*.

Did Thales believe there to be a vortex? West has argued that he did.[20] I disagree, and this brings us to a crucial divergence of interpretation which will recur through this chapter. West says that for Thales, 'A central fact about his lost cosmology can be recovered by simple inference.'[21] West then argues that Anaximander believed in vortices (so too Anaximenes), and that we can solve the problem of the genesis and current structure of the *kosmos* in Thales by attributing him a vortex as well.[22] Quite reasonably, West associates the formation of vortices with random motion. So he says that for Anaximander there is an eternal motion, and that:

> This motion probably consists of random currents running in no fixed direction. From time to time they result in a vortex which lasts for a while and then slows down and disappears.[23]

This fits well with the view that Anaximander believed in multiple *kosmoi*. In Thales' case, West points out that in many water-based creation stories, the first part of creation 'does not simply appear, as it were *ex nihilo*, but grows from some natural disturbance in the water'.[24] I do not wish to deny that this is a possible view of Thales, given the paucity of information we have on him.[25] The alternative is the panpsychism interpretation I have outlined above. These two alternatives will recur with both Anaximander and Anaximenes, and I will argue for panpsychism in each case. For Thales, we have no evidence for a vortex, for any form of random motion or for multiple *kosmoi*, co-existent or successive. We do have some small evidence for panpsychism. If Thales did adopt some form of panpsychism, and that psyche is able to steer the *kosmos* into existence, as I argue with Anaximander and Anaximenes, then Thales has no need of vortices, random motion or multiple *kosmoi*. With Anaximander and Anaximenes, we have no evidence for random motion, some evidence for multiple *kosmoi* and vortices and better evidence for panpsychism. This is not to say that I believe the panpsychism evidence to be indubitable, but on balance a more coherent case can be made for this evidence rather than for vortices in each Milesian.

IV. Anaximander: general principles

With Anaximander we have considerably more information, and considerably greater controversy about the nature of his cosmogony. Simplicius gives us Theophrastus' summary of Anaximander's views:

> Of those who say it is one, in motion and unlimited (*apeiron*), Anaximander, son of Praxiades of Miletus, was a follower and student of Thales. He said that the *archê* and element of existing things was the unlimited, being the first to give this name to the *archê*. He says this is not water, nor any of the other so-called elements, but some other unlimited nature, from which are generated all the heavens and the *kosmos* in them. The source of generation for extant things is that into which destruction occurs, **according to necessity. They pay penalty and retribution to each other for injustices according to the assessment of time**, as he says in a poetic fashion.[26]

The words in bold here are generally accepted as Anaximander's own.[27] I translate *apeiron* here as unlimited rather than infinite. This is philologically better:[28] it is doubtful that Anaximander had a conception of the infinite as opposed to the unlimited, and unlimited allows the supposition that the *apeiron* may be qualitatively unlimited as well as spatially/temporally unlimited.

V. Anaximander's steering principle

For Anaximander, change from the unlimited is a matter of separation. Simplicius tells us that:

> It is clear that Anaximander, observing the change of the four elements into each other, thought it proper to make no one of these the substratum, but

> something else besides these. He produces coming to be not through alteration of the element, but through the separating off of opposites through the eternal motion.[29]

Pseudo-Plutarch gives us something more specifically cosmogonical:

> He (Anaximander) says that which is productive out of the eternal hot and cold was separated at the genesis of this *kosmos*, and that a sphere of flame was formed around the air around the earth like the bark around a tree. When this was broken off and enclosed in certain circles, the sun, moon and stars were formed.[30]

Hippolytus also tells us that for Anaximander:

> The heavenly bodies are generated when a circle of fire is separated from the fire of the *kosmos*, and enveloped by air.[31]

So far we have seen the processes by which a part of the unlimited is converted into a *kosmos*. Why should this separation from the unlimited produce a *kosmos*, something well ordered, rather than any other configuration? Aristotle tells us something very interesting:

> The unlimited has no *archê* … However, this seems to be the *archê* of all other things, and it surrounds and steers (*kubernan*) all, as with all those who do not suppose other explanations, such as mind or love, beyond the unlimited. This is divine, for it is immortal and indestructible, as Anaximander and most of the physiologoi say.[32]

So it would seem that the unlimited has some power to steer. This is a translation of *kubernan*, which means simply to steer, as in to steer a boat, or more metaphorically, to guide or govern.[33] The latter may be important in suggesting some form of intelligent guidance. Whether Anaximander considered the unlimited to be divine is another matter. This may be an attribution by Aristotle, knowing that Anaximander considers the unlimited to have a steering function and to be immortal and indestructible. This passage would be enough for later commentators to consider the unlimited to be god, and given the last sentence, to attribute the same sort of view to Thales and Anaximenes. Furley suggests that the steering may occur once and once only, setting the *kosmos* on a course which it then follows.[34] Although that is a possibility, I see no evidence to support it. In the Aristotle passage, the unlimited presumably surrounds the *kosmos* at all time rather than just at an initial moment, but no distinction is drawn between surrounding and steering in this respect. Nor do any of the other passages concerning steering suggest that this is an intermittent rather than a persistent factor in the *kosmos*. Later we will see Plato's criticism of Anaxagoras for postulating a cosmic intelligence which withdraws after initially ordering the cosmos. Had any of the Milesians had a similar theory, we might expect to see similar criticism of it somewhere in the record. Plato does ask:

> Should we say that the whole universe is ruled by unreason, irregularity and chance, or on the contrary, as some of those who came before us said, say that *nous* and a marvellous organising intelligence steer (*diakubernan*) it.[35]

This is interesting not only for the attribution of a steering principle, but also in that Plato perceives a sharp bifurcation of explanation by 'unreason, irregularity and chance' and explanation in terms of a steering principle. Slightly later in the *Philebus*, he says, 'This supports those of old who believed that *nous* always rules the universe.'[36] This is significant in that it cannot be Anaxagoras whom Plato is referring to here. Plato is critical of Anaxagoras for allowing *nous* to withdraw from the *kosmos*, while these men of old have *nous* always ruling the *kosmos*.

Anaximander is not alone in supposing some form of steering principle. Heraclitus,[37] Parmenides,[38] the Hippocratics[39] and Diogenes of Apollonia[40] all make use of a steering function,[41] and below I will argue that Anaximenes does as well, in addition to Thales. It is unclear on the available evidence what the precise nature of the steering function is or what it steers for. A reasonable supposition though would be that it steers the separation process so that a *kosmos* rather than a chaos is formed out of the unlimited.

VI. Anaximander and multiple co-existent *kosmoi*

There has been considerable debate on whether Anaximander believed in one *kosmos* or many. There are three positions here, that he believed in co-existent multiple *kosmoi*, in successive multiple *kosmoi* or in a unique *kosmos*. Back in the nineteenth century, Zeller argued for successive *kosmoi*,[42] and at the beginning of the twentieth century Burnet argued for co-existent *kosmoi*.[43] Cornford demolished Burnet's version of co-existent *kosmoi*, and argued for successive *kosmoi*.[44] Kirk and Kahn have argued for a single *kosmos*.[45] More recently McKirahan has argued for co-existent *kosmoi* and Finkelberg for successive *kosmoi*.[46]

There is one argument against multiple *kosmoi* in Anaximander which we can dispose of very quickly. Cornford argues that 'There is nothing in nature to suggest innumerable worlds'[47] and Kirk, following on from Cornford, argues that there is 'nothing whatever in "the appearance of nature" to suggest *successive* worlds'.[48] However, that is going to be true for Leucippus and Democritus, and for Empedocles as well, and they certainly did believe in innumerable co-existent worlds and successive worlds respectively. In any case multiple *kosmoi* was not an empirical thesis, on the evidence we have, but rather a supposition aimed at explaining why our *kosmos* has the order it has, if we do not suppose any steering or design principle.

Kirk also says that 'The idea of different worlds in time would be, surely, an absolutely extraordinary one for an early Ionian thinker.'[49] However, there are at least two senses in which an idea may be extraordinary. If what is meant is that this idea is strange, or revolutionary, or brilliant, then the simple reply is that Anaximander had several strange, revolutionary or brilliant ideas so we can hardly debar multiple *kosmoi* on these grounds. If what is meant is that this idea does not fit with Anaximander's pattern of thinking, then the argument is sounder but needs to be fleshed out. The standard way to do this is to contrast him with the thinking of the early atomists. The chief considerations here are that Anaximander, unlike the atomists, does not employ an ontology of what is and what is not, and does not have the requisite conceptions of the void or the infinite. One might also add comparisons with Empedocles here, in that Anaximander does not have a cycle of love bringing everything together into a one and strife dispersing it again into a many. Of course, Anaximander may believe in multiple *kosmoi* for reasons other than those of Empedocles or the early atomists, but there is no evidence that he has any such views.

I would like to add four further considerations here. Firstly, Anaximander believes that the unlimited has the ability 'to steer' and at no point resorts to explaining phenomena as occurring by chance. Where Leucippus and Democritus, and Epicurus and his followers need to postulate many *kosmoi*, such that it is then plausible that the workings of chance and necessity alone will produce something as well ordered as our *kosmos*, Anaximander has no such need. The unlimited can steer the differentiated parts of itself (hot, cold etc.) into a single *kosmos*.[50] Secondly, Leucippus and Democritus, against the challenge of Parmenides, reject the principle of sufficient reason in certain aspects of their cosmogony. Where Parmenides asks why here, and why now, the early atomists deny that there are preferred times and preferred places, and employ a form of argument known as *ou mallon*. These arguments deny a preference for the alternatives, as in 'not x rather than y', or 'no more x than y'.[51] There is no indication in Anaximander that he ever abandons the principle of sufficient reason or employs *ou mallon* arguments. Thirdly, where we find a clear belief in multiple co-existent *kosmoi* in antiquity, unsurprisingly we find clear boundary conditions and individuation criteria for those *kosmoi*. So with Leucippus and Democritus, and with Epicurus and Lucretius, we find that there is a physical boundary for each *kosmos*, and they are keen to assert what makes a *kosmos* one *kosmos* and differentiates it from the rest of the universe, and allows us to count one *kosmos*, two *kosmoi* ... many *kosmoi*. This is absolutely fundamental to a co-existent *kosmoi* view, but we find nothing of this sort in Anaximander. Fourthly, there is Anaximander's account of the stability of the earth. Aristotle tells us that:

> There are some, such as Anaximander among the ancients, who say that (the earth) rests on account of its likeness. It is fitting that what is established in

> the centre and has equal relations to the extremes should not move up, down or to the side. It is not possible for it to move in opposite directions at the same time, and so necessarily it remains still.[52]

So too in the *Phaedo* Plato has Socrates say:

> I am now persuaded that, firstly, if the earth is round and in the centre of the heavens, then it requires neither air to prevent it falling nor any other necessitation of this sort, but the uniformity of the heaven itself in every way and the equipoise of the earth itself is sufficient to restrain it. For something which is equipoised and is placed in the middle of something homogenous cannot yield to being moved aside in any way, but will remain steadfast.[53]

This account of the earth's stability can only work if the surrounding heavens are indeed uniform.[54] There is the possibility that all the other *kosmoi* are equally spaced such that the earth has no preferred direction to move in, though I think it highly unlikely that Anaximander thought this, and the language of both Aristotle and Plato here is against it. These concerns would explain Aetius' comment that:

> Of those who claimed the *kosmoi* to be unlimited, Anaximander held that they were equally far away from each another.[55]

However, if these other *kosmoi* are coming into and out of existence, as in the early atomist account, unless pairs of them do this in a synchronised manner the balance for the earth will be upset. The simplest solution here by far is that Anaximander believed in one *kosmos* in the middle of the homogenous *apeiron*. So Aetius' comment looks like a rather desperate attempt to save a multiple co-existent *kosmoi* view of Anaximander. McKirahan argues that:

> Anaximander's tendency to see events as instances of general patterns that occur in other places and at other times and in different contexts might well have led him to hold that the process that generated our world is not unique and in consequence unintelligible. [56]

I agree with McKirahan that Anaximander (and indeed Thales and Anaximenes) tend to see events as instances of general patterns, and this, as part of a break from mythology and theology, is much to their credit.[57] I would put the consequences in a slightly different way though. The Milesians see cosmogony as a process by which the *kosmos* is formed over time, rather than an instantaneous (or short-lived) event. The processes by which the *kosmos* is formed are ongoing, and are a permanent part of nature. They see no need to invoke anything other than the processes they see in action around them in order to explain the origins of the *kosmos*. Again, against a background of mythology and theogony, this is greatly to the Milesians' credit. Is the cosmogony of a unique *kosmos* unintelligible for them? Not at all. As cosmogony is now a natural occurrence, and in a sense an ongoing process, it can be explained as part of the general pattern of the action of current physical processes. If Thales and Anaximenes too see events as instances of general patterns, McKirahan's argument would generate multiple co-existent *kosmoi* for them as well. McKirahan also argues that 'If the *apeiron* original substance generated a world in one place, it would generate worlds elsewhere too – without limit.'[58] There are several possible replies here. One would be slightly deflationary, that these sorts of consideration only come to the forefront of Greek thought after Parmenides. But that does not deny that they might still exist prior to Parmenides. More significant though is the reply that if *kosmos* might form elsewhere, why not everywhere else? Literally everywhere else would be absurd (i.e. at every point in space), but why would we get an appropriate spacing of multiple *kosmoi*? There is also a suppressed premise here that all places are equal. If so, how is there differentiation between them? The atomists achieve this by denying the principle of sufficient reason and invoking chance, but there is no evidence of that sort of thinking in Anaximander, especially if we construe his account of the earth's stability as a sufficient reason argument. This leads into a further consideration, which is that if Anaximander accounts for the stability of the earth by placing it in the middle, 'established in the centre and having equal relations to the extremes', as Aristotle puts it, then all places are not equal. The idea which seems to be problematic here is that if the *apeiron* is spatially infinite, then it has no centre. It is not clear that Anaximander considered his *apeiron* to be spatially infinite rather than simply unlimited. In either case, it is not straightforward to arrive at the conclusion that there is no centre to something unlimited or infinite.[59] If this seems harsh on Anaximander, the notion that an unlimited universe has a centre is a common and intuitive one. I teach history of astronomy and cosmology to both advanced undergraduates and postgraduates, and the class is made up those studying physics, astronomy, history of science and philosophy of science. A very common misconception among these students is that an infinite universe would be spherical (same in all directions!), with infinite but equal distances from the centre to each point on the periphery of that universe.

I will offer one further argument against multiple *kosmoi* in Anaximander, co-existent or successive. The Milesians appear to appreciate the value of philosophical parsimony. They make a small number of principles work hard to cover all of the phenomena, and they do not invoke further principles. What then would they gain through invoking multiple *kosmoi*? This is not an empirical thesis as no one can (or even claims to be able to) observe further *kosmoi* nor can they experience previous or future versions of successive *kosmoi*. What might the supposition of multiple *kosmoi* explain? In other hands, it is used to explain how a beneficial *kosmos* came about by chance or is used as a reply to Eleatic concerns.[60] None of this worries Anaximander. His unique *kosmos* comes together guided by the steering principle, he has a centre to his universe where the *kosmos* can exist and the puzzles of Parmenides are in the future. Multiple co-existent *kosmoi* are simply otiose for Anaximander.

VII. Anaximander and multiple successive *kosmoi*

There is also the view that Anaximander believed in single, successive *kosmoi*, the next one not coming into existence until the present one has perished. Kirk's argument that the material of the world is divine is well answered by Finkelberg, who points out that no one disputes that the divine material is eternal; the question is whether the *kosmos*, which has its birth out of the divine material, could be destroyed back into the divine material.[61] However, to establish the successive *kosmoi* view, it is not sufficient simply to argue that Anaximander is committed to 'everything is generated out of x and will be destroyed into x'. There are two reasons for this. Firstly, Thales is committed to an 'out of x ... into x' formulation, but there is no suggestion that he is committed to successive *kosmoi*. Secondly, it is not clear what is generated and what is destroyed, and in what sequence. Once the opposites have been generated, it is quite possible that some instances of those opposites are destroyed without, at any stage, all of those instances being destroyed. This would allow the *kosmos* to continue to exist even if every instance of the opposites which constitute it will ultimately be destroyed. It is important to recognise here that at no point does Anaximander say that the *kosmos* is destroyed back into the unlimited.

Where we find clear cut cases of successive *kosmoi*, as with Empedocles and the Stoics, we also find strong criteria for differentiating between successive cycles. In Empedocles there are the states of total association and total dissociation of the elements, in the Stoics there is the process of *ekpurôsis*. We find nothing of this sort in Anaximander.

There is some weak doxographical evidence that Anaximander believed in successive *kosmoi*, but before we come to that I want to ask why the *kosmos* in Anaximander should go into decline. It has been argued that this would be an apt symmetry of coming into being and passing out of being.[62] I disagree because the generation of the *kosmos* for Anaximander is something just. Here I reject the old idea that the elements 'pay penalty and retribution' to the unlimited, which could be construed as supporting the idea of a *kosmos* which goes out of existence. In line with most modern commentators, I see them as paying penalty and retribution to each other. So the generation of something with good order can be just, without any debt to repay. That Anaximander envisaged something with good order is beyond doubt, if we consider the centrality and stability of the earth, its suitability for life and the precise arrangement of the heavenly bodies. The destruction of this order would be unjust, and I cannot see any intelligent

Greek considering justice and injustice to be symmetrical, at least not in the sense that it is proper or natural for injustice to follow justice. What the symmetry consideration masks is that there is no reason, either ethical of physical, for Anaximander's *kosmos* to go into decline. That is especially so if the *apeiron* is steering the *kosmos* into being, and continues to steer. There is no indication that the *apeiron* ceases in this activity once the *kosmos* is formed, though of course it may be modified.

Most commentators envisage the reparation in the extant fragment occurring over time, perhaps in a cycle, like the passage of day and night or of the seasons. So in summer there would be an excess of heat, which would later give way to an excess of cold. There is an alternative here. If we separate out some hot from the *apeiron*, this either leaves the *apeiron* colder or also generates some cold, but in a different place. If the *apeiron* were colder as a whole, it would no longer be unqualified, so my view is that when some hot is generated here, some cold is generated there.[63] While there can be local changes in the amount of hot or cold, globally the balance between hot and cold does not alter. So for the examples above, while it is night here it is day somewhere else, and while it is summer here it is winter somewhere else. As to the destruction of opposites, this reintegration is also straightforward. One characteristic is paired with an equal amount of its opposite so that on reintegration the unlimited will remain uncharacterised. One consequence of this is that if the *kosmos* were to become hot and dry it could not be reintegrated into the unlimited, lacking the opposites to make a balanced reintegration. This interpretation has the advantage that the global balance between the contraries is kept of necessity, as in the fragment.[64] One consequence of this interpretation is that the *kosmos* could not become entirely dominated by one component, such as the dry. There might be such a separation of the dry and the wet that it could no longer be called a *kosmos*, but that is a different matter and no one is suggesting that for Anaximander.

Engemann has suggested that we might take the relation of fiery 'bark' to moist tree in the Pseudo-Plutarch passage as indicative of the relative proportions of fire and moisture at the inception of the *kosmos*, the earth and *kosmos* then becoming progressively drier.[65] I doubt that Anaximander meant any such proportion, especially in such a loose analogy, and we would also have to take into account the intensity of the 'hot' here. I find it implausible that Anaximander's *kosmos* would contain an imbalance of opposites in totality, however they are distributed within the *kosmos*. I would reiterate that an excess of hot over cold or of dry over moist would leave the *apeiron* cold or moist, even if such a thing could be done, as it may be contrary to the necessary production of both opposites together. I see no evidence in any other passage to suggest that the opposites are not separated off in a balanced manner.[66]

VIII. Anaximander and the drying-out theory

One reason that Anaximander is supposed to believe in successive *kosmoi* is that he is sometimes believed to hold the view that the earth is gradually drying up. If water is seen as a principle of life, then drying up might well be seen as associated with dying, and of course Anaximander has life originating from water.[67] The drying-up view originates from a passage in Aristotle, where he says that:

> At first the whole area surrounding the earth is moist, but due to the drying of the sun, that which is evaporated becomes winds and the turnings of the sun and moon they say, while that which remains is the sea. Thus they think it is becoming less through being dried up, and will at some point end up being entirely dry.[68]

It is significant that Aristotle does not name Anaximander here, or suggest anywhere else that Anaximander holds this theory, and that there is nothing else in Anaximander to support such an attribution.[69] However, in his commentary on this passage Alexander tells us: 'Of this view were Anaximander and Diogenes, according to the history of Theophrastus.'[70]

It has been suggested that the drying up of the earth is part of a cycle, perhaps part of a great summer to be followed by a great winter.[71] I do not think that will quite do, as it would seem unlikely that the earth would revert to being surrounded entirely by moisture. The drying out of the earth need not be a completely reversible process. The first part of this may be cosmogonic, in forming the fully fledged earth and its systems. Some water may be evaporated so that winds and the turnings of the sun and moon are produced, and to 'prime' the weather system, but when more water is produced it will condense, fall as rain and replenish the sea. Aetius tells us that for Anaximander 'Rain comes from the vapours which are raised from things below the sun.'[72] Anaximander may not have made any clear statement about the future of the *kosmos*, simply because the total drying up of the earth and its implications for successive *kosmoi* may not have been an issue for him, if the cosmogonic drying out was supposed to result in an equilibrium of evaporation and rainfall. Theophrastus may have interpreted the cosmogonic drying out as a terminal process when no such process was intended.

The other piece of doxographic evidence in favour of successive *kosmoi* in Anaximander is in Simplicius, who tells us that:

> Those who hypothesised worlds unlimited in their number, as with Anaximander, Leucippus, Democritus and later Epicurus, supposed them to be generated and destroyed indefinitely, always with some being generated and others being destroyed, and they said that motion was eternal.[73]

This is echoed by Aetius, who says:

> Anaximander said that the unlimited is the *archê* of existing things. From this all things are generated and into this all things are destroyed. This is why unlimited *kosmoi* are generated and are in turn destroyed back into what they were generated from. At any rate he says why it is unlimited, in order that generation should not fail.[74]

As is widely recognised, however, there was an assimilation of Anaximander to the atomists in the later doxographic tradition, clearly evidenced here by the quotation from Simplicius. Thus it is hard to tell whether Anaximander actually held this view, or it is being attributed to him owing to a supposed similarity between his *apeiron* and the unlimited space and matter of the atomists.

I reject then both pieces of doxographic evidence that Anaximander envisaged an end to the *kosmos*. Theophrastus misinterprets Anaximander, while the later doxographic tradition has assimilated Anaximander too closely to the atomists to give any reliable evidence. Many commentators are puzzled by the apparent discord between the extant fragment and the supposition that the *kosmos* will dry up and be reabsorbed by the apeiron.[75] There is no need. Anaximander simply does not believe in a *kosmos* which is terminally drying out.

IX. Anaximander and the doxography

There is a considerable amount of doxographical evidence in favour of unlimited *kosmoi* in Anaximander.[76] Why is the doxographical tradition in error? Let us first be clear that such a doxographical tradition can be wrong. The orthodox modern view of Heraclitus is that he supposed there to be one *kosmos* only, and not successive *kosmoi*. However, Aristotle, Theophrastus and the following doxographical tradition all attribute successive *kosmoi*, and in the later doxographers Heraclitus is given the Stoic *ekpurôsis* theory. It is also worth noting that while Stobaeus gives us a list of Anaximander, Anaximenes, Xenophanes, Diogenes, Leucippus, Democritus and Epicurus as those holding that there are 'unlimited *kosmoi* in the unlimited',[77] Pseudo-Plutarch's version of this passage mentions only Democritus and Epicurus.[78] Stobaeus places Empedocles with Plato and Aristotle as believing in one *kosmos*. Pseudo-Plutarch says that Thales and his followers (and it is highly likely that he includes Anaximander here) believe in one *kosmos* only. Those who argue for successive *kosmoi* in Anaximander argue that while Theophrastus attributes only successive *kosmoi*, the later doxographic tradition attribute co-existent *kosmoi*. They argue, in my view correctly, that the later tradition tends to assimilate Anaximander to the atomists and incorrectly associate him with co-existent *kosmoi*.

In my view, Aristotle does not attribute multiple *kosmoi* to Anaximander, but there is a passage which could be construed as doing so. Aristotle also makes some more general philosophical pronouncements which would allow one to make a case that Anaximander believed in multiple *kosmoi*. Theophrastus attributes successive *kosmoi* to Anaximander, but does so in a way that could be construed as supporting co-existent *kosmoi*. It is possible that he does this under the influence of these passages in Aristotle, and the later tradition then moves quite definitely to attributing co-existent *kosmoi*. The passage in Aristotle which might be construed as attributing multiple *kosmoi* to Anaximander is:

> Some hypothesise one alone, water, air, fire or something rarer than water but denser than air, and this they say is unlimited and surrounds all of the heavens.[79]

One possible reading of this is that this attributes a multiple *kosmoi* view to Anaximander. Certainly some ancient commentators took it in this way, though modern scholarship is against them.[80] The key consideration is that at *Physics* 187a12ff. Aristotle distinguishes between those who posit a definite primary substance (one of the usual three or an intermediate between fire and air) and produce all else by rarefaction and condensation, and Anaximander (who is specifically named) who produces opposites by separation from his one.

Even if we do take this as referring to Anaximander, there are several other possible readings here. Aristotle distinguishes three senses of *ouranos* (heaven) in *On the Heavens*, and it is possible here that he does not intend to attribute multiple heavens, only multiple rings.[81] It is also possible that Aristotle is quoting or paraphrasing Anaximander, who did refer to the heavens in the plural, but was referring to the rings of his own system (each one being a 'heaven') rather than to multiple heavens. One might also paraphrase Aristotle here as saying 'This they say is unlimited and surrounds the heavens of Thales, Anaximenes, Heraclitus and Anaximander.' Admittedly, if this is what Aristotle meant he has been far from clear, but this would be an odd way to attribute both a strange matter theory and multiple *kosmoi* to Anaximander.

I am suspicious of this passage attributing multiple *kosmoi* to Anaximander for several further reasons. Firstly, this would be the only place that Aristotle attributes multiple *kosmoi* to Anaximander. Aristotle is very keen to attack this idea in other thinkers, and mentions their views in several contexts. Secondly, Aristotle is one of our sources for the equilibrium theory of the stability of the earth in Anaximander. He is of course critical of this theory, but not on the grounds that what surrounds the earth is not uniform (as one would expect with multiple co-existent *kosmoi*) or that something that is undergoing the later stages of terminal decay could hardly be expected to be stable on these grounds. It is notable that Simplicius, who attributes co-existent unlimited *kosmoi* to Anaximander, also attributes to him the theory that the earth is supported by air.[82] Thirdly, we need to consider this passage in relation to some others in Aristotle:

> Everyone believes it to have been generated, but some believe it to be eternal once generated, others believe that it is destructible like any other formation in nature, while others believe that it alternates, at one time being as it now is, and at another time changing and going to ruin, such as Empedocles of Acragas and Heraclitus of Ephesus.[83]

Anaximander is here conspicuous by his absence, and Aristotle recognises that there are those before him who believed that the *kosmos*, once generated, was eternal. Aristotle also says:

> They say that nothing is ungenerated, and that everything is generated. Once generated, some things remain deathless while others again perish. Hesiod and his followers take this position, as do the first of the natural philosophers.[84]

This is very significant, as one must wonder who the 'first of the natural philosophers' might be. The most natural assumption would be the Milesians, especially as Aristotle uses a similar phrase at *Metaphysics* I/3, 983b6 to refer very clearly to the Milesians. I doubt then that Aristotle attributed multiple *kosmoi* to Anaximander. That Hesiod believed the *kosmos* to have a beginning but to be eternal is important in showing that this idea is quite possible just prior to the Milesians. One might also note that it is quite common around the time of Hesiod for the gods to be thought to have births but to be immortal. This provides a perfectly coherent alternative to the notion that the *kosmos* in the Milesians must die, having been born. It is also interesting that we have no information whatever on whether successive *kosmoi* for the Milesians would be similar or different. This was a major debate in later antiquity, after the Stoics formulated their theory that successive *kosmoi* would be identical, in contrast to the different *kosmoi* of either groups of atomists or Empedocles.[85] Aristotle had already asked some of the key questions though, wondering if a god was in charge of cosmogony, why the succeeding *kosmos* would be better, worse or the same than the preceding one, and rejecting answers to all three possibilities and so the notion of successive *kosmoi* as well.[86] One can reasonably ask these questions of the Milesians. If the steering principle produces a good *kosmos*, why would it allow it to die, given that the new *kosmos* would not be any better? The Stoics believe in an inevitable decline of the *kosmos*, so the regeneration of the best *kosmos* is a reasonable option. I do not see any sense of an inevitable decline in the *kosmoi* of the Milesians though, or any reason why the steering principle would or should allow the *kosmos* to become disordered.

X. Anaximander and unlimited generation

According to Aristotle:

> Only in this way will generation and destruction not come to an end, if that which generation is produced from is unlimited.[87]

> If what exists outside of the heavens is unlimited, then so is body, and so it would seem are *kosmoi*. For why here in the void rather than there?[88]

Reading these quotations, one might move to the conclusion that unlimited space and unlimited body entail an unlimited number of *kosmoi*, and that as generation and destruction need not come to an end with an unlimited supply, there will be an unlimited number of successive *kosmoi*. Certainly a good deal of the doxography will first state that Anaximander believed in an *apeiron*, then say that unlimited *kosmoi* were generated from this. Simplicius says:

> Anaximander was the first to suppose an unlimited, in order that it could be used abundantly in generation. It seems he hypothesised unlimited *kosmoi*, each of these *kosmoi* coming from the unlimited element.[89]

There are several reasons why Theophrastus may have gone astray in attributing successive *kosmoi* to Anaximander. He may have interpreted the drying-up process as ongoing, leading to the end of the *kosmos*. It is significant that Theophrastus also gives Heraclitus a theory of successive *kosmoi*, in this case with the *kosmos* becoming more fiery. Theophrastus may have thought that Aristotle attributed multiple *kosmoi* to Anaximander, either directly or as a result of some more general philosophical considerations.

A further question here is whether Aristotle has Anaximander in mind when he says that some posited an unlimited in order that generation should not come to an end. Aristotle has the right answer to this in mind slightly later, when he says:

> In order that genesis does not cease, it is not necessary for there to be an unlimited perceptible body. It is quite possible that the destruction of one thing is the genesis of another in a limited universe.[90]

In relation to this argument, there is an important difference between a single *kosmos* and an unlimited number of *kosmoi*. Aristotle is of course correct that for one, finite *kosmos* there is no problem with continual generation, because the material can be recycled. However, one might well worry that genesis will give out if there is a finite amount of matter but an unlimited number of co-existent *kosmoi*. I doubt that Aristotle attributes unlimited co-existent *kosmoi* to Anaximander, and I also doubt that he refers to Anaximander in the passage about supposing unlimited matter so that genesis does not give out. One can see why someone who thought that Anaximander believed in unlimited co-existent *kosmoi* might then suppose that the Aristotle passage referred to Anaximander and that Anaximander supposed an unlimited in order to prevent genesis giving out.

Finkelberg has argued that if one is to disagree with Theophrastus, then one has to show how Theophrastus must have gone wrong, not just how he might have gone wrong.[91] I disagree. Where we have good theoretical grounds for disagreeing with Theophrastus (the nature of Anaximander's thought entails that he has no reason for successive *kosmoi* and has several reasons against), and where we have a parallel for Theophrastus going astray (his treatment of Heraclitus), then it is sufficient to indicate reasonable possibilities for where Theophrastus' error may have sprung from.

XI. Anaximander and vortices

The evidence for a vortex in Anaximander is supposed to be found in this passage from Aristotle:

> If the earth now remains in place by force, it was also assembled in the centre by being carried there because of a vortex. For this is the reason which everyone gives, through what occurs in water and around air. In these the larger and heavier things are always carried to the centre of the vortex. So all those who say that the heavens are generated have the earth being assembled in the middle.[92]

Aristotle might well exempt Anaximander from this, because he recognises that Anaximander does not use force but equilibrium to explain why the earth does not move.[93] I also suspect that Aristotle is talking rather loosely here, and we can put no great weight on this evidence. Some Presocratic thinkers (Anaxagoras, Leucippus and Democritus) did indeed believe that the earth is held in place by force and is assembled there by the action of a vortex. But is this so for all those who believed the heavens to be generated? Surely this would implicate Thales, as someone who believed that the heavens were generated? Aristotle of course recognises that for Thales the earth floats on water.[94] For Aristotle, floating on water does involve forces, so Thales cannot be exempted in the manner we might with Anaximenes.[95] Yet we have no evidence whatsoever that Thales believed in a cosmic vortex.

More generally, Anaximander has no need of a vortex. He has an equilibrium theory for the earth and so no need of a vortex to hold it in place. If the unlimited steers all things there is no need of a vortex to initiate the cosmogonical process, and the instigation of the vortex itself would need some explanation. Others who employ a vortex do give an explanation (chance, according to Leucippus and Democritus, mind according to Anaxagoras) but we have nothing of this sort in Anaximander.

Vlastos has argued that *to gonimon*, 'the productive', of Pseudo-Plutarch *Stromateis* 2, may be a process rather than a thing, and that if it were a vortex it would enable the separation of opposites.[96] That is a possibility, but one possibility among many even granted we consider *to gonimon* a process. What is odd is that none of the later doxographers attribute a vortex to Anaximander. This is a well known idea with a well defined term, *dinê*. If Anaximander used a vortex, with the doxographers so keen to assimilate his views to the atomists, it is very surprising there is no mention of it. If *to gonimon* was a vortex, it is surprising that it is not mentioned as a vortex or the better known term *dinê* used instead.

Let us assume that the heavens in Anaximander somehow move in a circular manner. West's argument then is that 'such a system patently resembles a vortex'.[97] That the heavens move in a circular manner does not entail the existence of a vortex, however. Plato and Aristotle are two obvious contrary cases. If the unlimited is divine in Anaximander, then perhaps we could work up some association between the divine and regular circular motion to argue that the primordial motion in Anaximander is circular. Again, though, circular motion does not entail a vortex. We have no evidence that Anaximander associated circular motion with the unlimited, and Aristotle's complaint is that Anaximander does not specify a natural motion. The evidence in favour of Anaximander using a vortex is small and highly dubious; he had no need of a vortex and the doxographers make no mention of it. The simple conclusion here is that Anaximander never used a vortex. This in turn undermines West's attribution of a vortex to Thales.

XII. Anaximander's zoogony

Anaximander is the first, as far as we know, to give a zoogony. I have argued elsewhere that the proper translation of the key passage should be:

> Anaximander said that the first animals were generated in moisture and enclosing themselves in spine like barks, as they advanced in age they moved onto the drier and shedding their bark for a short time they survived in a different form.[98]

I have also argued that what Anaximander does here is give a description of the life-cycle of the Caddis fly. These flies are born in water, and as larvae they form a case around themselves of twigs, bark, leaves, etc. They attach themselves to the underside of a stone and undergo metamorphosis. They then swim to the surface and fly away, living for a much shorter time as flies than they did in the water. If so, this piece of zoogony is not myth or outright speculation, but is based, however loosely, on observational knowledge.

What Anaximander gets out of this, whatever the first animals may be, is that something generated in water can become an animal that lives on the ground. It is generally the case that the views of ancient thinkers on cosmogony are tightly related to their views on zoogony, and where appropriate stoichogony as well. I would point out here that Anaximander makes no use of chance in either his cosmogony or his zoogony, nor is there a sense that cosmogony and zoogony happen, and happen differently in many places.

It is common for commentators to draw tight parallels between Anaximander's cosmogony and his zoogony.[99] I do not wish to deny that early cosmogonies were often modelled on biological processes. It is a perennial problem in cosmogony to be able to say what cosmogony was *like*. With limited options available, it was perfectly reasonable for early thinkers to think of the origins of a complex, self-sustaining, self-moving *kosmos* in biological terms. Nor do I wish to deny that Anaximander was influenced by biological processes in his cosmogony. What I would question though is the extent to which he envisaged the processes of zoogony and cosmogony as being identical. KRS compare the Aetius passage on zoogony with the Pseudo-Plutarch passage on cosmogony (beginning of the Anaximander discussion) and say: 'The use of *phloios* here reminds one of the bark-simile in the cosmogonical account; both ball of flame and prickly shell broke away from round the core (here *peri* not *aporrêgnusthia*).'[100] This is why KRS have the animal barks being expanded and then split by heating, while I have the animals breaking out for themselves.[101] As the texts stand, we do have different verbs for what happens to the *kosmos* and to the first animals. While the first living creatures create their own cases, the earth has its sphere of flame formed for it. Perhaps we should give Anaximander some greater credit here then in differentiating between what passively happens to the *kosmos* and what living things are capable of doing for themselves.

There is also no need to attempt to tie Anaximander's zoogony with his theory that the earth is becoming drier.[102] The first life-forms in their cases are not forced to migrate onto the drier by changing climatic conditions, but do so naturally as part of their life-cycle. Moving to dry land is not then an adaptation to a changing environment, as Burnet and Kahn have suggested.[103] Nor need we take it that the first life-forms were amphibians.[104] A further consideration here is that it is the sea that is said to be drying up, not the freshwater sources important for Caddis flies.

One argument for successive *kosmoi* is that if the *kosmos* has a biological 'birth', then so too it should have a biological death. One argument against this is that it is quite typical in early theogony to find that the gods have biological births but are immortal, so there are at least some things which are born but do not die (at least of natural causes). The second counter-argument is that while Anaximander undoubtedly is influenced by biological analogues in his cosmogony, the *kosmos* is somewhat different in its origins to life-forms, and so may have a different fate.

It is also highly important that the origins of man are conceived of as natural as well. So we find that according to Anaximander:

> Life-forms are generated from moisture by evaporation from the sun. In the beginning, man was similar to another life-form, namely fish.[105]

> In the beginning, he says, man was generated from other life-forms. Others are swiftly able to sustain themselves, man alone being in need of nursing for a long time. Because of this, he would not have long survived if his had been the original form.[106]

I would emphasise again there is no sense of anything here happening by chance or that the events might happen otherwise elsewhere.

XIII. Anaximenes: general principles

Anaximenes believed the *archê* to be air, and unlimited. Simplicius gives us Theophrastus' summary of Anaximenes' views:

> Anaximenes son of Eurystratus, of Miletus, a companion of Anaximander, says like him that the underlying nature is one and unlimited, though unlike him it is not undefined, as he says it is air. It differs in its nature by being more rare or more dense. Becoming rarer, fire is generated, becoming denser wind, then cloud and denser still water, then earth, then stones, other things being generated from these. He makes motion eternal, and change is generated in this way.[107]

Change, and the cosmogonical process, is a matter of condensation and rarefaction of air. Pseudo-Plutarch tells us that for Anaximenes:

> All things are generated by a certain condensation of air, and again by its rarefaction. Motion has existed for all time. He says that when the air felts, firstly the earth is generated, entirely flat, and because of this, it rides on the air. So too sun, moon and the other stars have their origins in generation from earth. At any rate he declares the sun to be earth, as it acquires abundant heat by moving rapidly.[108]

Hippolytus says that for Anaximenes:

> The heavenly bodies were generated from earth through the moisture rising from it. When this is rarefied, fire is generated, and from fire raised to a height the heavenly bodies are constituted.[109]

XIV. Anaximenes and the steering principle

Does Anaximenes have some form of steering principle, as we saw with Anaximander? I believe he does, and the key passage is from Aetius:

> Anaximenes declared air to be the *archê* of existing things. From it all things come to be and into it all things are dissolved. He says **as our soul, being air, holds us in order (*sungkratei*), so wind and air envelop the whole *kosmos*.**[110]

We do not know whether the part in bold here is a direct quotation, a paraphrase of something Anaximenes wrote, or whether it continues the first sentence in reporting what the author believed were Anaximenes' views. Concerns here are that the passage is not in the Ionic dialect and that the 'He says' formula in these situations does not guarantee a direct quotation. One might worry about the use of the words *kosmos* and *pneuma* (for air) so early, though there are parallels for both at this stage among medical writers and the Pythagoreans.[111] As KRS argue,[112] *sungkratei* is not found in our sources until much later. *Sungkratein*, they say, 'is really a compendium for *sunechein kai kratein*',[113] and that could only have come about in the much later Koine Greek. It is unlikely then that this is an unadulterated direct quotation. However, it may be close to a quotation and may well preserve the essential meaning of what Anaximenes had to say. There are two possibilities. First, that this is virtually a direct quotation, but it is rendered or paraphrased into a later dialect and where Anaximenes wrote *sunechein kai kratein*, Aetius writes *sungkratein*. Alternatively, that this may be Aetius' report on what he thought Anaximenes meant, and *sungkratein* expresses what the commentator thought he meant. Attested uses of *sungkratein* indicate that it means to keep troops together or to hold in, keep under control.[114]

To make the case that this is a steering principle, let us begin by returning to a passage generally discussed only in relation to Anaximander, which in fact has broader implications for the Presocratics. Aristotle tells us that:

> The unlimited has no *archê* ... However, this seems to be the *archê* of all other things, and it embraces everything and steers all, as those say who do not suppose other explanations, such as mind or love, beyond the unlimited. This is divine, for it is immortal and indestructible, as Anaximander and most of the physiologoi say.[115]

How much of this should we attribute to 'most of the physiologoi' as well as Anaximander and should we include Anaximenes amongst 'most of the physiologoi' here? Anaximenes must be a good candidate, as he does not suppose explanations such as mind (Anaxagoras) or love (Empedocles) beyond the unlimited. He does suppose air to be unlimited, as we have seen.[116] Aetius bluntly tells us that for Anaximenes air is God.[117] One might try to argue that 'embraces and steers all' applies to Anaximander alone though it is hard to see on what grounds. Those who do not suppose other explanations are definitely plural, and the discussion both before and after this passage is about the unlimited in general, and not about Anaximander's unlimited in particular. Aristotle may pick out Anaximander either as the first to hold a steering principle, or perhaps as someone who gave the clearest or best known statement of it. Certainly Diogenes of Apollonia, a follower of Anaximenes, holds that air steers:[118]

> That which has intelligence is called air by men, and all men are steered (*kubernasthia*) by this and it has a power (*kratein*) over all things. This seems to be a God to me and to have permeated everywhere, to arrange (*diatithenai*) all things and to be in all things.[119]

This is an interesting passage for several reasons. Firstly, we have the direct statement that air steers, and also that air is intelligent. Secondly, air has power over all things. As the verb here is *kratein*, if Anaximenes had written *kratein* in similar circumstances one can see how a later summariser might have written *sungkratein*.[120] Thirdly, *diatithenai* means to arrange in its proper place, or to manage well. That is significant because it links the steering principle to the cosmogonical process of arranging everything in the first place, and continuing to have it well arranged. If that is the case, then there is no need for the current *kosmos* to deteriorate or to perish.

KRS canvass several possibilities for what Anaximenes wrote and meant in relation to the Aetius report.[121] For their version of the proposal argued for here, they are concerned with a lack of parallels for the soul holding the body together prior to Aristotle, though they do remark that this is not a complex idea. I am not overly concerned with a lack of parallels, not least because the notion of the unlimited having a steering function on everything is there in Anaximander, if not specifically in relation to humans.[122] The idea is not complex, and if I am right that it has a cosmogonical facet as well (how the *kosmos* is steered into arrangement, as well as how it keeps that arrangement), then one can see how it might well tie in with the cosmogonical concerns of the Milesians.

XV. Anaximenes and multiple *kosmoi*

It has been suggested that Anaximenes too held that there were innumerable *kosmoi*, but there is even less evidence for this than for Anaximander. The theoretical opposition to multiple *kosmoi* in Anaximenes follows that applied to Anaximander. Anaximenes, like Anaximander but unlike the atomists, makes no bifurcation of what is and what is not, and does not have the requisite conceptions of the void or the infinite, nor does he have a cycle of love bringing everything together into a one and strife dispersing it again into a many. Anaximenes says nothing about boundaries or individuation criteria for co-existent *kosmoi*, or about how to differentiate successive *kosmoi*. Anaximenes has a steering principle and has no need of chance and necessity, and makes no mention of them. There is nothing to suggest that on certain key cosmogonical issues Anaximenes abandons the principle of sufficient reason for *ou mallon* arguments. The processes which generate the *kosmos* are ongoing and intelligible and the idea of a single birth for the *kosmos* is perfectly intelligible in Anaximenes. In explanatory terms, Anaximenes has nothing to gain from multiple *kosmoi*, whether they be co-existent or successive. If he has a steering principle, such theories are otiose. That everything is generated from and destroyed into air is insufficient on its own to establish a need for successive *kosmoi*, as we have seen with Thales and Anaximander. There seems to be no reason in Anaximenes why the *kosmos* should deteriorate or be destroyed, so there is no need for the notion of successive *kosmoi*.

The only evidence that Anaximenes held multiple co-existent *kosmoi* is Aetius 1,1,3 which we looked at in relation to Anaximander. Aetius, though, as we saw, seems so unlikely in some of his attributions that we cannot put much weight on his evidence. Pseudo-Plutarch contradicts this view, and Simplicius believes Anaximenes to hold successive, but not co-existent *kosmoi*.[123]

> Those who have one *kosmos* generated and destructible say there is always a *kosmos*, though it is not always the same, but becomes different at different times, according to certain periods of time, as with Anaximenes, Heraclitus and Diogenes, and later the Stoics.[124]

Simplicius' report seems to be closely modelled on a passage in Aristotle:

> Everyone believes it to have been generated, but some believe it to be eternal once generated, others believe that it is destructible like any other formation in nature, while others believe that it alternates, at one time being as it now is, and at another time changing and going to ruin, such as Empedocles of Acragas and Heraclitus of Ephesus.[125]

Aristotle mentions Empedocles and Heraclitus but not Anaximenes, which must throw some doubt on Simplicius' assertion.[126] That it is unlikely that Heraclitus believed in successive *kosmoi*, as will be argued in the next chapter, throws doubt on them both. So as with Anaximander, the doxographic evidence for multiple co-existent *kosmoi* is dubious, and I reject it.

XVI. Anaximenes and vortices

The evidence for Anaximenes using a vortex is the same passage which led to the suggestion that Anaximander used a vortex. Perhaps the case is slightly stronger for Anaximenes, as he does have the earth held in place, and Aristotle classes him in this respect with the atomists. He tells us that 'Anaximenes, Anaxagoras and Democritus say that it is the flatness of the earth which is responsible for it remaining in place.'[127] However, in the subsequent argument it is the way that a flat earth settles on air 'like a lid' that keeps it in place, without any contribution from a vortex. So as with Anaximander, Anaximenes has no need of a vortex to hold the earth in place. If the parallels I have drawn between Anaximenes' 'holding and ordering' air and Anaximander's 'steering' unlimited are correct then Anaximenes has no need of a vortex to initiate the cosmogonical process. There is no process in Anaximenes which we might suppose to be a vortex (like Anaximander's *to gonimon*), and there is no mention in any of the doxographers of a vortex in Anaximenes. As with Anaximander, the simple conclusion is that Anaximenes did not employ a vortex. Assuming that the motions of the heavens are in some way circular in Anaximenes[128] does not entail the existence of a vortex. It may well be that for Anaximenes:

> The earth is flat and is borne on air, and likewise the sun, moon and all the other heavenly bodies, which are all fiery, are carried on the air because of their flatness.[129]

However, that may only explain why the sun, moon and stars retain their relative position to the earth. None of them falls because all of them are flat and are supported by the air. Even if Anaximenes likens the heavenly bodies to leaves,[130] this does not mean that they are moved by circular currents of air. We are told that 'According to Anaximenes the heavenly bodies make their turnings by being pushed by enclosed and rigid air.'[131] However, the heavenly bodies may move of their own accord through stationary, evenly distributed air and then only change their course when meeting these pockets of air. Even if there is circular motion of the air driving the heavenly bodies, that does not entail a vortex, for the same reasons as given above with Anaximander. Similarly, if there is something which guides the *kosmos* in Anaximenes as in Anaximander, as I have argued, we cannot move from there to a primordial circular motion without further evidence, or from a primordial circular motion to a vortex.

XVII. Aristotle and steering

Against the line I have argued for all three of the Milesians, that for each there is some form of steering principle which guides the *kosmos* into being, is the fact that Aristotle does not seem to recognise the function of this principle, even though it is on his evidence that we can attribute a steering principle to Anaximander and 'most of the physiologoi'.[132] Aristotle says that:

> If all generation and destruction comes from one entity or even many things, why does this happen and what is the reason? For the substrate surely cannot change itself. Neither wood nor bronze can change itself, wood does not make a bed nor bronze a statue.[133]

He goes on to say that the first to take up this enquiry (Thales, Anaximander, Anaximenes?) were not worried about this issue, as opposed to the Eleatics. Dismissing the Eleatics on the grounds that they could not account for generation, Aristotle then says:

> Fire, earth or any of these cannot be the reason why existing things are or become to be well and beautifully disposed, nor is it possible to suppose this. Nor could the beautiful arrangement of things be left to spontaneity and chance. So when someone said that mind existed, and that in nature just as in animals it is the reason for all good order and arrangement, he seemed sober in relation to the random accounts uttered previously.[134]

In the discussion that follows, Aristotle mentions a considerable number of Presocratics and their idea on how the *kosmos* was brought together – Hesiod on love, Empedocles on love and strife, Anaxagoras on mind, Leucippus and Democritus on atoms. He makes no mention though of any of the Milesians. His view seems to be that the Milesians did not recognise the need for some account beyond their principal substance to explain the genesis of the *kosmos*.

The problem here is the old one that Aristotle assimilates the views of the Milesians to his own views on matter. Aristotle cannot conceive of how wood or bronze could organise themselves, so cannot conceive of how water, the unlimited or air might organise themselves, and so does not attribute this view to the Milesians, saying they had no account. It is interesting here that Aristotle makes no attribution of chance or an external deity or mind to the Milesians, views he is ready to attribute to others. Nor does he recognise any of the trappings of a multiple *kosmoi* approach in them as he does with Empedocles and the early atomists. Returning to the question of why the doxographical tradition goes astray, here is another possible influence. Aristotle denies there can be self-organising elements and does not attribute them to the Milesians, and under this influence the doxographical tradition does not either. Aristotle is wrong here on two counts. Firstly, self-organising elements are a conceptual possibility, and secondly, the Milesians did believe in them.

XVIII. Hesiod and the Milesians

In the previous chapter I argued that Hesiod failed to produce the first philosophical cosmogony on two general considerations. The world he envisaged was not a *kosmos* in the proper sense, and the processes he supposed created this world were not proper to philosophical cosmogony. The key issues were parsimony, invariance, consistency and a rejection of the supernatural.

With the Milesians we have one principal substance, one invariant cosmogonical process and no external gods. The *kosmos* is produced by an entirely natural process, as in Anaximander are animals and humans. So too in Anaximander cosmic justice is a necessary, natural and internal matter as opposed to the whim of the gods in Hesiod. There is no interference in the *kosmos* from any external or supernatural entity in the Milesians. I would disagree with Cornford, who says of Hesiod, 'Only a very thin partition divides it from the early Ionian systems.'[135] There are some arguments which might be thought to show a closeness between Hesiod and the Milesians. One might argue that the move from personal gods to a divine hylozoism is not really so great a shift after all. I have already questioned whether the Milesians thought their principal substances were divine. One can see why Aristotle and the later doxographical tradition would think they were divine, but we have no direct evidence that the Milesians thought in this way. Whether the principal substances were divine though is not really the issue. Of much greater consequence is how they behave, and the justification for them. The gods of Hesiod are not invariant, the principal substances are invariant. Hesiod is profligate with his gods, the Milesians, if they consider their substances divine, have a single god. The gods of Hesiod are supernatural, the principal substances are natural.

The Milesians could all be classified as hylozoists, though I prefer to term them panpsychists. Hylozoists hold that matter is alive and capable of self-motion, but might also suggest variable and unintelligent behaviour, which is clearly not the case for the Milesians.[136] Panpsychism has connotations of intelligence, goodness and invariance which I believe are more fitting for the Milesians. Even with hylozoism though, there is a difference between accepting that matter has all the characteristics of animal life and accepting that it has some of these characteristics. It is common to hold that the *kosmos* grows like a life-form for the Milesians, and they do of course use biological analogies.[137] There is though this difficulty. If the *kosmos* grows from a seed or some similar biological starting point, where did that starting point come from? It can only have come from the principal substance acting to produce that seed. Again, the key issue is not terminology but behaviour. The principal substance for the Milesians behaves in a fundamentally different manner to anything in Hesiod.

We can arrange the order of events in cosmogony in a table as follows, to bring out some of the parallels between Hesiod, Anaximander and Anaximenes.[138]

Hesiod	Anaximander	Anaximenes
Chaos	Unlimited	Air
	The productive of hot and cold	
Chaos and Earth	Air surrounds earth	Air surrounds earth
Night, Erebos, Day, Aether	Sphere of flame	
Heavenly bodies from Earth	Heavenly bodies from flame	Heavenly bodies from earth
Mountains, Sea	Sea and air	Sea

Put in this way the parallels seem quite close, partly because this strips away a great deal of Hesiod's profligacy, in terms of what is created and how it is created. The parallels also seem close if, with Cornford, we restrict ourselves to considering how Hesiod and the Milesians answered the following question: 'How did a manifold and ordered world arise out of the primitive state of things?'[139] Rather, for the Milesians we should phrase this: How did a parsimonious, invariant, consistent and natural *kosmos* come to be in a parsimonious, invariant, consistent and natural manner? One of the limitations to Cornford's approach to Presocratic cosmogony in general is that he believes the question quoted above to be the only problem in cosmogony addressed by them.

I do not wish to deny that the Milesians drew on previous accounts in their cosmogonies. Thales may well have been inspired by Babylonian accounts even if he did have further reasons for adopting water as his principal substance, Anaximander and Anaximenes may well have been influenced by Hesiod, and doubtless one can find, however loosely related to the Milesians, previous ideas of creation out of something like the unlimited or air. The nature of the Milesian accounts though is fundamentally different.

Did the Milesians, and Anaximander in particular, produce the first infinite universe in Furley's sense?[140] Furley takes it to be 'an integral part of the theory of the Infinite Universe that a world is simply a mortal compound like any other, and there is no limit to the number of worlds that may be in existence, now or ever'.[141] As I have argued that for each of the Milesians there is one and only one *kosmos*, they are clearly not the originators of the infinite universe in Furley's sense. We may though draw a distinction between the closed *kosmos* of Plato or Aristotle, where there is no space beyond the *kosmos*, and the unlimited principal substances of the Milesians, as well as with the infinite universes of the atomists. I disagree with West when he says that 'Anaximander for his part was no vigorous rationalist'.[142] I do not believe that Anaximander 'Allowed divinity an important place in his universe',[143] in the same sense as his predecessors. Even if the unlimited is divine, it is invariant. Nor do I see that 'it is more meaningful to classify as theological' language such as 'They pay penalty and retribution to each other for injustices according to the assessment of time' which Anaximander uses in describing change.[144] He may construe what we would call physical law in civil law terms, but he is not the last to do that and it was perfectly reasonable to do so at the time, without any theological connotations. There may have been oriental influences on Anaximander's cosmology,[145] but I find these harder to locate for his cosmogony. As I disagree with West on whether Anaximander believed in multiple *kosmoi*, I cannot agree that Indian thought influenced a multiple *kosmoi* view. Other influences (use of light, heat, moisture, egg analogy) seem too diffuse and ubiquitous to establish any firm link between Iranian creation tales and Anaximander,[146] and I have argued above that we should not take the process of cosmogony to be identical to that of zoogony in Anaximander.

XIX. Physics and cosmogony

An issue I have touched on in this chapter is whether the Milesians saw the processes involved in cosmogony as part of the ongoing physics of the *kosmos*. The issue comes about in modern cosmogony because of the claims of the big bang theory. The key assumption of big bang theory is that if we go back arbitrarily far in time with a universe that is expanding, ultimately we will reach a stage where the *kosmos* began. This timescale can be calculated, and currently is thought to be in the order of 13-14 billion years.

If the BB happens only once, and science involves the repeated observation of phenomena, can we have a properly scientific study of the BB? If space and time are generated with the BB, can we have any causal account of the BB, if our causal accounts are always tied into spatiotemporal contiguity? What determines the nature of our *kosmos* if nothing (no space, time, matter or laws of physics) existed prior to it? Why does the BB produce a specific amount of matter which we can now calculate? Why does the BB happen only once? As we are unable to say why it happened, we cannot say why it does not happen again.[147]

In the middle part of the twentieth century, the idea that the universe began in the manner described by BB had serious competition from what was known as Steady State (SS) cosmology.[148] Indeed, 'big bang' was first used by steady state theorists as a derogatory term. SS cosmogony claimed to be more scientific then BB cosmogony on several issues. The basic cosmological or Copernican principle stated that the universe would appear the same, on a large scale, from whatever point the universe was observed from. The perfect cosmological principle went one stage further and stated that the universe looked much the same whatever moment in time you looked at it.

It was agreed that the universe was expanding. The presumption of big bang theory was that this led to a decrease in the density of the universe. Steady state theory denied this, and claimed that matter was being continually created such that although the universe expanded, its density remained constant. Matter was then created continually throughout the universe. There were no preferred times and no preferred sites. Steady state cosmology claimed the continual creation of matter to be a philosophical plus over the big bang theory. It claimed that the continual creation of matter could be brought under invariant scientific laws, while the creation of matter in the big bang was a one-off, inexplicable event. Laws of creation of matter/ energy should not be arbitrary, and could not be switched on and off. So the processes of cosmogony were a permanent part of the physics of the universe. SS cosmogony had now been largely superseded, as several critical observational issues were resolved in favour of BB.[149] The issues of whether BB is a fully scientific cosmogony in the sense outlined above have yet to be resolved.

The Milesians attempted to distance themselves from the theogony of Hesiod in several ways, through adopting parsimony, invariance, consistency and a natural *kosmos*. Another facet of this distancing lies in their attitude to cosmogony. For each of the Milesians, the processes involved in cosmogony are not unique events outside the natural physical order of the *kosmos*. The processes which bring about cosmogony are part of the ongoing physical fabric of the *kosmos*. So the principal substance continues to undergo transformation into the components of the *kosmos*, and the components of the *kosmos* are transformed back into the principal substance. One view is that, both on a global and a local scale, an equilibrium is reached, under the guidance of the steering principle. I have already argued for this in relation to the weather cycle and the drying-out theory in Anaximander. The earth may dry out until a weather cycle is set up, when the earth can continue in a beneficial equilibrium. Something similar may happen with the interchange of principal substance and the components of the *kosmos*. Firstly, there is a process of producing the components from the principal substance which ultimately, again under the guidance of the steering principle, sets up an equilibrium of things being generated out of x and being destroyed into x. This would allow a *kosmos*, once generated, to be eternal, precursors for this idea being Hesiod and the general notion that the gods are born but may still be immortal.

XX. Conclusion

The standard view is that the Milesians give us the first philosophical cosmogonies, and I see no reason to doubt that. I see nothing mythological or theogonical in the structure of their thought. That the content of Thales' cosmogony has something in common with preceding mythology should not be in the least surprising. What Thales does with that content is far more important.

My argument throughout this chapter has been that for each of the Milesians, their originative substance plays a key role in cosmogony in guiding the *kosmos* into being and then keeping it in order. For each of them, the *kosmos* does not simply fall into order, nor is there is any trace, in any of the Milesian fragments or testimonia, of any role for chance. The evidence for some form of steering principle is stronger in Anaximander and Anaximenes, but Thales too has the requisite belief in an active originative substance and it is at least a reasonable conjecture, in the absence of any other cosmogonic ideas, that he too held some form of steering principle.

There are concerns about all the evidence in relation to the Milesians, including the evidence which attributes a steering principle to each of them. I take the evidence on soul and steering to be indicative of what they may have thought rather than definitive of what they did think. We cannot reject all the evidence, as that would leave us nothing to say about the Milesians. Our options then are to accept both a steering principle and multiple *kosmoi*, or reject steering, or reject multiple *kosmoi*.

For each of the Milesians I have argued that there is one and only one *kosmos*, and that once in existence, it is eternal. There is no motivation for them to believe in multiple *kosmoi*. They do not have to explain how a *kosmos* could come to be from chance and necessity alone, as there is no evidence of a role for chance in any Milesian cosmogony, and each has a steering principle. They have no need of the early atomists' *ou mallon* reply to Parmenides, and none of the Milesians seek to deny the principle of sufficient reason. None of them has the boundary conditions and individuation criteria needed for a multiple co-existent *kosmoi* view, and none has the criteria to differentiate successive *kosmoi* from each other. The 'created out of x destroyed into x' formula, as we have seen, does not require there to be successive *kosmoi*. The doxographical evidence for multiple *kosmoi* for the Milesians is highly suspect, and can be explained in terms of Aristotle's slight misrepresentations of Milesian thought followed by some rather programmatic and inflexible interpretations of Aristotle. We shall have cause to question the doxographical evidence in relation to Heraclitus as well in the next chapter.

The Milesians sought to bring cosmogony into the province of philosophy, out of the murkier regions of myth and theogony. To their great credit, they saw cosmogony as in one sense unexceptional. The processes which bring about cosmogony, whether they be separation or rarefaction and condensation, are processes which we can see going on around us now. There was no need for any external interference in the universe for a *kosmos* to come into being. Whether they quite managed to keep those processes invariant, or when those processes altered they managed to explain why they altered is another matter and one which subsequent thinkers will have something to say about.

3

Heraclitus

It is now generally accepted that Heraclitus (*c*. 544-484 BC) did not have a cosmogony, and believed the *kosmos* to be eternal. This is quite different from how Heraclitus was perceived in the post-Aristotelian ancient world. Aristotle, and after him Theophrastus, believed that for Heraclitus everything in the *kosmos* periodically became fire. The Stoics, seeing a precursor of their own views in this, then interpreted Heraclitus as a believer in a cyclical cosmogony where the *kosmos* is generated out of fire, and degenerates into fire again, in a cycle without beginning or end. The generation of the *kosmos* from fire was called *ekpurôsis* by the Stoics, literally 'out of fire'. I argue that the evidence within Heraclitus is quite strongly against such an interpretation, and that we can see the genesis of the Stoic interpretation in the way that Aristotle tends to assimilate Heraclitus to Thales and Anaximenes (as positing one basic element) and the way in which he misconstrues Heraclitus' meaning about everything becoming fire.

I do not believe that Heraclitus, having no cosmogony, belongs with the Milesians. I also take Heraclitus to be pre-Parmenides, at least in the sense that his views are not a reaction to Eleatic concerns about change, being and sufficient reason.[1] Nor does he belong with Aristotle, as I believe his reasons for rejecting cosmogony are quite different, and are a reaction, at least in part, to the views of the Milesians. There are some interesting comparisons to be made between Heraclitus and Empedocles. Empedocles does have a proper cyclical cosmogony, and is the first to do so. It is useful then to look at what motivates Empedocles into proposing multiple successive *kosmoi*, and then to compare this with the views of Heraclitus. As I have also argued that none of the Milesians believed in successive *kosmoi*, there is an interesting comparison to be made here as well.

I. Heraclitus and cosmogony

The key passage for Heraclitus' views on cosmogony is Fragment 30:

> This *kosmos*, the same for all,[2] was not made by Gods or men, but has always existed and will always exist. It is an everliving fire, kindling in measures and going out in measures.[3]

If we read *kosmos* here in the usual sense of 'an ordered universe', then there is little scope for debate. Heraclitus does not have a cosmogony, as for him the *kosmos* has always existed and always will exist. While some fire may go out, there is no time when all fire goes out, and further fire will be kindled. I agree with Guthrie that if we interpret *kosmos* in this manner then this fragment is decisive against any successive *kosmoi* view and must 'override later testimony'.[4]

However, there is an alternative here, and along with the Stoics[5] we might take *kosmos* in the sense of an underlying order such that the physical universe might undergo periodical *ekpurôsis* but the underlying order persists. Heraclitus does say that 'a hidden attunement is stronger than an obvious one'.[6] So combined with the consideration that *kosmos* at this time has yet to have a fixed meaning, it is possible that this is what Heraclitus meant. Also in favour of this view is that it seems a little odd to deny that this *kosmos* was made by men, as no one would suppose that in the first place. So in this case 'Gods and men' has to be taken as a polar expression meaning 'no one at all'.[7] If we read 'underlying order' though, the contrast might be between that order and any moral/ social order that might be created by Gods or men. The contrast would then effectively be one of the natural underlying order of the *kosmos* and any conventions generated by gods or men.

These two interpretations have been extensively discussed in the literature, but I want to at least suggest a third view as well. Kahn, who has strongly supported the successive *kosmoi* view, speaks of 'the issues of cosmogony and cosmic conflagration, which hang together'.[8] I do not think, as a matter of necessity, that they do. There are two minor possibilities here which I shall mention but not pursue:

(1) A *kosmos* without a beginning in time that then undergoes *ekpurôsis*.
(2) A *kosmos* that has a beginning (from fire?) and an end (in fire?) but has no regeneration, so there is no unending cycle. If fire consumes fuel and does not generate anything burnable, then fire will ultimately be extinguished, with no possibility of a second cycle.

A third possibility here, which I do think is significant, is a *kosmos* which has a beginning but does not have an end. Here there would be cosmogony, but no successive *kosmoi*. This means adopting the 'underlying order' reading of *kosmos*, to allow an ordered world to come into existence. Nothing in Fragment 30 though (or in my view, anywhere else in Heraclitus) compels us to accept an end for the *kosmos* as well. One advantage of this position is that we can grant fire an important position of priority (it is the stuff from which the ordered world comes about) without being committed to successive *kosmoi*. This is important because Kahn's lead argument for successive *kosmoi* is, effectively, that fire must be granted some sort of physical priority. The easiest and most obvious way to do that is to give a role in cosmogony, therefore we must accept successive *kosmoi*.[9] If the new position I have sketched is viable though, then Kahn's conclusion does not follow and the case for successive *kosmoi* is severely weakened.

II. Evidence for *ekpurôsis*

What evidence is there in favour of the successive *kosmoi* interpretation? Aristotle tells us that:

> All say that it (the *kosmos*) had a beginning, though some hold that having begun it is eternal, others that it is perishable like any natural grouping, and others that it alternates, at one time being as it is now, at another otherwise [perishing], as with Empedocles of Akragos and Heraclitus of Ephesus.[10]

This would seem fairly clear in ascribing something like successive *kosmoi* to Heraclitus. The only question would be perishing, in brackets because it is reckoned to be a later addition to the text, probably but not necessarily by Aristotle. Aristotle also tells us:

> It is impossible that the universe, even if it is limited, should become any one of these (elements), as with Heraclitus who said that all things at some time become fire.[11]

This I think is decisive in establishing Aristotle's view, though he may have misunderstood whether all things become fire at the same point or at some point, but not all together. Diogenes Laertius, drawing on Theophrastus, is also clear in his belief of successive *kosmoi* in Heraclitus:

> Fire is the element and fire is an exchange for all things, which are generated by rarefaction and condensation, though he does not set this out clearly. All things are generated from opposites, and the whole flows like a river, it is limited, and all there is forms one *kosmos*. It is generated out of fire and undergoes *ekpurôsis* according to certain alternating periods which go on for all eternity. These events happen according to fate. Of the opposites, that which leads to generation is called war or strife, and that which leads to *ekpurôsis* is called agreement and peace, change he called a road up and down, and according to this is the generation of the *kosmos*.[12]

There are of course Stoics who interpret Heraclitus as supporting successive *kosmoi*. What is more significant is that there are Stoics and other philosophers around the time of the Stoics who recognise a difference between successive *kosmoi* and the views of Heraclitus. Plutarch has one of his characters comment that the Stoic notion of *ekpurôsis* is 'encroaching on the writings of Hesiod, as well as those of Heraclitus and Orpheus'.[13] Marcus Aurelius, who believes in successive *kosmoi*, describes a situation when there has been a dissolution of compound bodies into their elements, or there have been elemental changes, and then says:

> These things are taken back into the logos of the whole, either by a periodic *ekpurôsis*, or by a renewal by eternal exchange.[14]

While this does not mention Heraclitus by name, the word used for exchange here, *amoibê*, is quite distinctive and is the same word Heraclitus uses in a key fragment:

> All things are an equal exchange for fire and fire for all things, as property is for gold and gold is for property.[15]

If there is an eternal exchange, as Marcus Aurelius suggests, then there is no need for *ekpurôsis* in Heraclitus. There are of course fragments of Heraclitus which will bear the successive *kosmoi* interpretation, but none that demand it or even point strongly in that direction. The fragments deriving from Hippolytus all need to be viewed with caution, as he has a strong Christian agenda. The following fragment has been supposed to support *ekpurôsis*: [16]

> The onset of fire judges and overpowers all things.[17]

However, it is again ambiguous. Fire may well judge and overpower all things, the question is whether there is a time when everything is fire, or whether fire in kindling in one place also goes out in another.

III. Evidence against *ekpurôsis*

An important piece of testimony against successive *kosmoi* in Heraclitus is in Plato. In the *Sophist*, the stranger has been speaking of the disagreement between those who hold the world to be many and those who hold it to be one. He then says that:

> Later, certain Muses in Ionia and Sicily recognised that it was safest to combine the two accounts, saying that what exists is many and one, and is held together by hatred and love. The more severe of these Muses say that in drawing apart it is always being pulled together. The milder relax this idea holding permanently, and speak of phases where at one time everything is one and loving under Aphrodite, and at another time is many and at war with itself due to some form of strife.[18]

The milder, Sicilian Muse is clearly Empedocles, and this is a reasonably accurate description of his views. The stricter, Ionian Muse is Heraclitus. The key point here is that in contrast with Empedocles, Heraclitus does not have phases of dissociation and recombination, leading to a dominance of love or strife, but the drawing apart and pulling together happen simultaneously. There is here then no sense of a cycle of change that would lead to the dominance of fire and so to *ekpurôsis* and successive *kosmoi*. Heraclitus does say that:

> Taken together things are whole and not whole, gathered together and dispersed, consonant and dissonant. Out of all things comes one, and out of one all things.[19]

One might try to generate some sort of Empedoclean cycle out of the last sentence here, but Plato's evidence is very strongly against that. The author of *On the Kosmos* quotes this fragment and takes it to refer to the simultaneous mixing of opposites in the *kosmos*, rather than any cyclical changes.[20]

If Heraclitus believed in a great year,[21] this shows little as there can be natural cycles within the *kosmos* without the *kosmos* itself coming to grief. Plato believes in a great year in the *Timaeus*,[22] but has no difficulty in asserting the *kosmos* to be eternal. Heraclitus clearly does believe in some natural cycles:

> In relation to these periods, the sun is overseer and guardian, determining, arbitrating, displaying and illuminating changes, and brings all things to birth, according to Heraclitus.[23]

In terms of cosmogony or successive *kosmoi* though such cycles are harmless.[24]

IV. Why is there no cosmogony in Heraclitus?

One objection to Heraclitus having no cosmogony is that this would be a very radical departure from both the literary tradition of theogony and the philosophical tradition of cosmogony. According to Kahn this would be 'Strange, almost unintelligible',[25] and his view is that 'cosmogony was rejected by no one before Aristotle, not even Parmenides'.[26] What would be Heraclitus' motivation for such a move? Here are some possibilities.

If Heraclitus is a radical and iconoclastic philosopher (or at least perceives himself to be so), he might well be attracted to a radical, philosophically interesting idea which differentiates him sharply from previous thinkers. He is vitriolic in his criticism of other thinkers.[27] If he does reject cosmogony, then Fragment 30 does so boldly, bluntly and almost as a challenge.

We saw with the Milesians that the processes which bring about cosmogony are ongoing processes. They do not explain cosmogony by different processes to those currently operating in the world, and so attempt to give a natural explanation of cosmogony. However, those processes must change in some sense when the *kosmos* has been formed. In Anaximenes for example, there must be a change in the rate of formation of the other elements from air, or all the air will be used up. Intelligent steering may accomplish this, but a change will nevertheless be required in the steering once the *kosmos* is complete. Empedocles is very interesting in this respect as the processes which produce a *kosmos* do not alter once a *kosmos* has been formed, but go on to wreck that *kosmos*.[28] If Heraclitus is looking for something unchanging which explains the changing nature of the world about us, one option is to reject cosmogony and have the *kosmos* as an ever-living fire.

One might also draw the following contrast between Heraclitus and the Milesians. For the Milesians, the *kosmos* is a state of good order. For Heraclitus, the *kosmos* is a well ordered process.[29] One might reasonably ask how a state of good order has come about, and then give a cosmogonical explanation. It is slightly different to ask how a well ordered process has come about. It is perhaps more intuitive then to answer that the process has always been ongoing rather than to try to explain how an ordered process arose from a disorderly process.

V. The priority of fire: fire as *archê*

If Heraclitus does not have a cosmogony, then there is a question why Heraclitus gives some form of priority to fire. If we treat Heraclitus as an extra Milesian, someone who completes the sequence of water – unlimited – air – fire as hypotheses of the primary stuff, then his combined assertions about the priority of fire and the denial of cosmogony look slightly odd. If there is a cycle of change between the elements, why choose fire as primary if it is not what everything else ultimately came from? There are several possibilities here.[30] Of the four elements, fire is the most capable of changing itself, and so is perhaps suitable for a philosopher who emphasises change.[31] It is the stuff of the human soul.[32] Fire seems to be what steers the *kosmos*.[33] Perhaps Heraclitus had some empirical support for his choice of fire.[34] Fire may have some form of causal priority. Reeve has argued that Heraclitus rejected Anaximenes' notion of rarefaction and condensation because it introduces a new notion beyond the elements, and gives no reason for the primacy of air.[35] Further, Heraclitus saw a more elegant solution in making fire the cause of change. One part of this is reading Fragment 31 in a certain manner:

> Fire's reversals: first sea, though half of sea is earth, and half lightning. Earth is dispersed as sea, and measures up to the same ratio as before it became earth.[36]

The Greek translated as 'reversals' here is standardly used for solstices where the sun reverses from setting successively more southerly or northerly on the western horizon to setting more northerly or southerly respectively. So the sense is not that fire changes into water,[37] but at the end of a series of changes fire turns around and goes back through that series again. This then accords with two other fragments on elemental transformations:

> The death of fire is the birth of air, and the death of air is the birth of water.[38]

> For souls it is death to become water, for water it is death to become earth. Out of earth water is generated, out of water, soul.[39]

So we end up with a series of earth – water – air – fire. This Reeve suggests is both the upward path (earth to fire) and the downward path (fire to earth) of Fragment 60:[40]

> The road up and down is one and the same.[41]

This would also accord with the kindling and going out of fire of Fragment 30, and also with Plato's evidence suggesting that there is a simultaneous 'drawing apart and coming together' rather than successive phases as in Empedocles. According to Reeve, the path up, to fire, is one of progressive heat gain, the path down, to earth, one of progressive heat loss.[42] Against Anaximenes, who has to impose rarefaction and condensation on his elements extrinsically, Heraclitus has heat as intrinsic to fire. Reeve argues that Heraclitus reduces justice to strife and strife to fire, thus producing a monism.[43] This monism he considers to be more successful and of a different type to the Milesians, Heraclitus' basic entity not being a 'static stuff to which change is extrinsic, but a measured process'.[44] As Reeve puts it:

> Since we can explain the priority of fire without recourse to *ekpurôsis*, we lose the only compelling reason for attributing that Stoic doctrine to Heraclitus.[45]

VI. The priority of fire: fire not as *archê*

Alternatively, we might see Heraclitus as a more radical thinker, one who does not play the Milesian game of specifying a primary element. This is not to deny a cycle of change between the elements, but it is to deny that fire is temporally or ontologically prior to any other element. Graham, in a paper which does much to undermine the idea of what he calls a 'generating substance theory (GST)' in Heraclitus, suggests that fire is symbolic.[46] It is meant to emphasise that the process of transformation is the only stable reality, as none of the elements is stable, and so for Heraclitus 'the choice of fire will reveal the utter failure of GST'.[47]

Much as I am sympathetic to liberating Heraclitus from GST, I don't think this does enough to explain why Heraclitus accords fire a leading role in Fragments 30, 64, 66 and 90.[48] I agree with Graham's reading of

Fragment 90, that gold does not produce property, nor does it constitute property and while it remains in circulation, it does not become property, and so is not temporally or ontologically prior to property.[49] I would add that the trade of fire/ gold for other things and vice versa is simultaneous and without change in the amount of gold/ fire. While this reading may free us from GST in Fragment 90, it does not tell us any more about why Heraclitus chooses fire instead of another element here or in other fragments. I suggest two possibilities as to why Heraclitus might single out fire for some form of priority without treating it as a Milesian element.

Graham objects that Reeve's account leaves Heraclitus with a form of GST.[50] That is so, but it is possible to reformulate the idea of fire having some form of causal priority without giving it any ontological priority. Fire need not exist prior to the other elements, and the other elements need not come from fire, but change is caused by fire. Here one can resist Reeve's reduction of justice to strife and strife to fire by giving a causal account instead. Strife may be produced by fire, but cannot be reduced to it, and similarly justice may be produced by strife but cannot be reduced to it. This is a more plausible reading of Heraclitus, who does not use reductionism as a strategy in other aspects of his thought. The fragments might be argued either way though. Fragment 94 might be thought to be non-reductive about justice:

> The sun will not overstep his measures. If he does, the Erinyes, the defenders of justice, will find him out.[51]

However, Heraclitus also says:

> It is necessary to know that war is common, and strife is justice, and that all things occur by strife and necessity.[52]

Either side might reasonably argue that the opposing evidence is expressed too loosely and figuratively to be entirely reliable. It would be odd if Heraclitus did not have a source of change in mind. However, he does not specify anything extrinsic to the elements as an agent of change. In part this makes Heraclitus like the Milesians, in giving one of the elements a philosophical, and not merely a symbolic priority. It does not give him a GST though. Fire is the cause of change in the cycle of the elements, but is not what the elements come from. Secondly, even if we deny that fire is primary in cosmogony for Heraclitus, it may still have an important cosmological role. We saw with the Milesians that the steering principle may continue to play a role after the *kosmos* has been formed, keeping the *kosmos* in order and preventing any degeneration. Heraclitus too is concerned about the possibility of degeneration and indeed may have a more radical view of what would happen to the *kosmos* if what keeps it in order is removed. He says that 'even the *kukeon* separates unless it is stirred'.[53]

The *kukeon* was a drink made of barley meal, grated cheese and wine,[54] which would separate out very quickly unless it was stirred. If we take this cosmologically, there is a need for continual strife to stop the *kosmos* from separating out. What agency might provide that? Two important fragments in relation to this are:

> The thunderbolt steers all things.[55]
> All things are steered through all.[56]

If we take 'thunderbolt' to be an allusion to fire, then, as with Anaximander and Anaximenes, we have the prioritised element steering the *kosmos*. There is also the association of fire and intelligence in that for Heraclitus:

> The dry soul is wisest and best.[57]

So if fire generates strife, and what is more directs this intelligently, there will be no need for the *kosmos* to separate out. That is a significant role for fire, as it was for Anaximander's unlimited and Anaximenes' air. This does not pick out fire merely because it is associated with soul or because it has some unspecified steering function.[58] Fire has a specific, and very important cosmological role. Again, this gives fire a philosophical and not merely symbolic priority without introducing a GST.

There are then two important ways in which fire can be said to be primary. It is the cause of changes in the cycle of the elements. It has a steering function to keep the *kosmos* from separating out. Either of these can be formulated with the other elements being ontologically dependent on fire or not.

VII. Motivation for successive *kosmoi*?

Returning to the *kukeon* fragment Kahn comments:

> This may serve as an image for the celestial rotation of the sun and stars by which the cosmic order is maintained, and which is itself the residue of a primordial vortex or whirlpool (*dinê*) from which this order was generated.[59]

I see no allusion whatsoever to a vortex here. Certainly physical stirring may involve circular motion, but it is a long way from that consideration to a vortex. Rather more likely it is strife that metaphorically stirs up the contents of the *kosmos* and prevents any separation. So if strife keeps stirring, then the *kosmos* will continue. There is no suggestion in Heraclitus that strife ought to cease or will cease at any point, rather the opposite, as Aristotle tells us:

> Heraclitus criticised the author of 'Oh, that strife should be utterly destroyed among Gods and men.'[60] For there would be no musical scale without higher

> and lower notes and no life-forms without female and male, which are opposites.[61]

In commentary on this passage, Simplicius says that strife cannot cease for Heraclitus: 'otherwise everything would be destroyed, he says'.[62]

What motivation might Heraclitus have for successive *kosmoi*? Leucippus and Democritus have a very strong motivation for co-existent and successive unlimited *kosmoi*. Empedocles has a proper cyclical cosmogony and has very strong motivations for adopting successive *kosmoi*. Any notion that Heraclitus has a mortal *kosmos* to mirror mortal human beings seems fairly thin, especially as we are neither born nor die in fire. In short, I do not see Heraclitus' motivation here. As we saw above, Heraclitus has the idea of fire being intelligent and able to steer the *kosmos*, and so has an agent that is capable of keeping the *kosmos* in order, as with Anaximander and Anaximenes. If so, as with the Milesians, I do not see the need for any degeneration of the *kosmos*. There may of course be limited cycles of change (such as the seasons) without the need for the end of one *kosmos* and the beginning of the next one.

VIII. The value of testimony, Aristotle and after

Those who argue that Heraclitus did not have a cosmogony point to the absence of any evident cosmogony in the extant fragments, and the fact that the ancient commentators do not allude to any cosmogonical passages that we are not aware of. Those who argue that Heraclitus did have a cosmogony point to the fragments on the interchange of the elements,[63] from which it would be relatively easy to construct a Milesian type of cosmogony, and the following in Diogenes Laertius:

> By thickening earth turns into moisture and by compaction, water comes about, and by congealing water turns into earth. This is the path down. Earth is liquefied, and from this comes water, and from this comes the rest, as he derives nearly everything from exhalations from the sea. This is the road up. Exhalations occur from both earth and sea, some are bright and pure while others are dark. Fire grows from the bright, and moisture from the others. It is not clear what the nature of that which surrounds is.[64]

This attributes a cosmogony to Heraclitus, and indeed we have here as much testimony on cosmogony as we have for the Milesians. However, a good deal of this is unattested by any other source on Heraclitus. In particular, there is the worry that there is no reference, in any of the recognised fragments, to:

(i) Rarefaction and condensation, or any similar processes in relation to changes between elements.

(ii) A role for peace.
(iii) Change being the road up and the road down.
(iv) The road up and the road down in relation to cosmogony.
(v) Bright and dark, etc. in relation to the elements.

The later tradition is to a large extent reliant on Theophrastus, who is in turn reliant on Aristotle. How reliable though is Aristotle's evidence? As we saw with the Milesians, Aristotle tends to see their *archê* in terms of his own material explanation. Aristotle tends to assimilate Heraclitus to the Milesians, as he groups together those he thinks believe in:

> one underlying body, either one of the three or something else which is denser than fire but lighter than air, and they produce plurality by having everything else generated by rarefaction and condensation.[65]

This clearly includes Thales (water), Anaximenes (air) and Heraclitus (fire).[66] We have already seen Aristotle's carelessness when it comes to attributing ideas to groups of people when in all likelihood only some of that group subscribe to that idea. Here, clearly Anaximenes does believe in rarefaction and condensation, but we cannot extend this to Heraclitus on this type of assertion from Aristotle. He assimilates Heraclitus to the Milesians on the issue of flux as well:

> They say that everything is in a state of generation and flux, and nothing is stable, and there is one thing which persists, out of which all the changes are produced. This seems to be the view of Heraclitus of Ephesus and many others.[67]

One can easily see though how the doxographical tradition following on from Aristotle might then attribute rarefaction and condensation to Heraclitus. Similarly, one source of the successive *kosmoi* attribution to Heraclitus may well be Aristotle, who may have misunderstood what Heraclitus had to say, or assimilated him with someone else who believed in successive *kosmoi*.

IX. Cosmogony and chance

Heraclitus does have one (apparently) paradoxical thing to say about the *kosmos*, which is that:

> The beautiful *kosmos* is a massing together of refuse.[68]

Kahn translates this as 'The fairest order of the world is a heap of random sweepings.'[69] Against taking there to be any connotation of random order here, there is no suggestion in any other fragment of a role for chance in the composition of the *kosmos*.[70] On the contrary, we have the notion of steering, and indeed if 'all things are steered through all' a very strong notion of steering which would appear to exclude chance. We also have the notion of measure, as in Fragment 94. So too we have fire going out and kindling by measure in Fragment 30, fire being an equal exchange for all things in Fragment 90, and everything being subject to fate in Fragment 80. I simply do not see a role for chance events in Heraclitus, certainly not on the basis of Kahn's interpretation of Fragment 124 against other evidence. This is important because it removes another possible motivation Heraclitus might have had for successive *kosmoi*, and forms an important contrast with Empedocles, who I will argue does have successive *kosmoi* but also has a significant role for chance in his cosmogony as well as his zoogony.

One further point here is that Kahn wishes to take *kosmos* here in the usual sense of an ordered universe rather than as underlying order, as he did with Fragment 30, and as is critical to the successive *kosmoi* interpretation. The underlying order of the *kosmos* could hardly be a collection of refuse. If we want to read *kosmos* univocally in Heraclitus, then Fragment 124 virtually forces us into taking *kosmos* in the usual sense of ordered universe.

Ultimately, I don't see anything particularly problematic about this fragment, certainly no more so than with some other fragments. It may, in one relatively superficial sense, be paradoxical that our beautiful *kosmos* is a collection of refuse, but this is not going to raise questions about radical flux, the identity of opposites and the law of non-contradiction.

X. Conclusion

No one has ever suggested that Heraclitus believed in co-existent multiple worlds, and indeed there is no evidence whatsoever to suppose he did. The evidence is also strongly against Heraclitus believing in successive *kosmoi*. Fragment 124 requires us to read '*kosmos*' as 'ordered universe', and if we read *kosmos* in Fragment 30 similarly, this appears to rule out successive *kosmoi*. Some fragments are consistent with successive *kosmoi*, but all of these are amenable to other interpretations and none demands successive *kosmoi*. I do not see that Heraclitus has any motivation for successive *kosmoi*. There is no role for chance and no vortex in Heraclitus, instead we have strong roles for fate, measure and equal exchange.

Plato's evidence is very much against successive *kosmoi*, in that it implies simultaneous 'drawing apart and coming together' rather than a succession, as in Empedocles' phases. Aristotle assimilates Heraclitus to the Milesians, which makes his evidence to some degree suspect, and so too the evidence of the doxographical tradition from Theophrastus on. The Stoics have their own reasons for attributing *ekpurôsis* and successive *kosmoi* to Heraclitus. The successive *kosmoi* interpretation of Heraclitus is untenable.

Whether Heraclitus had a cosmogony, but did not believe in successive *kosmoi* is a trickier question. It does allow the most obvious sense of the priority of fire, the Milesian one of the other elements and the *kosmos* being generated out of fire. There are other viable senses in which fire may have priority though, and this interpretation too runs up against the problem of what is meant by *kosmos* in Fragments 30 and 124. Giving Heraclitus a cosmogony out of fire may also make him too Milesian, where there are more interesting possibilities of seeing him as a more innovative and critical figure. It is important though that we recognise that the cosmogony/ no *ekpurôsis* interpretation is a possibility, as it shows how much more work the successive *kosmoi* interpretation has to do. It is not enough simply to argue for a cosmogony, there is a need to demonstrate that Heraclitus envisaged an end to the *kosmos*, and, if we are to rule out the one cycle interpretation, that he believed there to be a new cosmogony following the destruction of the *kosmos* as well.

If the standard interpretation of Heraclitus on cosmogony is correct then, he is the first person we are aware of who openly rejects a beginning for the *kosmos*. One assessment of Heraclitus I would disagree with is that of Barnes, who comments: 'Fire is the primary stuff of the world. The thesis has a traditional Milesian ring; and on his monism Heraclitus constructed, perhaps not in conscientious detail, a physical science of a standard Milesian type.'[71]

The rejection of cosmogony, though, is original and radical, as are the reasons for it if I have divined them correctly. What steers the *kosmos* should do so in a consistent manner, and not arbitrarily change once the *kosmos* has been formed. This consideration may have been highly influential on Empedocles. Heraclitus may also see the *kosmos* as an ordered process rather than a state. We can give a strong sense to the priority of fire whether or not we consider Heraclitus to be a monist. Fire may have causal priority, and in steering and keeping the *kosmos* in order, it may have a cosmological priority as well, without having any temporal or ontological priority. If fire does have causal priority, there is another significant difference with the Milesians in that Heraclitus has the cause of change intrinsic rather than extrinsic to his element theory.

Heraclitus was a bold and innovative thinker, distancing himself from the Milesians in many respects. Although his rejection of cosmogony was not taken up by any other Presocratic, the reasons for this may have had a more significant influence than is generally recognised, especially on Empedocles.

4

Parmenides and Eleaticism

Parmenides (*c*. 515-450 BC) is rightly famed for having formulated an important principle in cosmogony, that no thing can be generated from nothing. This is a clear example of a perennial problem in cosmogony. Once formulated, any subsequent cosmogony has to be able to deal with this principle, either finding reasons to deny it or means by which to accommodate it. Modern cosmogony has yet to provide a fully satisfactory answer. As ever with Eleatic problems, the more literally we take nothing, no matter, no energy, no time, no space, no laws of physics, the more difficult and interesting the problems become. I argue that Parmenides produces several more problems for cosmogony. Some are influential on immediately subsequent thought on cosmogony, some have a more lasting influence.

I. Nothing from nothing

Parmenides' way of truth has been highly influential. Having rejected any discussion of what is not, he tells us that:

> There remains only one tale of a path, that it is. On this path there are a great number of signs, that being ungenerated and indestructible it is, whole and of one kind, unmoved and complete. It never was nor will be, as it is now, all together, one, continuous.[1]

He then continues:

> What birth will you seek for it? In what way, from what source did it grow? I will not allow you to think or say from not being, for it is not to be thought or said that it is not; and what necessity would have driven it later rather than sooner, beginning from nothing, to grow?[2]

The problems Parmenides sets here are quite general, for anything coming to be, not just a *kosmos*, though this is taken seriously by Greek cosmogonists. The principle that no thing can be generated from nothing is accepted quite generally up to the later antiquity debates about the nature of God's creation in Christianity. I believe that Parmenides is quite correct to assert that no thing can come from nothing. I would only add the hypothetical puzzle, if something were to come from nothing, what would it be and how much of it would we get? As there is nothing, there is nothing which can determine this. This is a problem for modern cosmogony, as our universe has determinate values. There is a determinate amount of matter/energy, the universe is expanding at a determinate rate and there are determinate values for the fundamental constants. It is entirely unclear why this should be so if our universe came from absolutely nothing.[3] Parmenides makes his conclusion abundantly clear:

> Therefore it must be entirely, or not at all. Nor will strength of opinion allow anything beside it to be generated from what is not. Thus Justice has never let slip her bonds to allow it to be generated or destroyed, but holds it fast.[4]

II. Sufficient reason

This passage above also poses a sufficient reason problem.[5] If something should begin to grow, what reason is there that it should happen at one time rather than any other time? If what we begin with is homogeneous, and time is homogeneous, then how can one moment in time be preferred to another?

Apart from being an interesting general principle, this can also be seen as a criticism of the Milesians.

It is almost certain that the Milesians did not have their primordial substances generated out of nothing. However, it is also the case that they almost certainly envisaged the primordial substance as entirely homogeneous before the beginning of any cosmogonic process. If this is controlled by an intelligence, why does the intelligence choose any particular time for this process to begin? If the process begins by some necessity, why does that necessity apply at one moment as opposed to any other?

A similar problem, though not expressly formulated by Parmenides, flows from these considerations. Where would a *kosmos* begin to grow? What reason would there be for a *kosmos* to begin to grow here rather than there?[6] This may be easier for the Milesians to answer, as they may think there is a preferred point in space where it is difficult to see that there are preferred points in time for them. Even if the Milesians considered space, or their primordial substance to be unlimited, they may well have believed there to be a centre to that space. So Anaximander can have the earth stable as it is equidistant from the extremes, and may have the *kosmos* begin to form around the one significant point in his space, the centre. Of course, they may be in error in believing that an unlimited space has a centre, but as long as they believe there is a centre the sufficient reason argument about where the *kosmos* begins to be formed can be answered.

The early atomists suffer from both of these problems as they believe in a centre-less unlimited space. They have quite a radical answer to these problems, effectively denying the principle of sufficient reason on these issues, as well as on the sizes and shapes of the atoms.

III. The initial state dilemma

Parmenides is insistent that what is is one, is continuous and is homogenous.

> It is not divided, since it all exists in a similar fashion. It is not more here than there, as this would prevent it from being continuous, but it is all full of being. Everything is continuous, for what is draws near to what is.[7]

> Since there is a further limit, it is complete, resembling the bulk of a sphere, well rounded in all aspects, equivalent in all ways from the centre. It is not greater or lesser here or there. Neither is it non-existent, preventing it reaching its like, nor does it exist with more being here, and less there since it is all inviolate. Equal to itself in all respects, it falls uniformly within its limits.[8]

There is a dilemma in cosmogony which can be expressed like this. If the original state of the universe is homogeneous, and it obeys laws which are place invariant, it is then difficult to see how anything interesting will develop in that universe. If the original state of the universe is not homogeneous, we require some further explanation of why the original state is not homogeneous and why it has this particular state as opposed to any other. Modern big bang cosmogony has this problem. If we assume that what expands from the initial state is entirely homogeneous, and that it expands isotropically, then all that will result will be a larger homogeneous universe, which only differs from earlier stages in being less dense and cooler. The modern solution to this difficulty is to invoke quantum mechanics. At an early stage of the universe, quantum effects create inhomogeneities in the distribution of matter/ energy throughout the universe (the famous 'ripples' in space/ time). The areas of greater energy density will then act as gravitational attractors. Depending on one's model of cosmic evolution, stars or galaxies will then form around these areas of higher energy density. [9] In reply to the question of why here rather than there, the answer will be that quantum mechanics is fundamentally probabilistic in its nature, and can only give probabilities of energy distribution. Leucippus and Democritus and Epicurus and his followers adopt this type of solution, albeit to slightly different problems.

The second arm of the dilemma is that if we do not suppose a homogeneous beginning to the *kosmos*, then we require some explanation of the initial distribution of the contents. If we can give no sufficient reason for that distribution, then we are driven back to the homogeneous beginning. Whether we should consider homogeneity as a special case which does not need further explanation is another matter.

A further aspect of Parmenides' thought is that what is is one and is not many. In this he was supported by Zeno and Melissus.[10] This view was not of lasting importance in itself, but was certainly significant for Parmenides' immediate successors. The early atomists see the need to assert not only chance but also plurality, and Empedocles has the universe cycling between being one and being many.

IV. Parmenides' cosmogony

Parmenides also produces his own cosmogony, and the relationship between the way of truth and the cosmogony is a standard problem. Parmenides effects the transition from truth to opinion by saying:

> With this I end my trustworthy account and thought concerning truth. Now learn the opinions of mortals, and listen to the deceptive order of my words.[11]

In Fragment 8 he then introduces the opinions of mortals:

> They decided to name two forms, of which they ought not to name one, and in this they have wandered. They distinguished them as opposite in nature and set up signs separate from one another, to one the aethereal flame of fire, rare and very light, in every way equal to itself but not to the other. The other in itself is opposite, dark night, dense in form and weighty. The entire ordering I tell you as is fitting, in order that no thought of mortals will overtake you.[12]

In Fragment 9 Parmenides opts for light and dark as his two forms:

> Because everything has been named light and night, and according to their powers each goes with this or that, all is full of light and invisible night together, both equal, as nothing shares in neither.[13]

His explanations in cosmogony seem to have been quite extensive, as Plutarch tells us:

> Parmenides made an ordering, mixing the elements of light and dark, producing all of the phenomena out of them and through them. He has said much concerning the earth, the heavens, sun, moon and the generation of mankind.[14]

We do have some information on his cosmogony and cosmology.[15] Aetius tells us that:

> Air is separated off from the earth due to its being vaporised by strong compression. The sun is an exhalation fire, as is the milky way.[16]

In relation to the 'steering principle' view I have argued for the Milesians, it is interesting that Parmenides says:

> The narrower rings are full of unmixed fire, those close by are full of night but with some measure of flame. In the middle of this there is a goddess, who

> steers all things, ruling the hateful birth and mixture of all things, sending female to have sex with male, and conversely male with female.[17]

There are a variety of views on the relation between the way of truth and the cosmogony which Parmenides puts forward. Now discarded is the old view of Zeller and Burnet, that the cosmogony is not Parmenides', but is a summary of the views of some of his contemporaries.[18] As Long comments, it has not been satisfactorily explained why Parmenides would expound someone else's view at this point, it is not clear whose cosmogony this might be, and the ancient commentators all took this to be Parmenides' own cosmogony.[19]

A second view is that the cosmogony is in some ways an extension of the way of truth. Parmenides, so some say,[20] cannot have denied the existence of phenomena entirely, and so the discussion of mortal opinion in some way reinstates the phenomena or mitigates against taking the conclusions of the way of truth in their most radical forms. As Owen put it, there is 'the problem of whether, as Aristotle thought and as most commentators still assume, Parmenides wrote his poem in the broad tradition of Ionian and Italian cosmology.'[21] A third view is akin to some interpretations of Plato's views of the physical world, that the cosmology is the best that can be managed by mortals. If we take *eoikota* in Fragment 8,60 as 'most likely' rather than 'fitting', and look to a similar usage by Plato in the *Timaeus*, then perhaps the cosmogony is the most likely account we can have, the best account of the physical world possible. Finally, there is the view that Parmenides makes no claim at all that the cosmogony is true.

V. The aims of Parmenides' cosmogony

The view I am going to put forward here is that the cosmogony is a demonstration of some of the problems which face mortal cosmogony. As with my treatment of the way of truth and its relation to cosmogony, my emphasis is on the principle of sufficient reason. Let us begin with Fragment 8,53-61. The two forms (*morphai*) which are named by mortals may be:[22]

(1) Light and dark
(2) Being and not-being
(3) Light/ being and dark/ not being
(4) Any pair of opposites.

I am reluctant to rule out any of these categorically. Studies of Plato have greatly benefited from the idea that some of his works are not simple repositories of doctrine, but may attempt to challenge the reader to question their own views and to provoke some form of dialectic with the reader. The goddess's injunction to 'learn the opinions of mortals, and listen to the deceptive order of my words' may be meant for us, mortal, opinion holding readers as much as for any character in the poem. The first challenge then may be to think about what the forms named by mortals may be. Similarly, there has been considerable diversity of opinion on what is meant by 'of which they ought not to name one'. As the goddess has warned us specifically about the 'deceptive order of my words', we should at least be open to the possibility that this is an issue that we either cannot or ought not to settle by grammatical or syntactical considerations. The intention may be to be ambiguous, and throw the emphasis on the reader to sort out their own mortal opinion, and realise the shortcomings of it. There are three main views on what should and should not be named here:

(1) Two forms should be named, and not one.
(2) One form should be named, but the other should not.
(3) No form should be named.

Again, I am reluctant to rule out any of these categorically, especially as an answer here may be related to an answer to the previous question about what the forms are. So naming one alone makes better sense on the view that the *morphai* are being and not-being rather than opposites. Then, in line with the way of truth, being may be named but not not-being. Simplicius' view is that the logic of opposites means that we cannot name only one. However, who, prior to Parmenides (or indeed at any other time) has named only one opposite? There are several thinkers who have clearly named two or more, but no one who has named one alone. Naming only one of a pair of opposites would be fairly thin as an analysis of why all mortal cosmogony goes astray. As I wish to follow the thread of the *morphai* as opposites in general, and it would make little sense to name only one opposite, the question will be whether mortals should name two forms or none.

It is worth paying attention to the order of what is introduced next. First mortals distinguish opposites, which at this stage are non-specific, then they attribute 'signs' to them, turning them into light and dark. This would seem to lend support to the idea that the *morphai* are opposites in general rather than any specific opposites. The ultimate upshot of this, according to Plutarch, is that Parmenides is able to reproduce all of the phenomena.[23] So schematically, the procedure goes:

(1) Set up a non-specific pair of contraries, x and the contrary of x.
(2) Assign them contrary qualities.
(3) Reproduce all the phenomena.

There are two sufficient reason considerations that we might raise here.

(1) What sufficient reason is there for choosing light and dark as opposed to hot and cold (Anaximander) or dense and rare (Anaximenes) or anything else as the primary opposites?
(2) What sufficient reason is there for the proposed distribution of qualities, or indeed any distribution of qualities?

The second question here may look a little strange, but consider it in relation to the Pythagoreans, who use contrary pairs such as right and left, male and female, etc.[24] Why should we associate one of each pair in one group, and the other in another group? What sufficient reason is there for that? Within Parmenides' poem we might ask why fire should be associated with mild and light and why dark should be associated with dense and heavy.

There is also the important point here that Anaximander and Anaximenes cannot then justify their choice of opposites on the grounds that they alone can produce the phenomena. Depending on how radically we want to take this, either Parmenides shows that the phenomena can be reproduced by the alternative pair of light and dark, or he shows that this can be done with an entirely arbitrary choice of opposites, and an arbitrary choice of qualities to characterise those opposites. No other mortal account could outstrip this, as it has done all the mortal account could reasonably be expected to do. This interpretation has one significant advantage over many of its competitors. It is not clear, on many interpretations, why Parmenides gives any sort of cosmogony/ cosmology at all, let alone a detailed one. If there is some fundamental philosophical error in talking about cosmogony at all, that is immediately evident at the beginning of the way of seeming, e.g. that cosmogony involves talking about non-being, there is no sense in giving a detailed cosmogony that covers all the phenomena.

The cosmogony of the way of opinion is then no better and no worse than any other mortal attempt at cosmogony. We can take the goddess's statement that no mortal will ever overtake someone who knows this cosmogony in two ways. Firstly, this cosmogony covers the same phenomena and makes the same types of judgements as other cosmogonies and so cannot be worse than other cosmogonies. Secondly, if one picks out some of the 'deceptions' in this account, one will know how to undermine any other account. I take a major deception to be that mortals appear to have made proper judgements in selecting opposites and selecting their qualities, when in fact they have not. A lesson from the way of truth might well be that a sound judgement is one that is based on the principle of sufficient reason. A further deception, and distinction from the way of truth, is that mortals set up signs where in the way of truth the signs are there independently.[25]

VI. Conclusion

Parmenides had a huge impact on subsequent cosmogony. As we shall see, Empedocles and the early atomists devise their systems and their cosmogonies specifically to counter some of Parmenides' ideas. Parmenides is most famous in the history of cosmogony for his rejection of the idea of creation *ex nihilo*. While this is of course important, it is not all that he deserves to be famous for, nor perhaps was it the most influential thing he had to say in relation to the development of Greek cosmogony. Did anyone prior to Parmenides believe in an *ex nihilo* creation? Not as far as we are aware. Would those coming after him have accepted creation *ex nihilo* if Parmenides had not so decisively rejected it? The first real thrust toward creation *ex nihilo* comes from the early Christians. Parmenides presents a challenge to any theory of change, and this is taken up in many ways, perhaps most radically by the early atomists. There are some important problems raised more directly with cosmogony though. Questions of sufficient reason, of why this time and this location for cosmogony will also loom large. They will be problematic until the Christians introduce the radical idea that god generates time and space along with matter.

5

Empedocles

Empedocles of Akragas (*c.* 495-435 BC) is significant in the history of cosmogony for introducing the idea of cyclical *kosmoi*. A *kosmos* is generated, it then degenerates and after this process is complete, a new *kosmos* is produced. This process has neither beginning nor end. I have argued that the evidence is against any of the Milesians or Heraclitus having a cyclical conception of *kosmoi*. I have also argued that they lacked the philosophical motivation that might generate such a theory. The evidence for Empedocles having some form of cyclical cosmogony is conclusive, and he has clear motivation.[1]

This philosophical motivation for cyclical *kosmoi* is the main question I wish to address in this chapter. What is it that makes him put forward a cycle of everything coming to be one under the growing influence of love while strife wanes, and then becoming many under the growing influence of strife as love wanes? Clearly Eleatic questions loom large here, but I shall argue that for a full understanding of Empedocles we need to see him in relation to Heraclitean ideas on cosmogony as well.[2] I also argue that chance plays a very important role in Empedocles' cosmogony. It has long been recognised that chance plays a significant role in Empedocles' zoogony, but the extent to which it plays a crucial role in cosmogonical processes has been greatly underestimated. Empedocles introduces an influential new pattern of explanation into cosmogony, based on this role for chance.

There has been considerable debate about the nature of the cosmological cycle in Empedocles, particularly concerning the role of strife. In broad outline, while all are agreed that under increasing love there is cosmogony and zoogony, the nature of the period of strife's dominance and whether there is cosmogony and zoogony under strife have all been contested. I do not propose to examine all the evidence and alternatives here in detail, as this would be the work of another book in itself. I am going to set out my own view (which is the orthodox view with a small variation) and what I see as the main arguments for that view, which I believe have been considerably bolstered by the publication of the Strasbourg papyrus, which gives us new Empedoclean material. We can then move on to the more intriguing question of the role of chance in Empedocles' cycle, and whether successive cycles will be identical or different, and if different whether they will be radically or marginally different.

I. Empedocles, Parmenides and Heraclitus

I will begin by linking Empedocles to Parmenides, and to a lesser extent to Heraclitus. It has long been recognised that there are allusions to Parmenides' poem in Empedocles' work, often with Empedocles differentiating himself from Parmenides. So Parmenides says that:

> With this I finish my reliable account and thought concerning truth. Learn now the beliefs of mortals, listening to the order of my words.[3]

Empedocles seems to allude to this directly when he says:

> Listen to the undeceitful passage of my words.[4]

I believe that Empedocles makes the same sorts of allusions to Heraclitus' work as well. So Heraclitus says:

> Much study does not teach understanding. Otherwise it would have taught Hesiod and Pythagoras, Xenophanes and Hecataeus.[5]

While Empedocles says:

> Come, listen to my tale, for learning increases wisdom.[6]

KRS argue that Empedocles engages with the epistemological challenge of Parmenides, and that he replies in a Heraclitean manner, mourning the limitations of the senses but claiming an intelligent use of empirical information will help us,[7] and Trepanier has brought out some stylistic and intellectual affinities.[8] I would hold that Fragment 3, 'Not holding sight to be more trustworthy than hearing', is a comment on Heraclitus Fragment 101, 'Eyes are more accurate witnesses than ears'.[9] As we saw in Chapter 3, the following is cosmologically very significant for Heraclitus: 'even the *kukeon* separates unless it is stirred'.[10] The mix of wine, barley and cheese will separate very quickly unless there is strife to keep the mixture in motion. In the context of the actions of love, though, Empedocles talks of 'gluing together barley-meal and water'.[11] The allusion here may be to baking and to love being able to put together and hold together a much more homogeneous mixture than can strife.

II. Empedocles and creation *ex nihilo*

To a certain extent, Empedocles agrees with Parmenides in that there can be no generation from nothing. So Empedocles says:

> There is no birth of mortal things, nor do any end in some unhappy death,

> but there is only mixing and interchange of the mixed, birth being the name given to these things by men.[12]

> It is not possible for something to be generated from the non-existent and that something which exists should be destroyed cannot be accomplished and is utterly unheard of.[13]

On issues concerning the nature of the universe as a whole, Empedocles is also in agreement with Parmenides. He says:

> No part of the All is empty or overflowing.[14]
> None of the All is empty, so how could anything come into it?[15]
> What could increase this all? From where would it come?[16]

Empedocles clearly reacts to Eleatic concerns, but relating Empedocles to the Eleatics will only take us so far. In reply to puzzles about the one and the many, Empedocles has love bringing a one out of a many, and then he has strife bring a many out of a one. That is a little excessive as a reply to Eleaticism. If we accept Plato's contrast here,[17] that Heraclitus considers the *kosmos* to be one and many at the same time, while Empedocles envisages a succession of states progressing between one and many, then Empedocles could rationalise his scheme. Empedocles could have a beginning with his four elements in different clumps, and have love associate them into a *kosmos*, and eventually into a one, or he could have a beginning with a perfect mixture of the elements dissociated by strife in to many. As long as the time span is long enough to accommodate many generations of humans, either will account for the phenomena and give Empedocles his answer to the one and many problem.

III. The history of the cosmic cycle

There has been a considerable history of different interpretations of Empedocles' cosmic cycle. This is at least in part the result of Empedocles' style. He wrote in verse, like Parmenides, and many of his poetic allusions are difficult to give a precise philosophical sense to. For example, take this line from Fragment 17:

> Strife apart from these and equal in every way, and Love in amongst them, equal in length and breadth.[18]

One could try to infer from this something about the shape of the *kosmos*,[19] but there is no guarantee that Empedocles meant anything to be inferred about the shape of the *kosmos* here. I am not convinced that Empedocles spelt out the full details and chronology of all of aspects of the cycle.[20] I would not follow Osborne in suggesting that Empedocles did not give a great deal of thought to the cycle, but the fact that he appears to have worked out some parts in great detail (notably aspects of the zoogony) does not entail that he did this for all. I would agree with Osborne though that we need a broader approach than simply seeing Empedocles as providing one sort of answer to Parmenides.[21]

In the modern era, what might be called the orthodox interpretation of the cycle stems from the work of Panzerbieter in the mid-nineteenth century.[22] I take the orthodox interpretation to be that there is cosmogony and zoogony in times of both increasing love and increasing strife. This is also sometimes termed the symmetrical interpretation, as it gives symmetrical roles to love and strife. The evidence of Aristotle supports the orthodox interpretation.[23] According to Aristotle, we live under increasing strife:

> At the same time he says that the *kosmos* is in a similar condition now under the influence of strife to that which it was in under the influence of love.[24]

If that is so, and at the end of the reign of love there is a complete association of the elements such that there is no *kosmos* and no living beings, then there must be a cosmogony and a zoogony under strife. Aristotle's evidence has been questioned, but it is interesting to note that Plato is of the same view. While Plato's evidence is tricky to handle, with the exception of O'Brien it is not discussed in the Empedocles literature.[25] While Plato does not refer to Empedocles by name, it is clear that some passages are in direct criticism of Empedocles, and are usually satirical in nature.[26] Panzerbieter was supported by Zeller and Bignone,[27] and although the orthodox view was criticised by Tannery, von Arnim and others it remained the dominant interpretation.[28] In 1965, Bollack, Solmsen and Holscher all produced challenges in different ways,[29] while in 1969 O'Brien published his *Empedocles' Cosmic Cycle* which firmly supports the orthodox view on the back of an exhaustive analysis of the fragments and testimonia.[30] Long produced the most challenging opposition to the orthodox view in 1974,[31] and Osborne has been consistently sceptical about many aspects of our reconstruction of Empedocles' thought. Of the important recent histories of Greek philosophy, Guthrie and Barnes support the orthodox interpretation while KRS oppose it.[32]

There is no single alternative view to the orthodox interpretation, and none without difficulties. Dissenting views deny either a cosmogonic or a zoogonic role for strife, or both, and some deny a phase where the role of strife increases, such that there is a leap from the complete dominance of love to the complete dominance of strife. One motivation for alternative interpretations has been that there is a disproportion of passages describing the cosmogonic and zoogonic activities of love and strife. However, one might point out that there is also a disproportion of passages describing the cosmogonic and zoogonic effects of love compared to the *kosmos* and life destroying effects of love as love continues to associate the elements into a homogenous one. Does that mean that we should seek to excise or radically revise the role of increasing love once the *kosmos* has been formed? In what follows I support the orthodox interpretation, as I think it makes best sense of the evidence, though I am more concerned with the role of chance and the relation of successive cycles than the internal details of the cycle.

IV. The nature of the cosmic cycle

It is traditional within the orthodox interpretation to break up the cycle into four phases, though it must be emphasised that Empedocles makes no such divisions himself. As this is a cyclical process, the numbering of the phases, and indeed any point at which one starts an exposition of them is entirely arbitrary. The key processes are that love associates the four Empedoclean elements into an homogeneous mixture, and then strife dissociates the elements into four disjoint groups. While some of the literature has been happy to refer to love and strife as 'forces', I decidedly am not.[33] They are entities on their own rather than properties of matter, they do not generate attractive or repulsive forces between matter, their effect varies over time and their effect discriminates between different types of matter.[34] It is better to refer to love and strife as principles of association and dissociation respectively.

With the total success of love, the universe can be said to be one, with the total success of strife it can be said to be many. I also associate rest and motion with the polarity of total love and total strife, such that there is rest when there is total love and maximum motion with total strife.[35] This gives us four phases:

(1) The complete domination of strife, where the four elements are entirely dissociated and there is no mixing.
(2) Increase in the influence of love, such that the elements become increasingly mixed with one another.
(3) The complete domination of love, such that there is a perfect mixing of the elements.
(4) Increase of the influence of strife, such that the elements become increasingly dissociated from each other.

There is cosmogony and zoogony under the increasing influence of love, working by association of the elements, but there is also a destruction of living things, and to a great extent of the *kosmos*, as love nears complete domination. Similarly, there is cosmogony and zoogony under the increasing influence of strife, working by dissociation, and again a destruction of living things and of the *kosmos* as strife nears complete domination. When love is completely dominant there is rest. It is not clear how long this lasts for, but at least long enough for Aristotle to complain about the conceptual difficulties involved in restarting motion in the universe from a state of total rest. Aristotle tells us that:

> Empedocles posits strife as an explanatory principle for destruction, as strife is able to produce everything but the one, for all things except god come from it. He says 'From these has come everything that was and is and will be. Trees, men and women have sprung up, beasts, birds and water-nurtured fish and the long living gods.' This is clear even apart from this. If there had not been strife in things, all things would have been one, he says.[36]

When strife is completely dominant there is maximum motion, although it is unclear what that might amount to. This is momentary and love begins to increase its influence directly after the high point of strife. Technically this gives us only three phases as the dominance of strife is only momentary, but it has traditionally been counted as a phase in the orthodox interpretation.

There have been widely diverging conceptions of what might happen in total strife. O'Brien has argued that strife ultimately produces a universe with a central sphere of earth and concentric shells of water, air and fire.[37] Tannery on the other hand considers the rule of strife to be one of chaotic, disorderly motion.[38] I believe O'Brien's account to be too deterministic while Tannery's is too chaotic, so I offer something in between as a suggestion.

Imagine a spherical 'lava lamp', with four immiscible fluids, each with a different colour dye, representing the four Empedoclean elements. As the state of total strife approaches, any separate groups of any one liquid/element join together and do not split apart until after the moment of total strife and love begins its process of association again. The groups of the elements will not have any specific shape, and what shape they do have will be in a perpetual state of change, as will the position of the elements.

In the lava lamp the fluids are kept in motion by a small heat source, the hot fluids rising to the top where they cool and then fall back down again. This does not produce a stratification of the fluids, but rather a flux. In relation to Empedocles, the contrary actions of love and strife, if we consider these as acting from the centre and the periphery respectively, may generate a flux of the elements rather than a stratification if the differences between the elements in respect of weight are relatively small.

The complete domination of strife might produce a stratification of the elements as love would not be able to produce any contrary effects. However, if the complete domination of strife is only a moment between the decreasing and increasing influence of love, it may not have the time to produce that stratification, and so the 'changeover' may occur with any shape and distribution of the four elements as long as there are four disjoint masses of them. Within that, the concentric shells model is one possibility among many, and within that the earth, water, air, fire order is one possibility among many. It seems somehow fitting that in Empedocles' system there should be a single state for the universe when it is one and many states for it when it is many. The lava lamp analogy would fit with Plutarch's description of strife for Empedocles:

> The first principles of the whole are unmixed, without affection and alone, not forming a mixture or a community with each other, but fleeing and turning, carried by separate and remorseless motions.[39]

This also avoids the problem that Fragment 35 poses for any rigidly stratified account of total strife. When love begins to increase its influence again, it does so from the centre, associating the elements together. However, if there is only earth at the centre, love's influence will be at least delayed, and its first mixing will be of earth and water, air and fire having to wait even longer. On the lava lamp account love would have at least the possibility of immediate access to all four of the elements.

I don't wish to push the lava lamp analogy too hard as it is obviously anachronistic and Empedocles is unlikely to have seen anything that functions in a similar way, though he may have seen some recirculating oil and water effects in a domestic context, which might give him the basic ideas. It does give us another way of conceiving of what a state of near total strife might be like for Empedocles, and one that avoids the rigid stratification of O'Brien and the chaos of Tannery. The importance of this will become more apparent later in this chapter, where I discuss the role of chance within the cycle and the relation of one cycle to the next.

Empedocles' vision of total strife has some important successors. Plato in the *Timaeus* considers what the pre-cosmic state would be like, and one of his suggestions is the complete dissociation of the elements into four separate groups.[40] Aristotle too has the separation of the elements into four concentric rings as a consequence of what would happen to the terrestrial realm if it were not for the action of the sun stirring and mixing the terrestrial elements. The vision of total love is rather easier to conceptualise:

> Concerning the form of the *kosmos*, on what it is like when it is ordered by Love, he speaks in this manner, 'no double branches come from its back, it has no feet, no swift knees, no genitalia' but 'it was a sphere', and equal to itself.[41]

This too has a correlate in the *Timaeus*, where Plato is emphatic that the *kosmos* is perfectly spherical and has no arms or legs, etc., though for Plato that is the case in perpetuity once the *kosmos* is formed.[42]

V. Fragment 17

The orthodox interpretation is the most natural reading of Empedocles Fragment 17:

> I will give a double account. At one time, they grew together to be one alone out of many, at another time they grew apart to become many out of one. In two ways[43] are mortal things generated, and in two ways do they fail. They are born and perish through the coming together of all things, and they are nourished and vanish as they are drawn away once again.[44] These things never cease their continual interchange. At one time all things were brought together through Love, at another each being drawn apart by Strife's hatred. In that, they have learned to grow in to one from many, and the one is again drawn apart into many. In this way they are generated and themselves have no stable life, in that they never cease their continual interchange, they are forever unchanged in the cycle.
>
> Come, listen to my tale, for learning increases wisdom. As I said earlier in declaring the limits of my tale, I will give a double account. At one time, they grew together to be one alone out of many, at another time they grew apart to become many out of one. In two ways are mortal things generated, and in two ways do they fail. Fire, water, earth and the great height of air, and destructive Strife apart from these and equal in every way, and Love in amongst them, equal in length and breadth. You must perceive her with your mind, and not sit with amazed eyes. It is she who is considered to be implanted in mortal limbs, and so they think in a friendly fashion and complete peaceful deeds, calling her Joy and Aphrodite by name. She is perceived by no mortal man as she circles amongst them, but listen to the undeceitful passage of my words.
>
> These are all equal and contemporaneous, though each is ruler in a different realm and each has its own character, and each prevails in turn as time passes. Apart from these, nothing is generated or destroyed. What could increase this all? Where could it have come from? In what way could it be utterly destroyed, as nothing is empty of these things? There are just these, flowing through each other they become some things at some times and other things at other times, and yet ever and always the same.[45]

VI. Literary form and philosophical content

Even in the ancient world, there was a dispute about the nature of Empedocles' poetry. Plutarch tells us:

> It is not Empedocles' habit to adorn his topic with vacuous descriptions, akin to some bright flowers, to create a beautiful style, but rather he gives a precise illustration of all substances and powers.[46]

Aristotle, on the other hand, is rather more disparaging:

> Homer and Empedocles have nothing in common except metre, so while it

would be just to call the former a poet, we should rather call the latter a physical philosopher.[47]

The essential question for modern interpreters is whether the structure of Empedocles' fragments is part of the philosophical message he wishes to convey. There are alternative readings of Fragment 17, but I would agree with Graham that 'B17 must be read as more than an argument or a programme for an argument: it is also a mimetic structure which portrays the world condition which it describes.'[48]

Given the success of relating literary form and philosophical argument in other ancient thinkers, notably Parmenides, Heraclitus, Anaxagoras and Plato, this is an approach that at the very least needs to be explored. The central consideration here is the way in which Empedocles repeatedly uses the formula of one line describing association under the influence of love followed by another line on the influence of strife.[49] These pairs are grammatically structured so as to bring out the parallels between the functions of love and strife, and stress the symmetry of their roles.[50] The overall effect of this use of line pairs is to emphasise the theme of eternal recurrence and the symmetrical functioning of love and strife in associating and dissociating the elements. So I would agree with Graham that 'The verbal structure of the poem is after all not dispensable in the way that Aristotle thought it was, for it is in poetic form that Empedocles reveals the eternal recurrence of the cosmic cycle and its fearful symmetry.'[51] These considerations weigh very strongly against any non-cyclical or non-symmetrical interpretations of Empedocles.[52]

VII. The unity of Empedocles' thought and the Strasbourg papyrus

Whether Empedocles wrote one poem or two, and what the relation is between those fragments which might be reckoned to be 'physical' and those which might be reckoned to be 'religious' are highly contested topics.[53] I do not believe we have enough evidence to decide the one/ two poem issue, and I am suspicious of any criteria we use for distributing the fragments between two works (or two parts of one work) when we have so little knowledge of how Empedocles distributed them.

The discovery of the Strasbourg papyrus has shifted the grounds of the debate somewhat. As Trepanier has argued,[54] ensemble d from the papyrus shows a much tighter integration of physical and religious themes than previously known fragments, and calls into question any sharp division of Empedocles' work into physical and religious categories. If there is this mixture of themes, then as Trepanier points out, there are passages which previously were considered to be solely religious in nature which we might now consider to have a physical import as well. The most important of these is Fragment 124:

> O miserable, wretched collection of mortals, you were born from such strife and lamentation.[55]

This would appear to show that there is a zoogonic function for strife, if men are born from strife and lamentation. Fragment 62 is also significant:

> Come and hear now how fire as it was separated out raised the nocturnal shoots of men and pitiable women, for the account is neither far from the mark nor ignorant. Whole-natured forms sprang from the earth, having a share of both water and heat. Fire sent these up, willing to come to its like.[56]

The Strasbourg papyrus gives further evidence in favour of the orthodox interpretation. I agree with Trepanier that ensemble a, which continues directly on Fragment 17, carries on the general description of the cosmic cycle.[57] It gives us further evidence of cosmogony and zoogony under strife, the poetic structure thesis of Graham and the integration of themes in Empedocles' writing.

VIII. Successive *kosmoi*

There can be no question that Empedocles believes there to be multiple successive *kosmoi*. There is no suggestion in Empedocles of any of a beginning or end to the cycle, and of course there is no beginning nor end to the existence of what is, in the Eleatic fashion. There is no specific statement of unlimited time in Empedocles, but Fragment 16 does talk of 'countless ages'.[58]

In Chapter 2 I argued that for Anaximander there is no vortex, no role for chance, no co-existent multiple *kosmoi*, no successive multiple *kosmoi* but that there is a steering principle. I argued a similar thesis for Anaximenes, only with air as the steering principle. For both, one *kosmos* is generated and then continues indefinitely. Heraclitus too has no vortex, no role for chance, a steering principle and a single *kosmos*, this time without beginning or end. Leucippus and Democritus have vortices and a significant role for chance, no steering principle and have co-existent and successive multiple *kosmoi*. Where does Empedocles fit into this?

Empedocles does not have a steering principle, or at least not in the same sense as the Milesians. Certainly the actions of love and strife bring about the formation of the *kosmos*. The actions of love and strife also bring about the dissolution of the *kosmos*, and as such could hardly be said to be steering it in the manner of Anaximander. In one respect, Empedocles sides with Heraclitus in making the key cosmological processes consistent and unvarying. The steering of Anaximander and Anaximenes, as we have seen, has to alter so that the processes which form the *kosmos* do not go on to destroy it. While Heraclitus abandons cosmogony, Empedocles allows love and strife to act in a consistent manner and both to create and then to destroy the *kosmos*.

Neither love nor strife in Empedocles could be said to act in an intelligent manner, unlike the steering principles of his predecessors. Nor does Empedocles seek to portray love and strife as in any way intelligent. That in many respects is also in sharp contrast to Anaxagoras, who will seek to impose a cosmic intelligence on his primordial vortex. Empedocles has a greater affinity with Leucippus and Democritus, where a *kosmos* is not the culmination of a process that will alter on the completion of that *kosmos*, but is rather a by product of the processes which govern all that exists. These processes will, unchecked, bring about the downfall of a *kosmos* as well. Empedocles does not seem to favour any one of the four elements or suggest that any one or any group of them play a steering role. If neither love nor strife act in the steering role, nor do the elements, then that seems to exhaust Empedocles' ontology.

Empedocles does talk of gods, but these do not appear to have any cosmological or cosmogonical function. According to Fragment 21, line 12 the Gods are 'long living' but in sharp contrast to both his literary and philosophical predecessors, they are not immortal. Love, strife and the four elements though do seem to be immortal. Empedocles does say, 'All things have thought and some understanding.'[59] However, it is by no means clear what 'all things' refer to here and Empedocles does not seem to link this to any steering function.

IX. Chance: the evidence of Aristotle

We need now to consider what role chance might play in Empedocles' cosmogony. Aristotle tells us:

> Empedocles says that air is not always separated out upwards[60] but according to chance – He says in his cosmogony 'Thus at one time it ran by chance, but many times it was otherwise' – and he says that the parts of animals are for the most part generated by chance.[61]

> Although strife dissociates, it was not by strife that aether was carried upwards, but sometimes Empedocles speaks as if this were due to chance, saying 'Thus at one time it ran by chance, but many times it was otherwise', but he also says that it is the nature of fire to be carried upwards, and aether, he says 'sank beneath the earth with long roots'.[62]

If this evidence is reliable, it is of considerable significance because the separation of air is absolutely fundamental to Empedocles' cosmogony.[63] It is also significant because Guthrie has attempted to play down the role of chance in Empedocles.[64] Guthrie draws a perfectly reasonable distinction between chance in the sense of random, uncaused or inexplicable events and the association of chance and necessity. For teleologists such as

Plato and Aristotle, the results of events that are necessary might be described as chance occurrences if there is no purpose involved. So Empedocles could have a fairly strongly deterministic system and still have Aristotle criticising him for his use of chance. What goes strongly against this though is that, in Empedocles' own words according to Aristotle, 'many times it was otherwise'. Thus this looks to be chance in the uncaused, inexplicable sense rather than the necessity without purpose sense. It is also important, should one want to try to differentiate between the role for chance in Empedocles' zoogony and his cosmogony, that this quotation is in a specifically cosmogonical context.

As I have been very critical of Aristotle as a witness of Presocratic views in previous chapters, we need to consider carefully how reliable Aristotle is here. That the doxographers also attribute a significant role for chance to Empedocles is by no means clinching, for as we saw with the Milesians and Heraclitus, Aristotle's views had a considerable influence on the doxographers. In favour of this being reliable is that the types of errors that Aristotle commits with the Milesians and Heraclitus (seeing material cause everywhere, assimilating Heraclitus to the Milesians, etc.) do not seem to be relevant here. A different concern though would be this. If Aristotle cannot detect any theory of the natural motion of the elements in Empedocles, or considers what he finds to be inadequate, and if he considers the roles of love and strife to be association and dissociation only, then Aristotle is likely to conclude that Empedocles' elements can only move by chance, whether that is what Empedocles believed or not. Aristotle does seem to think that Empedocles' reasons for change are inadequate:

> Strife is no more the reason for destruction than it is for being. Similarly, nor is love in relation to being. For drawing everything into one it destroys everything else. At the same time he gives no account of the reason for change, except that it is natural – 'When great strife had grown strong in the limbs, and sprang to its realm as the time came, which was appointed for them in turn by a broad oath' – as if change were necessary. But he gives no reason for this necessity.[65]

Aristotle backs his view on Empedocles and chance with a direct quotation. What is more, and to some extent this is proof against Aristotle's carelessness in reporting the Presocratics, he quotes precisely the same fragment in two different places. Aristotle has a further consideration here, which is that in his view, entities for Empedocles are not random ensembles, but have specific ratios of constituents. Aristotle may have Fragments 96 and 98 in mind here, where entities (such as blood, flesh and bones) are constituted from ratios of fire, water, air etc. Aristotle says:

> Empedocles says that nothing is generated through coming together by chance, but according to a certain ratio. What is the reason for this? It is not

> fire or earth. Neither is it love or strife, as these are the reasons for association and dissociation. The reason is the substance of each, and not only, as he says, 'The mixing and the separation of things mixed', as the name for this is chance, and not ratio. For it is possible for things to be mixed by chance.[66]

Also important for the reliability of Aristotle's evidence is that he links it to Empedocles' zoogony, where it is quite clear that chance does indeed play a role:

> Is it possible that disorderly motion produces some mixture, from which naturally constituted bodies come about, such as bones or flesh, as Empedocles says happened under the rule of love. He says 'Many a head sprouted up without a neck.'[67]

Aristotle's Greek here for 'disorderly motion' is *kinoumena ataktos*, literally 'motion without order' so there is little doubt about what he is attributing to Empedocles. Aristotle also says:

> Wherever everything came together as if it were for the sake of something, there they survived, having been accidentally brought together artfully, but where this was not the case, they died and are dying, just as Empedocles says of his man-faced ox progeny.[68]

There can be no doubt about the position which Aristotle adopts, which is that Empedocles has a significant role for chance in both his zoogony and his cosmogony.

X. Chance: the doxographic evidence

The doxographic tradition strongly supports Aristotle in attributing a significant role to chance in Empedocles. While Philoponus gives us an excellent description of the role of chance in Empedocles, one has to wonder how much such evidence is based on Aristotle's reading of Empedocles alone:

> Empedocles says that it is by chance that air seized the upper position, all things having at one time been previously mixed together in the sphere, having been separated out by strife, each was carried to the place where it is now, not by providence, but by chance. Indeed, concerning the upward motion of the air he says 'Thus at one time it ran by chance, but many times it was otherwise.' Water is now on top of the earth, but at another time, if it chanced to happen in another *kosmos* making, when again the *kosmos* is generated from the sphere, a different order and place would be obtained. He also says that most of the parts of animals were generated through chance as though through forethought.[69]

Aetius tells us:

> Empedocles says that the places of the elements are not set throughout everything, nor are they defined, but all exchange places with each other.[70]

Achilles Tatius also informs us:

> Empedocles did not give defined places to the elements, but he says they exchange places in turn with each other, with the result that earth is carried to high places and fire down to low.[71]

Philoponus gives Fragment 53 exactly as in Aristotle. It is also repeated by Themistius and Simplicius. The latter is the most significant, for while the former two may have Aristotle as their only source of this fragment, Simplicius quotes a great deal of Empedocles. Simplicius in his commentary on Aristotle's physics gives seven passages to illustrate Aristotle's contention that Empedocles gave a significant role to chance and goes on to say, 'It is possible to find many such expressions in Empedocles' physics.'[72] He compares Empedocles with the atomists, and states:

> Empedocles is not as worthy of consideration, as he seems to employ chance in minor matters without explaining its nature.[73]

One of these seven is Fragment 53, the others we will look at in the next section on zoogony.

XI. Chance and zoogony

There can be no doubt that Empedocles has a significant role for chance in his zoogony. In what follows I am not going to be overly concerned with the details of the zoogony, or what happens under love and strife respectively. I am more concerned to establish the pattern of explanation that Empedocles uses in zoogony. There are several sages to Empedocles' zoogony, beginning with the generation of body parts, through their chance association to the formation of creatures which are viable in themselves and are able to reproduce. In the first stage of generation:

> Earth came together by chance with these in roughly equal quantities, Hephaestus, rain and shining air, anchored in the perfect harbours of Kupris, either a little more or in more of them. From these were generated blood and the types of flesh.[74]

> And kindly earth received in its broad melting-pots two parts of the glitter of Nestis out of eight, and four of Hephaestus; and they became white bones, marvellously joined by the gluing of Harmonia.[75]

> Of those which are compact within and loosely formed outside, having chanced on this flaccidity in the hands of Kupris.[76]

Here one would reasonably suppose that there were many other occasions where these ingredients came together, but not in the right proportions, and so blood, flesh and bones were not formed. Now Aetius tells us that:

> Empedocles believed that the first generation of animals and plants was not generated complete in all parts, but consisted of parts not joined together, the second of parts joined together as in a dream, the third of wholes, while the fourth no longer came from homogenous substances like earth and water, but by mingling with each other.[77]

So we then have the dream (nightmare?) stage of:

> On the earth there burst forth many faces without necks, arms wandered bare bereft of shoulders, and eyes wandered needing foreheads.[78]
>
> Many sprang up two faced and two breasted, man-faced ox progeny, and conversely ox-headed man progeny.[79]
>
> These things fell together, encountering each other by chance, and many other things were constantly being produced.[80]

So we have further accidents. Aristotle then says:

> Whenever everything happened as if it were generated for the sake of something, then they survived, having come together accidentally in a suitable manner. Where this did not occur, however, they perished and are still perishing, as Empedocles says concerning his 'Man-faced ox progeny'.[81]

Presumably, those species capable of having viable offspring survived once the first male and female had come together by chance. The other types of creatures would have perished.

> There by the desire of chance all things are sentient.[82]
> The best have fallen together by chance.[83]

What is the nature of the explanation here? What needs to be explained are the origins of something complex, which appears to have been designed. Empedocles shares the view that it is not plausible to explain this by the action of chance in a single instance or a small number of instances. Instead, we have a multiplicity of accidents, giving us a multitude of differing results. The generation of something complex can then be explained as one of a series of accidental happenings.

This is the first instance of this type of explanation. It is taken up by several thinkers in the ancient world, most notably the early atomists, and forms a stark contrast with the guiding principles of the Milesians, and especially with the intelligent design theories of Plato and his followers. Empedocles' pattern of explanation is still going strong in zoogony where it is effectively the dominant idea. One might also note that in line with later evolutionists (cf. the nineteenth-century debate about the age of the earth),[84] Empedocles feels the need for there to have been a great amount of time for the process of the formation of species. Aetius tells us that for Empedocles:

> When the human race was first generated from out of the earth, the day was of the length that ten months are now, owing to the slowness of the sun's advance. As time went on, it became the length of seven months.[85]

XII. Chance and cosmogony in Empedocles

Is there further evidence in Empedocles that he has a significant cosmogonical role for chance? In the Strasbourg papyrus, ensemble d, line 11 we have:

> When there chanced to be an unyielding flame.[86]

The Greek translated here as 'chanced to be' is *sunetungchane*. We might compare this with Fragment 85:

> The gently shining flame chanced upon short-lived earth.[87]

The Strasbourg papyrus ensemble a also gives us:

> They never cease from persistently shooting in all directions, in tight vortices [88]

which might suggest not only chaotic but random motion, and would tie in with Aristotle's assertion of disorderly motion. The evidence for a significant cosmogonical role for chance in Empedocles is then very strong. Aristotle gives us a key fragment twice, repeated without question in the doxographic tradition. Simplicius, who knows a good deal of Empedocles, accepts this fragment and gives us six more attesting the role of chance. There is further evidence in Empedocles concerning the role of chance, which fits well with the fragments produced by the doxographers.

How significant then is the role for chance in Empedocles? Returning to Aristotle *Physics* II/4, 196a20ff. and the context given there, it is absolutely critical, because of the way Empedocles describes his cosmogony:

> Air was separated off from the original mixture of the elements and crowded around in a circle. After this, the air/ fire, having no other space, ran up and out under the solidified air.[89]

Similarly, Aetius tells us that:

> Empedocles held that first aether was separated off, secondly fire and after

> this, earth, from which water sprang out because of the tight constriction due to the speed of the rotation. From this water came air, by evaporation, and the generation of the heavens was from aether, of the sun from fire and by compression of the others came the things around the earth.[90]

Furthermore, Eusebius tells us that for Empedocles:

> There are two hemispheres rotating around the earth, one wholly of fire, the other a mixture of air and a little fire. This he supposes to be night. The origin of the motion is due to the chance happening that fire pressed tightly together into a mass.[91]

The role for chance then goes right to the heart of Empedocles' cosmogony. Some of the key processes begin with chance occurrences, which might well have happened otherwise. Here I am in agreement with Trepanier, who comments:

> Empedocles' cosmic cycle did not consist of a completely identical pattern of recurring events, as in the later Stoic doctrine of the cosmic cycle.[92]

One objection to a significant role for chance in Empedocles might be that there are identical recurring states. That will certainly be so under the dominance of love, with total mixture, and would also be so in the concentric sphere view of strife. This might then give the appearance of a determinism operating between these states, but this need not be so. As long as love and strife ultimately produce total association and total dissociation, some events on the way, such as zoogony and on my account cosmogony, can come about by chance. As there is no question that zoogony occurs by chance, there can be no objection in principle to cosmogony occurring at least partly by chance as well, even with two identical recurring states. Of course, as I have argued above, I think there is only one recurring identical state, that of total love. There are many dispositions of the elements at total strife, though of course two identical states could occur accidentally.

XIII. Modern cyclical cosmos theories

There are modern theories of a cyclical universe. That the universe may expand and then contract is a straightforward possibility in big bang cosmogony. If the contractive effect of gravity (determined amount of matter and the value of the gravitational constant) is greater than the expansive energy of the universe, then the expansion will slow and stop, and a contraction will result. At present, we believe our universe to be expanding slightly too quickly for this to occur.

What happens should the universe contract is of course a matter of conjecture. It may collapse to a singularity, which either goes out of existence in a converse way to the big bang coming into existence, or the singularity may exist as such for all time. Thus we would have only one cycle to the *kosmos*. Alternatively, the universe may reach a singularity, or some point close to it, and then begin to expand again. If so, then there would be an endless cycle of universes. If the fundamental constants are in some way reset, perhaps randomly, each time the universe contracts to a singularity, then there is a way of explaining the nature of our own universe. Over time, there are a multiplicity of universes with different values for the fundamental constants, so our universe with its apparently fortuitous values for the fundamental constants is no longer a surprise, but is one of many.

Empedocles' problem is slightly different, in that he needs to explain how a *kosmos* can occur without intelligent guidance, both from a complete association and a complete dissociation of the elements. His answer is structurally the same as the modern answer though. If there is an element of chance, and if we have enough goes, then the occurrence of a *kosmos* can be explained as one among a multiplicity of accidents. Empedocles gives the same sort of answer in zoogony, and here too he finds a modern counterpart. Species evolve through a multiplicity of accidents leading to viable species. Whatever we may think of the details of the explanation that Empedocles gives us, the explanatory structure he uses is important.

XIV. Empedocles and teleology

Does Empedocles use teleology? If he does, it will be a very odd sort of teleology, for several reasons.

Firstly, the *telos*, the end aimed at, will not be a good organisation of the world, a *kosmos*. Love and strife, if they aim at anything, aim at total association or total dissociation. They may generate a *kosmos* as the universe moves from association to dissociation and back again, but the *kosmos* could hardly be said to be the goal of love or strife.

Secondly, teleology is generally associated with intelligence and the good. The *telos* envisaged is good, and there are intelligent actions to bring about that end. That would mean though that the destruction of the *kosmos* under love would be both intelligent and good, contrary to usual teleology. If strife is an agency for the bad,[93] aiming at total dissociation, then the construction of the *kosmos* under strife is bad, again contrary to usual teleology.

Thirdly, strife would seem to have a very odd role. Is it possible to call the actions of strife intelligent when its end goal is bad? Yet do we have any reason to suppose that strife is any less clever than love, when it can accomplish cosmogony and zoogony, and its goal of dissociation? Do we have reason to believe that strife is any less purposive than love?

Fourthly, however anthropocentric it may be, humans are typically seen as the good result of intelligent design by teleology. However, as we live under increasing strife, we are, at least to some extent the products of strife and everything that entails. The same point can be made about the *kosmos* which we live in. One can then draw a dilemma. If strife does involve teleology, the problems noted above all occur. If strife does not involve teleology, then neither humans nor the current *kosmos* are explained teleologically.

Fifthly, Plato and Aristotle do not seem to recognise Empedocles as a teleologist. Given that Plato criticises Anaxagoras, and is keen to criticise both cosmogony and zoogony in Empedocles, that is odd. One interesting perspective on the *Philebus* is to read the criticism of hedonistic love as in part an attack on Empedocles. Hedonistic love, on its own and without intelligence, will not produce a good life, nor will it produce a sustainable *kosmos*.[94]

Finally, those who employ teleology hold that our *kosmos* is unique, this best *kosmos* having been brought about by intelligent design, and that animal species have come about through intelligent design as well. Empedocles though employs a different explanatory structure, animal species having come about through a multiplicity of accidents, and our *kosmos* being one of many successive *kosmoi*, where chance plays a significant role in their formation. Empedocles may be an exception to this, but I argue there is another explanation for what Empedocles is doing.

The best evidence for teleology in Empedocles is in three fragments: 86, 87 and 84, taken in succession:[95]

> From these Aphrodite produced untiring eyes.[96]

> Aphrodite having given them bonds of affection.[97]

> As when someone intending to go forth on a journey prepares a lamp for himself, a fire burning through the night of winter, fitting linen screens as protection from assorted winds, which stop the blast of the blowing winds, but the finer light escapes, and shines across the threshold with unfailing beams. In the same manner elemental fire, wrapped in membranes and delicate flesh, was concealed in the round eye.[98]

We might also consider Fragment 95, which is also in the context of the generation of the eyes:

> When they initially grew together in the hands of Kupris.[99]

Sedley has argued that:

> Even if one strips from this the figurative personification of Love as a divine artisan, one is left with the impression of an intelligent and purposive creative force. The architectonic role of Love in Empedocles' cosmic cycle makes it a very hard task indeed to portray him a pure mechanist.[100]

One can commonly find in modern biology texts explanations which look as if they imply purpose or intelligent design when no such implication is intended. One might reasonably say, in the context of explaining the function of the eye, that 'the eye is designed such that ...' while remaining a staunch evolutionist. I am not convinced then, on the basis of this evidence alone, that Empedocles employed teleology. Wright comments: 'The whole passage is Homeric in vocabulary and rhythm, as well as in the simile form.'[101]

If so, we must have further doubts as to whether we can take Fragment 84 as a direct assertion of teleology, especially as this is the only passage where teleology appears to be directly asserted. Fragment 95, while not contradicting a teleological interpretation, does not do a great deal to support it either. One could read it simply as the eyes growing under the influence of love (association occurring) without any design or purpose on the part of love. It is significant in relation to Fragments 75 and 98, where components do come together by chance, 'in the hands of Kupris' or 'anchored in the perfect harbours of Kupris'. Empedocles does make considerable use of a craftsmen analogy for the actions of love. Does that imply intelligent, purposive behaviour on the part of love or does that merely draw a parallel between the way that craftsmen and love are both able to associate things together? If, prior to Anaxagoras, teleology as such is not a prominent or sensitive issue, Empedocles may employ analogies which are intended to be simple craft analogies without the intention that they be interpreted as teleological analogies.

We need not take Empedocles as someone who envisages a mechanical picture of the world if we reject him as a teleologist. I am not sure that anyone in antiquity had a fully mechanical world view, at least in the sense of the mechanical philosophy put forward by Descartes.[102] Descartes had pieces of matter/ extension which had the properties of length, breadth, depth and motion only. There were no attractive or repulsive forces, and interaction was by contact alone. Most supposed mechanists in the ancient world at least have a 'like-to-like' principle in addition. The ancients also lack the mechanical analogues typical of the mechanical philosophy, most notably the idea that clockwork is the key analogue for natural processes, including biological processes.[103] Ancient technology does not lend itself easily to mechanical analogues.[104] In contrast to Descartes, who explained gravity in terms of the contact action of particles alone, the Newtonians had gravity as an attractive force which acted at a distance.[105]

Empedocles is clearly not a mechanist in the Cartesian sense, nor can we really call him a materialist. Love and strife are clearly active principles which cannot be reduced to matter and motion. Active principles though do not need to be teleological principles. They can act on matter, be irreducible to matter and indeed act at a distance without being intelligent or purposive. In a static universe, if gravity is the only significant force then all the matter will be attracted together into one large lump.[106] No one would suggest though that is the aim of gravity, or that gravity is an intelligent or purposive force. So between mechanism and teleology, there is space for Empedocles to have some active but non-intelligent, non-purposive principles.

Earlier in this book I discussed the question of what cosmogony is *like*. As it is beyond our experience, we have to model it on something within our experience. We can also ask what forces are *like*. Not surprisingly, in the early days of philosophy and science, this was a difficult question to answer. The modern conception of force is not at all intuitive and was very hard won. This is a further reason why I prefer to call ancient attempts in this area principles rather than forces. Love and strife are clearly not forces in any modern sense, as I remarked earlier.

XV. Some Empedoclean weaknesses

Empedocles' cosmogony is original and influential. As one might expect from something this original and this early in the history of cosmogony, it also has its weaknesses. Aristotle is surely correct to be concerned about how motion can begin again once there is total love and unity has been achieved. If, like total strife, total love was a single moment as love reaches its zenith and strife its nadir, then there would not be a difficulty.[107] Both Aristotle and Plato, as teleologists, of course have their criticisms of Empedocles which we shall come to in due course.

Although this is a more modern concern, one might wonder why love and strife have the degree of efficiency which Empedocles attributes to them, rather than some other value. This is similar to the modern concern as to why fundamental constants have specific values. If love had a stronger associative effect, then it could bring about total mixture and unity much more quickly. So quickly, perhaps, that the *kosmos* would be relatively short lived and there would be no time for the evolution of sentient beings. If love were a weaker effect, then it might not complete its work of association, nor produce a *kosmos* or living things, before strife began to increase again. The modern analogue to this is the balance between the expansive effect of the initial big bang balanced against the contractive effect of gravitation. If this balance is off, either the universe expands too quickly for star and galaxy formation, or it collapses too quickly for life to begin. As with the value of the gravitational constant and its balance with other forces, the values and balance of love and strife, this looks fortuitous and unexplained. The balance of contrary effects/ qualities is something that is often assumed in Presocratic thought, but ultimately it does need some justification.[108]

Of course there are weak points in Empedocles' account of zoogony as well, again as one would expect in such an original theory so early in the history of science. Even in an ancient context though, the explanation of the origins of flesh, bone and blood as the chance event of elements coming together in certain ratios looks rather thin. He also appears to make a significant leap, from the formation of bones and flesh to the formation of well formed biological entities such as arms, legs, necks and eyes. One issue I have left untouched here is the nature of the zoogony under strife. A simple reversal of the order of zoogony under love, although that is what some of the evidence points to, would seem an extremely implausible train of events in the *kosmos* that we live in.

XVI. Empedocles' motivation

Why does Empedocles adopt this cyclical conception with cosmogony and zoogony both under love and strife? In the words of Kirk and Raven, he seems to have 'set himself a task which might well overtax the most fertile imagination'.[109] Clearly a reaction to Parmenides is at least part of the explanation here, but will not take us all the way. If Plato is right, and Empedocles' answer to the one and many problem is a progression of states, rather than Heraclitus' simultaneous state, then Empedocles could simply give us a progression from one to many or from many to one. As long as the timescale is long enough to accommodate many generations of humans, this will serve as an answer. So why does he do otherwise?

Empedocles accepts Heraclitus' criticism of the Milesians. Cosmogonical processes cannot arbitrarily change once a *kosmos* has been formed. Against Heraclitus though, Empedocles wishes to reaffirm the possibility of cosmogony. An important idea which I take to be original to Empedocles is that the processes which produce the *kosmos* continue to act in precisely the same manner even when the *kosmos* has been fully formed. So the *kosmos* is a by product of the actions of love and strife and not the end goal, and love and strife will both wreck the *kosmoi* they create. It should be clear from this that one option is ruled out for Empedocles, that of love bringing a *kosmos* together and then strife destroying it. It is significant that the turning points for Empedocles are not in any way cosmocentric or anthropocentric. Physically and philosophically, there is nothing else that the universe can do once it has become entirely one, other than become many again, and so too once it has become entirely dissociated by strife it can only begin to associate again. That love and strife change their effect (or their relative balance) over time is not problematic, as long as that change is continual and predictable. As Aristotle points out, there may be other problems here associated with the state of rest under the rule of love. How is it that from total rest the universe can begin to change again? Empedocles at least tries though to give both cosmogony and consistency in physical processes.

Empedocles seems to be the first to make significant use of chance in cosmogony and zoogony. As love and strife are blind principles of association and dissociation, which ultimately destroy what they create, Empedocles rightly recognises that to make his views plausible, he needs multiple accidents in order to produce living things and a *kosmos*. So in his zoogony we have the grotesque tale of how body parts come together by accident. In cosmogony, Empedocles opts for many instances of the cosmic cycle, each of which may turn out differently, even if they have the period of total love in common.

Empedocles differs from Leucippus and Democritus in that he postulates only successive *kosmoi*, and not co-existent *kosmoi* as well. Why should that be so? Essentially this is a question of how Empedocles and the early atomists react to Parmenides. Parmenides raises sufficient reason issues for cosmogony. He poses two fundamental questions. Why here? Why now? Empedocles can answer that there is only here for it to be and each *kosmos* is part of a determinate cycle of change. Although he is not explicit on this issue, I strongly suspect that Empedocles would say that there is no first cycle of change and the cycle continues as long as there is love, strife and the elements. Empedocles seems to envisage a limited *kosmos* which becomes one at repeated intervals. Aetius is in my view completely mistaken when he says that for Empedocles:

> There is one *kosmos*, however, the *kosmos* is not the all, but is a small part of it, while the rest is inert matter.[110]

Empedocles of course makes no mention of any inert matter, and it would be a nonsense to claim that the universe is one, and completely mingled, if the inert matter took no part in that mingling.[111] This brings out an important point though, which is that Empedocles envisages a unity for the universe in a way that the early atomists do not. For them, space is unlimited and the number of *kosmoi* within it are unlimited too. Empedocles' conception of the state of total love perhaps does not determine a single *kosmos* view (he could have a finite number), though his limited conception of the universe would rule out an unlimited number of *kosmoi*. It does lead him in a single *kosmos* direction though, and he has no philosophical need of more than one *kosmos* as long as he has an unlimited succession of *kosmoi* generated by the cycle. I will argue that the early atomists take the radical approach of rejecting the principle of sufficient reason for the why here, why now questions and say that a vortex, and hence a *kosmos*, can form at any place and at any time, by chance. Thus Empedocles also uses chance in a different way to the early atomists.

Why is there cosmogony and zoogony under strife as well as love? One answer to that is simple efficiency. We get two *kosmoi* for each cycle. We also get a destruction of the *kosmos*, and life, under both love and strife. If love both creates and destroys, it is hard to see why strife should not do the same, especially if we strip both down to principles of association and dissociation respectively. If it is correct that Empedocles, contra the Milesians, wants processes that do not alter once a *kosmos* has been formed, and, contra Heraclitus, consistent processes which allow cosmogony, then love and strife have to be indifferent to the *kosmos*. That they appear to be, as their critical points are total association and total dissociation. If love and strife are indifferent to the state of the *kosmos* in this sense, then there is no reason why strife should not have symmetrical cosmogonic and zoogonic roles to love.

XVII. Conclusion

Empedocles produced the first cyclical cosmogony, with multiple successive *kosmoi*. Love and strife, by their roles of association and dissociation, both have a cosmogonic and a zoogonic role, and both, by their continued action, destroy what they have created, on the way to a complete association or dissociation of the elements. According to Aristotle, Empedocles was the first to use two principles in cosmogony.[112] Empedocles was the first to produce a non-*kosmos*-centred cosmogony. The *kosmos* is a by product of the actions of love and strife, which carry on regardless once the *kosmos* has been formed, unlike the guiding principles of the Milesians. Empedocles was the first to give chance a significant role in cosmogony, and so too zoogony. In doing so he introduces a new pattern of explanation in cosmogony. The *kosmoi* generated by the cycle are not identical, and the differences are down to chance. Our *kosmos* can now be explained as one of an unlimited series of *kosmoi*. Empedocles was the first to use two processes for cosmogony, and to have those processes subject to regular change. In all of this, he is radically different from Heraclitus and the Milesians.

The motivation for this radically new cosmogony comes from several sources. Empedocles' reaction to Parmenides is clear and well documented, and one can readily see the agreements and differences between the two. Less evident, but equally important is his agreement with Heraclitus against the Milesians that physical processes must be consistent and not arbitrarily *kosmos*-centred, and his attempt to reinstate cosmogony contra Heraclitus with consistent, non-*kosmos*-centred processes.

Our knowledge of Empedocles' cosmic cycle is far from complete, and it may well be that Empedocles did not think through all aspects of it with the thoroughness we moderns would like. Even if we interpret Empedocles fairly generously, as I have done here, there are serious weak points and lacunae. What is more important than these flaws though is that Empedocles was both highly original and highly influential. Leucippus and Democritus build on Empedocles' new pattern of explanation, and although their reply to Parmenides is different, and in some ways even more radical, chance has a highly significant role to play in their cosmogony. The later atomists too draw on Empedocles' thinking, and the basic structure of explanation is still alive in some cosmogonical theories today. While the questions modern cosmogonists seek to answer are slightly different, the idea of a cyclical cosmogony, where each cycle is different from previous cycles is still a possibility to be explored.

6

Anaxagoras

Anaxagoras of Clazomenae (500/497-428/427 BC) was the first Presocratic to give a cosmic intelligence an independent existence and a role in cosmogony.[1] While the Milesians, as I have argued, were effectively panpsychists, Anaxagoras has intelligence separate from and unmixed with matter. Aristotle tells us:

> Someone said that *nous* was present in animals and in nature, and was the reason for the arrangement and order of everything, and he seemed a sober man compared with the accounts that preceded him. Clearly we know that Anaxagoras held this account, though Hermotimus of Clazomenae stated it first. Those who thought in this manner supposed there to be a first principle which was the reason for the beauty of things, and also gave rise to motion.[2]

It is interesting to note that Aristotle contrasts this view both with the Milesians, on the grounds that an element on its own cannot organise itself into a *kosmos*, and with anyone who employs spontaneity or chance. There is no indication in any of the ancient evidence that Anaxagoras gave any role to chance.

I. Anaxagoras and the Eleatics

Anaxagoras, like virtually everyone who came after Parmenides, both agreed with him on some issues and reacted against him on others. So we can find Anaxagoras saying:

> The Greeks were not correct in recognising generation and destruction. Nothing is generated or destroyed, but is brought together or separated off from the things that exist.[3]

On the other hand, Simplicius tells us that the following passage was the opening of Anaxagoras' book:

> All things were together, unlimited both in number and smallness. The small too was unlimited. All things were together, but none of them was visible because of their smallness. Air and aether restrained all things, both being unlimited. These are the greatest components of the mixture of all things, in number and size.[4]

So Anaxagoras sets out in defiance of Parmenides. In his first sentence, he uses a past tense and asserts the plurality of things, in contrast to Parmenides who says:

> It never was nor will be for it is now, all together, one and continuous.[5]

Anaxagoras also seems to allude to Zeno, wishing to disagree on divisibility and plurality:

> There is not a smallest part of the small, but there is always a smaller. It is not possible that what is should cease to be in the future. So too there is always something larger than the large.[6]

Zeno, on the other hand, says:

> If there is to be plurality, of necessity there must be small and great. So small as to be without magnitude, so great as to be unlimited.[7]

II. Cosmic *nous*

The starting point for Ananaxagoras then is a homogeneous mixture of different ingredients. The simplest summary of Anaxagoras on cosmogony is given by Diogenes Laertius:

> All things were together, when *nous* came and set them in order.[8]

The key cosmogonical statement, however, is Fragment 12:

> Other things have a portion of everything, but *nous* is unlimited, rules itself and is unmixed, being alone by itself. If it were not by itself, and mixed with other things, it would have a share of all things, if it mixed with any. In everything there is a part of everything, as I said previously. Mixed things would hinder it, with the result that it would be able to rule nothing in the same way as it does now, alone and by itself. *Nous* is the finest and purest of all things, it has all knowledge concerning everything and it has the greatest power. *Nous* controls all things that have life, both the greater and the smaller. *Nous* controlled the revolution of the whole, such that it revolved in the beginning. At first it revolved in a small region, but now it revolves in a greater, and will revolve in a greater still. The things which are mixed and separated and divided are all known by *nous*. *Nous* ordered them all, this revolution in which the stars and the sun and the moon and the air and the aether which are being separated off. This revolution produced the separation. The dense is separated from the rare, and the hot from the cold, the bright from the dark and the dry from the moist. There are many parts of many things. Nothing is entirely separated or divided from anything else except *nous*. *Nous* is entirely alike, both the greater and smaller parts of it. Nothing else is like anything else, but what is most in each single entity is most clearly what it is and was.[9]

Nous, then, is the critical factor in Anaxagoras' cosmogony. In my translation of Anaxagoras' work I shall render *nous* as intelligence throughout. *Nous* could also be translated as mind, but Anaxagoras does not have the cosmic *nous* perceiving or feeling, or at least this is not emphasised. The tasks of cosmic *nous*, as far as we are aware from the fragments, are to know, to order and to initiate motion. It initiates motion in the primordial mixture, and as differentiated objects are separated off, it brings them into order, creating a *kosmos*. *Nous* is distinct from all other things, being unmixed. Anaxagoras makes use of a vortex in his cosmogony. Vortices do occur naturally as whirlpools or tornadoes. So Anaxagoras appeals to something that is within our experience. In whirlpools, objects are dragged into the centre, so the sorting function of vortices in cosmogony is a reasonable extension of observation. Vortices were still in use as a model for gravity into the eighteenth century, until experimental results worked in favour of Newton's action-at-a-distance theory and against Descartes' vortices.[10] Against other ancient uses of the vortex, Anaxagoras has his vortex initiated by *nous*. Along with other ancient uses though, the vortex is an ongoing feature of the universe.

III. Anaxagoras on cosmogony

We can supplement the main account of cosmogony in Fragment 12 with a few more fragments:

> As these things rotated in this way they were separated out by the force and speed. It is the speed which generates the force. The speed is unlike any speed which now exists among mankind, being in every way many times more rapid.[11]

> Air and aether are being separated off from the many which surround, and the surround is unlimited in number.[12]

> The dense, the moist, the cold and the dark came together here, where the earth now is, while the rare, the hot, the dry and the bright went out into the further reaches of the aether.[13]

> From these things, as they are separated, earth is condensed. Water is separated out from the clouds, earth out from water, and from this stones are solidified by the cold, these stones having a greater tendency to move outward than water.[14]

The last fragment appears to lean heavily on Anaximenes for its process of transformation.[15] Why stones should be an exception (in physical terms) in their greater tendency to move out is unclear, though the result of the production of sun and planets is clear enough.[16] Finally, Aetius tells us:

> Diogenes and Anaxagoras say that once the *kosmos* had been formed and had

produced living things from the earth, it in some way, of itself, tilted to the south.[17]

Anaxagoras' views on zoogony, at least on the initial generation of life, are fairly straightforward. Hippolytus tells us that for Anaxagoras:

> Life-forms were generated from moisture initially, but later on from each other.[18]

Theophrastus tells us:

> Anaxagoras says that the air contains the seeds of all things and these are brought down together with water and generate plants.[19]

IV. Plato and Aristotle on Anaxagoras

Plato and Aristotle are highly critical of Anaxagoras in relation to teleology. Plato, in the *Phaedo*, has Socrates say:

> One day I heard someone reading from a book, which, he said, was by Anaxagoras. This said that it is intelligence/ *nous* which sets everything in order and is the aitia for everything. I found this theory of *aitiai* (explanations) pleasing, and it seemed to me to be something good that intelligence should be the aitia of everything, and I supposed, if this is so, then when intelligence brings everything to order it establishes each thing as it is best for it to be. So if someone wishes to find the aitia for each thing, the manner of its generation, destruction or existence, it is necessary to seek concerning this, in what manner it is best for that thing to be, or otherwise to suffer and to act. On this theory, then, a man should seek to investigate nothing else, whether concerning himself or anything else, but what is good and best. The same man will necessarily know the worse, for the knowledge is the same. Considering these things, I was delighted to think I had found in Anaxagoras a teacher of the reason for things in accordance with my own way of thinking. I thought that firstly he would tell me, whether the earth is flat or round, and having told me this, he would explain the reason and the necessity, stating the nature of the better and why it is better for the earth to be as it is.[20]

Socrates is disappointed, though, and goes on to say:

> At no time did I think that, having said that things are ordered by *nous*, that he would bring in any other reason than it being best for them ...[21] My wonderful hopes came to ruin, since proceeding with my reading I perceived a man neither making use of (his) intelligence, nor assigning to it any reason for the order of things, but giving the reason as air, aether, water and many other absurd things.[22]

I take it that Plato has two objections here. Firstly, Anaxagoras gives explanations couched in the same inadequate (to him), entirely physical terms as those of the physiologoi. Secondly, Anaxagoras postulates an entity which does no explaining. Anaxagoras is reputed to have remarked that, though all men have intellect, not all of them use it, which may make Plato's comment all the more cutting.[23] As Fragment 12 makes it clear that intelligence does have a role in cosmogony, the problem here may be that Fragment 13 suggests that intelligence retires from an active role in the *kosmos*. Anaxagoras tell us that:

> After *nous* initiated motion, it became separated from all that was moved, and all that *nous* moved was divided. As what was in motion was divided, the rotation considerably increased the process of division.[24]

Plato is critical of Anaxagoras at *Laws* 967bff. Hippolytus tells us that for Anaxagoras:

> The sun, the moon and the stars are burning stones which are carried around by the rotation of the aether.[25]

Plato objects on the grounds that only ensouled, intelligent bodies could act in the perfectly regular manner we associate with the heavens.[26] Anaxagoras may begin on a teleological account of the *kosmos*, but according to Plato he does not see it through far enough. Aristotle too is critical of Anaxagoras. He says:

> Anaxagoras makes use of mind as a device in cosmogony, though whenever he is at a loss to explain something as from necessity, he drags *nous* in, and he uses anything other than *nous* to explain generation.[27]

> Those who speak of *nous* or love think of these reasons as the good, though they do not say that anything exists or is generated for them, but that motion comes from these things.[28]

Plato, in the *Phaedo*, may underplay how much Anaxagoras explains with intelligence and overplay what can be done with teleology for dramatic effect. Given the role that Anaxagoras assigns to intelligence in Fragment 12, in ordering everything into a *kosmos*, Plato can hardly deny that Anaxagoras uses intelligence in some way. Simplicius points out that Plato himself, in the *Timaeus*, after discussing the good and the role of the demiurge, makes use of physical explanations.[29] Whether Anaxagoras thinks of cosmic intelligence as the good, or the reason for beauty, as Aristotle informs us, is open to some doubt. Anaxagoras does not say this explicitly, though he may take it for granted that an intelligent ordering of matter would be both beautiful and the best organisation of matter. Aristotle may be interpreting Anaxagoras in terms of his own thought here. What is clear though is that Anaxagoras does not use the cosmic intelligence in a sufficiently thoroughgoing manner to satisfy either Plato or Aristotle.

V. The withdrawal of *nous*

Anaxagoras' position has some interesting similarities with those adopted by some of the mechanical philosophers in the seventeenth century. There a Christian God created the world in a ready ordered state (in line with the book of Genesis) and then took no further part in the running of the universe. Here *nous* provides the initial impetus and ordering before withdrawing, and it is interesting to note that like a Christian God, *nous* knows all, is immortal and is all powerful.[30] The critical difference here is that in the seventeenth-century conception God also creates a framework of physical law and forces to which matter is subject. Given the initial ordering, these forces and laws are supposed to be sufficient to explain the ongoing order of the world. As in the ancient world, there was a religious debate around this idea. Newton and his supporters argued that removing god from any part in the running of the world would lead to atheism. Leibniz on the other hand argued that to suggest that god was an incompetent craftsman and would produce a mechanism that needed 'winding up and cleaning' was derogatory of the power of God and would lead to atheism.[31]

It is at least a plausible view that without the direction of intelligence and without some framework of laws and forces the *kosmos* will not be able to keep its initial order, especially in a Greek context. Some have taken the view that even with such a framework the *kosmos* will not be able to maintain order. Newton, for instance, believed that without the active intervention of God, the stars would move into one giant fireball under mutual gravitational influence,[32] and that the gravitational attraction of the planets for each other would gradually disorder the solar system until it was no longer a system.[33]

Anaxagoras, however, provides no such framework for the explanation of why the *kosmos* has continuing order. It is not the ongoing work of *nous*, yet nor does he give anyone in an ancient context sufficient grounds to believe that matter will be able to sustain or develop the initial ordering. In particular while Anaxagoras allows that *nous* controls living things, he does not make the *kosmos* itself a living thing controlled by *nous*. The *Timaeus* provides a rather different picture, as after the initial ordering the *kosmos* has a soul, and so do the celestial bodies, which is why they move in an intelligent, ordered and regular manner. That of course is in particular contrast to Anaxagoras' view, of which Plato was well aware, that the sun and the stars were merely hot stones.[34] If *nous* withdraws its influence, then there is no framework for the maintenance of cosmic order. Necessity will continue to prevail, but is that enough? Anaxagoras might reply yes, if the initial ordering has been precise then necessity will do the rest. Plato of course will reply no, necessity is incapable of keeping the planets in complex orbits. That requires the consistent application of intelligence, not just an intelligent initiation.

VI. Multiple *kosmoi* in Anaxagoras?

On the basis of the following passage, a part of Anaxagoras Fragment 4, there has been a debate as to whether Anaxagoras believed in multiple *kosmoi* or not:

> If this is so, it is right to believe that there are many things of all types in all that has been separated, seeds of all things and all sorts of shapes, colours and tastes and men have been formed and animals have souls, and that men have lived in cities and established farms, as with us, and have sun and moon and the others, as with us, and the earth grows for them a great amount of many sorts of things, of which they harvest the useful and take into their houses to consume. This then I have said concerning separation, that separation would occur not only with us, but elsewhere too.[35]

Aristotle at least does not appear to recognise multiple *kosmoi* in Anaxagoras:

> Empedocles and Anaxagoras separate everything out from a mixture. They differ in this, that the former has these things occur periodically, the latter has them happen only once, with unlimited numbers of homeomeres and opposites, while the former separates out only the so-called elements.[36]

This would seem to count against the idea of successive *kosmoi* in Anaxagoras. Aristotle also says:

> Those who say there are unlimited *kosmoi*, these *kosmoi* being generated and destroyed, say there is always motion, as the generation and destruction of these *kosmoi* necessarily involve motion. Those who believe in one *kosmos*, or that there is not always a *kosmos*, hypothesise opinions concerning motion consonant with their view. If at some time there is no motion, there are only two ways of conceiving of this. As Anaxagoras says, all things were at rest and at the same place for an unlimited amount of time, until *nous* initiated motion and separated them out. Alternatively, as Empedocles says, they were alternately changing and again at rest. They move when Love is making a plurality out of unity or when Strife is making a unity out of a plurality, but they rest in the time in between.[37]

This would seem to count against the idea of co-existent *kosmoi* in Anaxagoras. Simplicius does not place Anaxagoras in his list of those who believed in co-existent *kosmoi* or his list of those who believed in successive *kosmoi*.[38] Simplicius tells us:

> Those who say that the *kosmos* is unique, like Anaxagoras and Empedocles,

> have no need of an infinity of elements, so that Empedocles does better to make them limited.[39]

Simplicius does discuss whether Anaxagoras believed in other *kosmoi*, based on an analysis of the wording of Fragment 12, but comes to no firm conclusion.[40] Aetius is confused on whether Anaxagoras believed in multiple *kosmoi*.[41] Anaxagoras does say:

> The things in one *kosmos* are neither separated from each other nor hewn off with an axe, nor is the hot from the cold or the cold from the hot.[42]

However, this cold be taken as an assertion of the unique nature of the *kosmos* or it could be taken as an assertion of the unity of one *kosmos*, which could be one among many.[43]

VII. Successive or co-existent *kosmoi*?

There is no evidence, and no suggestion that Anaxagoras supposes there to be successive *kosmoi*. It is true that Anaxagoras does not seem to envisage an end to the process of separating off, and that he believes there is an unlimited amount of matter to draw on.[44] However, this proves little as much the same could be said for any of the Milesians. As Simplicius points out, Anaxagoras refers to other civilisations in the present, not the past tense. One might add that they are not referred to in the future tense either. There is nothing in Anaxagoras which requires successive *kosmoi* or suggests successive *kosmoi*. There is no sense that the current *kosmos* has any predecessors, or that there is any process of decay which will lead to the end of this *kosmos*.

There have been supporters of the view that Anaxagoras believed in multiple co-existent *kosmoi*.[45] As Aristotle does not recognise this, and as Simplicius, who sees more multiple *kosmoi* supporters than there really were, does not recognise this either, this seems unlikely. Certainly, Anaxagoras cannot believe in something like the early atomist pattern. While he has an unlimited amount of matter,[46] there is no void,[47] no role for chance[48] and no use of the *ou mallon* principle.[49] While Leucippus and Democritus had an unlimited number of vortices, coming into existence at unspecified times, Anaxagoras has only the one vortex to produce *kosmoi* from. Anaxagoras does not specify any boundary to the *kosmos* (no membrane of atoms)[50] which would separate it from the rest of the vortices and other *kosmoi*. So Anaxagoras has no means of individuating different *kosmoi* within his one vortex. If the rotation of the vortex keeps the heavenly bodies in motion, there is no provision for this in other *kosmoi*, assuming our *kosmos* is central.[51]

If Fragment 4 is meant to introduce multiple *kosmoi*, it is odd that Anaxagoras emphasises that in other *kosmoi* everything is 'just as with us'. A key point of multiple *kosmoi* is to be able to explain our *kosmos* as one of a large, or an unlimited array of other different *kosmoi*. So in Hippolytus' report of Democritus' views, it is the differences between the *kosmoi* (how many and what size suns and moons, whether they have water, etc.), which is emphasised.[52] To have many *kosmoi* all the same would only exacerbate the difficulty of explaining their nature. If Anaxagoras did want to introduce a multiple *kosmoi* view in Fragment 4, he does not do so clearly or obviously when there is ample chance for him to do so. In particular, he says for these other civilisations that 'the earth grows for them' rather than 'their earth grows for them'. The grammatical difference between the two is very small and simple, and if Anaxagoras postulated more than one earth, it is hard to see why he did not simply draw this contrast between our earth and other earths. It is most odd, if Anaxagoras did hold that there were multiple *kosmoi* that there is no mention, and no intimation of them in the long and important Fragment 12. If there are supposed to be multiple *kosmoi* in one vortex, it is hard to see how a separating out process based on a revolution could work for multiple *kosmoi*.

VIII. Other interpretations of Fragment 4

I must disagree with KRS when, concerning the number of *kosmoi* in Anaxagoras, they say this:

> Since the question was anyhow not for Anaxagoras (as it was for the atomists) one which arose inevitably from first principles and consequently demanded a definite answer, it seems wisest to follow the guidance of Simplicius and leave the question open.[53]

I do think this is a matter which stems from basic principles for Anaxagoras, and I also think we can come to a reasonably definite solution. If Anaxagoras does not believe in multiple *kosmoi* in the sense rejected above, what does he believe in Fragment 4? Cornford and Guthrie have defended the view that Anaxagoras is referring to other civilisations on the earth,[54] perhaps comparable to the 'hollows in the earth' of Plato's *Phaedo*, where the civilisations are unaware of each other's existence.[55] The key objection to this, and similar proposals,[56] is that surely, on both grammatical and philosophical grounds, Anaxagoras is referring to cosmogonical separation in general, not just the generation of humans.[57]

A rather different line was taken by Fränkel, who suggested that Anaxagoras was indulging in a thought experiment. So we should read Fragment 4 as saying that if there were to be another world, it would be exactly similar to our own. This in many ways is a highly attractive proposal. There is no plurality of civilisations, and no plurality of *kosmoi*. We get an interesting contrast with the early atomists, in that while

Anaxagoras seems to say that if there were any further *kosmoi* (which there are not), they would all be similar to the current one, Leucippus and Democritus say that there are further *kosmoi* and they are different. We get a similar contrast with Empedocles, if different cycles produce different *kosmoi*. The great difficulty with Fränkel's account, as Schofield points out, is with how *dokein*, to suppose, functions in this passage. In the Greek, *dokein* is used once and governs the rest of the long first sentence. Fränkel's account requires us to take *dokein* first as 'it is right to believe', down to where men are formed, where we then take *dokein* to be counterfactual.

Furley gives *dokein* the sense of 'it is only reasonable to expect'.[58] Anaxagoras is then explaining what we can expect given the initial conditions. The phrase 'as with us' uses the 'as' of exemplification rather than comparison, so there is no question of further *kosmoi*.[59] The main difficulty with Furley's account is with how the final sentence fits with this analysis. Why can Anaxagoras claim, on Furley's interpretation of what the first sentence says, that separation would occur in the same manner as it does here?

An altogether different approach is the one favoured by Leon,[60] Strang,[61] Mansfeld,[62] Schofield[63] and Sisko.[64] Critical for this interpretation is Anaxagoras Fragment 6:

> Since there are parts equal in number of the large and the small, in this way all things will be in everything. It is not possible for them to be separate, but all things have a share of everything. As there is no smallest part, it is not possible for anything to be drawn apart or to be generated by itself, but as it was in the beginning so things are now. In all things there are many of the things which are separating off, equal in number to the small and the larger.[65]

Schofield argues that we can derive from this (and other passages on Anaxagoras)[66] 'the thesis that complexity is no function of size.' [67] So there could be another *kosmos*, with a full arrangement of parts and all the necessary complexity, on a scale far smaller than our own *kosmos*. Our own *kosmos* could be replicated at a level far below our perceptual thresholds. Fragment 4 is then to be understood as proposing at least one of these *kosmoi*. I see no reason though to restrict this view to one extra *kosmos*. If complexity is not a function of size, and there is unlimited divisibility, there is no reason why a 'micro' *kosmos* should not in turn contain another micro *kosmos* ad infinitum. So too in the other direction, for larger *kosmoi*, and it is noticeable that Anaxagoras couples the idea of 'always a smaller' with 'always a larger'.[68] We can hardly bar larger *kosmoi*, for several reasons. If perception of larger but not smaller *kosmoi* is an issue here, it is possible that larger *kosmoi* are imperceptible to us.[69] There are two logical considerations here as well though. Our *kosmos* is bigger than a micro *kosmos*, yet according to this theory there are humans in the micro *kosmos* as well, for whom our *kosmos* is bigger. Unless there is something arbitrary about us as humans (why, if size is irrelevant?) then there can be larger *kosmoi* than ours. Secondly, if there is an infinite series of *kosmoi*, it can hardly be the case that we are the last member of that series, and any assertion that the series is infinite in one direction only will be entirely arbitrary (again, why, if size is irrelevant?).

There are two versions of this multiple sized *kosmoi* theory. One version, which can be called the Leibnizian version,[70] has *kosmoi* inhabiting every piece of matter in our *kosmos*, [71] while in the other version, which I shall call the nested version, all the *kosmoi* are concentric, so our *kosmos* exists in the centre of a larger one and a smaller *kosmos* exists in the middle of ours. As Sisko argues, the latter is the more plausible, especially when we consider the way in which the vortex expands over time from a small beginning for Anaxagoras.[72]

I don't find either plausible, for two reasons. Firstly, if we do interpret Fragment 4 in this manner, then it is odd that such a radical and important thesis is not alluded to in any other extant fragment, and in particular it is not alluded to in Fragment 12, the key cosmogonical fragment. So too there is no trace of this theory in the doxography, a tradition very keen to see multiple *kosmoi* theories wherever it can. Secondly, Anaxagoras is desperately in need of some way of individuating his *kosmoi*, if this is what he believes. What is it that marks off a microcosmos from a macrocosmos? The early atomists are very specific on this point, there being a membrane which separates a *kosmos* from the surrounding void, matter and other *kosmoi*.[73] I see nothing along these lines in Anaxagoras, yet the theory is likely to fall flat without such criteria.

IX. Anaxagoras' anthropic argument

I take a rather radical line on the interpretation of this fragment. To begin with, I think it is important to consider the whole of Fragment 4. The modern view has taken the part we have yet to look at, here labelled (1), to follow on from the part we have been analysing. Here I propose to go back to the original order that Simplicius and Diels-Kranz have, which is that this part (1) precedes the section we have been analysing (2).[74] So Fragment 4 runs:[75]

> (1) Prior to these things being separated off, when all things were together, no colour was clear in any way. The mixture of all things prevented this, of the moist and the dry, the hot and the cold and the bright and the dark, as there was a good deal of earth in the mixture, and an unlimited multiplicity of seeds in no way like each other. Since this is so, we must suppose all things were in the whole.
> (2) If this is so, we must further suppose that there are many things of all types in all that has been separated, seeds of all things and all sorts of shapes, colours and tastes, and men have been formed and animals have

> souls, and that men have lived in cities and established farms, as with us, and have sun and moon and the others, as with us, and the earth grows for them a great amount of many sorts of things, of which they harvest the useful and take into their houses to consume. This then I have said concerning separation, that this separation would occur not only with us, but elsewhere too.[76]

What I take section (1) to establish is the homogeneity of the mixture prior to any separation beginning. That I take to be uncontroversial, as it is implicit in the idea that everything is in everything else. What I believe to be going on here is a piece of anthropic reasoning. The anthropic principle in modern cosmology was introduced to limit the Copernican principle. Its initial statement was:

> Our location in the universe is necessarily privileged to the extent of being compatible with our existence as observers.[77]

The anthropic principle can be used to 'explain', or at least reduce our surprise at some of the more astonishing features of the *kosmos*. It does this by taking as basic that we are intelligent carbon-based life-forms, and then asking what is necessary in cosmological and cosmogonical terms for the existence of such life-forms. If the process of star formation, heavy element formation, supernova, solar system formation and evolution took twelve billion years, it is no surprise that the universe is as old as it is or as big as it is. There are two main variants on the anthropic principle:

The weak anthropic principle (WAP)
The observed values of all physical and cosmological quantities are not equally probable but they take on values restricted by the requirement that there exist sites where carbon-based life can evolve and the requirement that the universe be old enough for it to have already done so.

The strong anthropic principle (SAP)
The universe must have those properties which allow life to develop within it at some stage in its history.

I do not want to suggest, of course, that Anaxagoras is following any modern version of the anthropic principle. However, if we put the anthropic principle in general terms, it is that we take the existence of humans as basic and then ask what is required for them to come into existence, then it is reasonable to suppose that Anaxagoras is following anthropic reasoning. Here I would agree entirely with Furley that the 'as with us' phrase is not a comparison between our *kosmos* and others, but simply describes our *kosmos*. One advantage of this view is that we can give a natural and consistent reading to *dokein* all the way through. At the end of what I take to be the first part of Fragment 4 'we must suppose' that all things are in the whole, a simple supposition based on Anaxagoras' general principles. At the start of the second part of Fragment 4, 'we must suppose' the separation of the whole leading down to the existence of humans.

A further aspect of this fragment is the relation of what I take to be the final line to what I take to be the first section. On other readings, it is unexplained why Anaxagoras should conclude that separation would occur similarly if it happened anywhere else. As noted above, what the first section of Fragment 4 does on my reading is establish the homogeneity of the initial mixture. So if *nous* were to act in the same way on any different part of the mixture, as the mixture is homogeneous the same *kosmos* would result.[78] One might take it that *nous* would always act in the same way in the same circumstances. That sort of invariance seems fundamental to Presocratic philosophy. However, Schofield objects that 'Mind could surely have devised a *kosmos* which was *not* providentially constructed, or providentially constructed in some other way.'[79]

Earlier I looked at Plato and Aristotle's criticisms of Anaxagoras on teleology. According to them, Anaxagoras attempted teleological explanation, but did not go far enough. What we need to consider here though is what Anaxagoras thought he could do with the idea that *nous* orders everything from the initial state. Did that consideration generate for him a unique, most intelligently ordered *kosmos*? I doubt, on the current state of the evidence that this could be settled conclusively, but I would say that the most likely answer is yes, for Anaxagoras there was one most intelligently ordered *kosmos*. He may not explicitly make the claim that this is the best organisation for the *kosmos*, but that may be because, pre-Plato, an explicit analysis in terms of the best is not really at issue, and all that is required is that this is the most intelligent initial organisation of the *kosmos*.[80] That, and the fact that intelligence does not have an ongoing role in the *kosmos* may not satisfy Plato and Aristotle, but may satisfy Anaxagoras. Doubtless *nous* could have conceived of an unintelligent ordering, or of a less intelligent ordering. Why though would it wish to instantiate it?

One final comment on multiple *kosmoi* in Anaxagoras. Leucippus and Democritus do employ an unlimited number of *kosmoi*. As with modern multiple universe theories, they seek to explain something about the nature of our own *kosmos*. The structure of our own *kosmos* is accidental rather than providential, but we can understand why it has this structure as it is one of an unlimited array of other accidental *kosmoi*, which instantiate an unlimited number of other structures. Anaxagoras cannot use this mode of explanation for two reasons. Firstly, he uses *nous* as an explanation of the structure of our *kosmos*. Secondly, all the *kosmoi* are supposed to have the same structure for Anaxagoras. As noted above, this exacerbates rather than solves the explanatory problems. It is difficult to see what motivation Anaxagoras might have for multiple *kosmoi*.

X. Why an anthropic argument?

Why would Anaxagoras use something akin to the anthropic principle? A possible answer here is that this is a reaction to Parmenides. As McCabe has argued, Parmenides' way of truth is highly paradoxical, in that its conclusion denies the existence of any *nous* which might either construct or think about the way of truth.[81] In short, if reality is one and unchanging, human beings and human minds are not part of reality. Anaxagoras might then, quite reasonably take some different starting points, asserting in the first place the existence of plurality and the existence of human beings. A question for him then would be how we can get from a homogeneous but plural starting point to the existence of human beings, in other words an anthropic analysis of cosmogony. Another important contrast here may be with Empedocles. Rather than treating the total association and total dissociation as goals, with the *kosmos* and humans as accidental and transient by products, Anaxagoras has humans as a planned product of cosmogony and zoogony. If there were other *kosmoi*, humans would be generated just the same, perhaps a sharp contrast to Empedocles where different cycles may produce radically different results. Humans, for Anaxagoras, are not an accidental product of some *kosmoi*, but a necessary product of any *kosmos*.

In relation to the why here and why now questions set by Parmenides, I suspect Anaxagoras would answer that the vortex forms in the middle of the infinite matter, and that the earth will form in the middle of the vortex. On the question of when, Sisko puts forward the very interesting suggestion that for Anaxagoras there is no first event.[82] Sisko emphasises that for Anaxagoras the vortex begins to revolve from a small area, and then makes the point that for Anaxagoras, there is no smallest area. This is a highly ingenious and very attractive solution in many ways, and one that finds several correlates in modern cosmogony. However, in my view it founders on Anaxagoras' insistence that there is an original, perfectly mixed state and that *nous* initiates motion. *Nous* need not initiate motion in a smallest part of matter, or indeed in any one part of matter as opposed to several at once. One might object here that *nous* then has to make a choice of what size of matter (and how much matter) it will initiate motion with. If Anaxagoras is truly indifferent to size, how will *nous* choose? This sort of objection though has to be faced by all interpretations of Anaxagoras, including the Leibnizian and nested *kosmoi* theories. Unless Anaxagoras proposes to instantiate all sizes of *kosmoi*, *nous* has to choose which sizes to instantiate. It also has to choose the difference in size between the *kosmoi*.

It may be the case that with the principle of infinite divisibility, size is no indication of complexity for Anaxagoras, as a philosophical notion. That does not establish that he is indifferent to size, however, because certain structures can only function within specific parameters of scale. The

Greeks were well aware of scale problems in civil construction, with e.g. the difficulties involved in generating a structure which will support a roof of twice the surface area. All Anaxagoras need recognise is that size matters in the construction of many things. The cosmic *nous* can then make an intelligent choice of the appropriate size and amount of matter to initiate the vortex with such that the most intelligent arrangement, in terms of scale, eventually comes to be as our *kosmos*.

If there are such discriminations of size for Anaxagoras, then he has no need of larger and smaller *kosmoi*. It is simplest by far if Anaxagoras has just the one *kosmos*. Admittedly, this leaves us with the difficulty of why *nous* would choose a specific moment to initiate motion. Certainly one possibility would be to say that time is inextricably linked to motion and so there is no time before *nous* initiates motion. If that was Anaxagoras' view though, one would expect that to have left a trace in the doxography, especially given Plato's views on time and cosmogony and his willingness to criticise Anaxagoras by name. It also runs counter to Aristotle's report that for Anaxagoras everything was at rest for an unlimited amount of time before *nous* initiated motion.[83]

XI. Conclusion

Anaxagoras then believes in just one *kosmos*. If we go back to Simplicius' order for Fragment 4, it can be seen as the first example of anthropic reasoning. Where Parmenides' way of truth reaches a conclusion which appears to deny the existence of thinking entities, Anaxagoras boldly asserts the existence of humans, and a pluralistic starting point for the universe, and then asks how it is that a *kosmos* and humans could come to be from the primordial state of matter. It is no surprise then that there is no mention of the supposed multiple *kosmoi* in any fragment of Anaxagoras, or that Aristotle and the doxographers do not attribute him multiple *kosmoi*, whether successive, co-existent or nested.

Anaxagoras was the first to employ a vortex in cosmogony. While he has only one vortex, and that is initiated by *nous*, we shall see that Leucippus and Democritus are quick to suppose there are an unlimited number of vortices in the universe, initiated by the motion of matter in the absence of a controlling *nous*. Anaxagoras was also the first to make a cosmic, controlling intelligence separate from the matter of the universe, unlike the Milesians.

There is a good deal that is original and interesting in Anaxagoras' views on cosmogony. Where he is perhaps slightly disappointing is in his responses to Parmenides' question about when and where the *kosmos* might be supposed to begin.

7

Leucippus and Democritus

Leucippus (fl. 440 BC) and Democritus (fl. 420 BC) gave a new turn to cosmogony in the fifth century. I will not attempt to separate the views of Leucippus and Democritus. Leucippus was earlier and is recognised by some sources as the inventor of atomism.[1] It is unclear what, if any, differences there were between the two on the issue of cosmogony.[2] Early atomism is often viewed as a reaction to the views of the Eleatics.[3] Parmenides is supposed to have denied that there can be a multiplicity of things, denied that there can be change, and denied any sort of existence to that which is not. In contrast the atomists assert that there are a multiplicity of atoms, that there is a void, that atoms can move within that void, and that while the atoms do not themselves change, the compound bodies they make up do. So too against Parmenides, they assert that there can be cosmogony at least in the sense of a *kosmos* coming to be, even if the ultimate constituents of that *kosmos* are eternal. Indeed, they assert that not only one *kosmos*, but many *kosmoi* come into being, and that the process of *kosmos* formation is an ongoing one, and *kosmoi* can pass out of existence, being dissolved back into their components again.

I. The *ou mallon* principle

Critical to the atomist view is the *ou mallon* principle. Translated literally, it means 'not rather' as in a sentence such as 'not x rather than y'. This does not mean a preference for y, though. It rather asserts an indifference, or lack of preference between x and y.[4] Leucippus and Democritus use this principle in a wide variety of situations.[5] What is exists no more and no less than what is not.[6] Perceptions are no more true than other perceptions.[7] It is also used to explain why there are an unlimited number of atoms with an unlimited number of shapes. Simplicius tells us:

> Leucippus supposed there to be an infinite number of atoms that are always in motion and have an infinite number of shapes on the grounds that nothing is such rather than such (*dia to mêden mallon toiouton ê toiouton einai*).[8]

Parmenides' argument led to the conclusion that what exists is one and is spherical. There is no real question of its size, at least in relation to anything else.[9] If there are to be multiple things, though, what are their shapes? Here the early atomists assert that there are no preferred shapes, and so with an unlimited number of atoms there are an unlimited number of shapes. If we accept that Leucippus and Democritus postulated any size of atom (as opposed to Epicurus, who specifically limits atoms to very small sizes),[10] then the *ou mallon* principle may have been applied to the sizes of atoms as well. Allowing for the possibility of huge atoms may not have been an oversight, corrected by Epicurus, as is generally thought, but an integral part of early atomist thought.

II. *ou mallon* and early atomist cosmogony

The *ou mallon* principle is also significant in relation to other challenges to cosmogony that Parmenides provides. If cosmogony is to begin, whether from nothing or from a homogeneous substance, what sufficient reason is there for when it should begin? Parmenides does not discuss place in the same manner, but one can easily see how his sufficient reason argument could be extended to where a *kosmos* is generated as well, given the requisite assumptions about space.[11] How do Leucippus and Democritus react to these issues? The first stage of early atomist cosmogony is vortex formation. The fullest statement of early atomist cosmogony we have comes from Diogenes Laertius:

> Leucippus holds that the whole is infinite … part of it is full, and part void … from these innumerable *kosmoi* come to be and are dissolved into these again. The *kosmoi* are generated in this manner. By 'cutting off from the infinite' many bodies of all shapes move into a great void, where they are crowded together and produce a single vortex, where colliding with each other and circulating in all manner of ways, they separate out like to like. When, because of their great number they are no longer capable of moving around in equilibrium, those that are fine spread out into the outside void, as if sifted, while the rest 'hold together' and becoming entangled, they unite their motions and create the first spherical structure. This stands apart like a 'membrane', containing in itself all kinds of bodies. As they whirl around, due to the resistance of the middle, the surrounding membrane becomes thin, and the close packed atoms flow together due to touching the vortex. In this way the earth came into being, the atoms which had been borne in to the middle remaining there together. Again the surrounding membrane itself is increased, due to the influx of external bodies.[12] As it moves around in the vortex, it takes in whatever it touches. Some of the bodies which become entangled form a structure which is firstly moist and muddy, but which dries out as it revolves with the vortex of the whole, and then ignites to produce the constitution of the stars.[13]

The first phase of cosmogony, then, is the 'cutting off from the infinite' (*kata apotomên ek tês apeirou*). This is a strange phrase and its meaning is far from clear. The literal meaning of *apotomên* is 'cutting off' (LSJ),[14] but that does not take us far in understanding this process. Presumably it does not denote the generation of matter *de novo*, but rather a gathering together of atoms which then form a vortex.[15] As ever, there is great difficulty in saying precisely what the cosmogonical process is like. What is important is that this 'cutting off', however that is supposed to work in detail, is different from the 'separating out' (*diakrinesthai chôris*) that occurs once the vortex is formed. It is notable that as with Leucippus and Democritus, Epicurus uses 'to separate' (*apokrinein*)[16] and its cognates when he talks of the principles by which matter is arranged into something coherent.[17]

Once the atoms have gone through the cutting off process, they 'move into a great void'.[18] This seems odd as there is a sense in which all the atoms are already in a great void, i.e. the expanse of not being. If we take great void in a different sense, as a volume completely devoid of matter, it would seem enormously fortuitous that there is a local great void for the newly cut off mass of atoms to move into. From the *ou mallon* principle, we might reasonably expect the atoms to be randomly distributed through the void prior to the beginning of the cosmogonical process. If there is then a cutting off of some of the atoms, which gather together, that may create a volume of complete void with a tight mass of atoms in it, rather than the usual diffuse spread of atoms throughout the rest of the void. There is then no need for a separate great void which they move into. Epicurus has a different account of *kosmos* formation, but it is interesting that he says that:

> This will occur in a place with much void, but not in a place of complete void, as some have said.[19]

It is generally accepted that his target here is Democritus, and if so, this would seem to lend strength to the idea that 'great void' in Democritus should be read as 'complete void'. Another distinctive feature of this cosmogony is that it happens an unlimited number of times. It is interesting that Metrodorus of Chios, a pupil of Democritus, says that:

> It is strange for one ear of corn to be produced in a great plain, and for one world in the boundless.[20]

From a modern perspective, we might be somewhat suspicious of the biological aspects of this analogy, but the point in favour of multiple *kosmoi* is clear enough.[21]

In relation to the origins problem, of whether the initial state is homogeneous and so tricky subsequently to differentiate, or inhomogeneous but inexplicable, Leucippus and Democritus have two sorts of answer. Firstly, as they deny an origin to the universe, there is no original state. However, it seems likely there is a random (*ou mallon*) distribution of particles and random motion of those particles leading to the formation of *kosmoi*.

III. Cosmogony and chance

According to Aristotle, Leucippus and Democritus viewed the formation of a vortex as a matter of chance:

> There are some who make chance the reason for the heaven and all of the *kosmoi*. For from chance arose the vortex and the motion which by separation brought the universe into a state of order.[22]

There is ample evidence in the doxographical tradition to support this attribution of chance, e.g. Simplicius tells us:

> When Democritus says that 'A vortex of all shapes is separated off from the all' (how or by what cause he does not say), it appears that this occurs spontaneously or by chance.[23]

There is a difficulty though, in that Leucippus says:

> Nothing happens at random, but everything for a reason and by necessity.[24]

One can resolve the apparent contradiction here in two ways. Perhaps Leucippus and Democritus say something happens by chance when actually they mean it occurs by necessity, but the causes are unknown or unknowable to us.[25] Alternatively, perhaps Aristotle calls these chance events because they have no teleological purpose,[26] and the doxographical tradition then follows him, while Leucippus and Democritus consider them necessary events.

The problem with this is that it leaves Leucippus and Democritus with a rather feeble reply to Parmenides' when and where questions. They accept the principle of sufficient reason, but are unable to give any sufficient reason for when and where *kosmoi* form. Out of necessity from a previous state of distribution and motion of matter in the universe will not do, as Parmenides will simply ask for a sufficient reason for the previous state.

The alternative is to allow Leucippus and Democritus an *ou mallon* reply to the when and where questions, that there are no preferred times or places. One version of this would be that there is necessity outside of the vortices, but there is no reason for when and where vortices form. The second version is more radical in supposing there to be genuine chance events outside the vortices. Aristotle does recognise chance as distinct from necessity in some contexts, as we saw with Empedocles. Where purposeless necessity would repeat itself, a chance happening will occur differently on different tries, as with air being separated off upwards or otherwise in Empedocles.[27] Aristotle contrasts the chance formation of the vortex for the early atomists with the supposed subsequent formation of animals, which he says happens 'by nature'.[28] This would accord with the evidence of Diogenes Laertius, who says that for Democritus:

> Everything occurs by necessity, the vortex being the cause for the coming into being of all things, and this he calls necessity.[29]

So perhaps outside the vortices there is no (or limited) necessity, and the fragment we have from Leucippus is missing a context or a qualification. Eusebius certainly attributes random motion to the atoms outside of the vortices for the early atomists and random motion for Epicurus too:

> Those who name the unlimited number of small bodies atoms, and who suppose a void unlimited in extent, say the atoms move by chance in the void. With their disorderly motion they collide with each other ... Epicurus and Democritus held this view.[30]

If there are no preferred places, this would also accord with the evidence of Hipploytus, who says:

> The spaces between *kosmoi* are not equal, in places there are more and in others less.[31]

Leucippus and Democritus have a radically different conception of the vortex to Anaxagoras. Where Anaxagoras has a single vortex initiated and controlled by *nous*, Leucippus and Democritus have many vortices forming by chance, with no control over them. Leucippus and Democritus might reasonably point to the unpredictable and uncontrollable occurrence of vortices (whirlpools) in rivers and the sea.

IV. The like-to-like principle

The next phase is the separating out of atoms according to the like-to-like principle. Sextus Empiricus tells us:

> There is an old view which, as I said previously, has long been prevalent among the *phusikoi*, that like recognises like. Democritus confirmed this opinion and Plato spoke of it in his *Timaeus*. Democritus founds his argument on both animate and inanimate things. For animals, he says, flock with animals of the same kind – doves with doves, cranes with cranes, and so with the other irrational animals. Similarly in the case of inanimate things, as can be seen from seeds that are being winnowed and from pebbles on the sea-shore. For in the one case the whirling (*ton dinon*) of the sieve (*tou koskinou*) separately arranges lentils with lentils, barley with barley, wheat with wheat; and in the other case, by the motion of the waves, oval pebbles are pushed into the same place as oval pebbles, and round pebbles as round as pebbles, as though the similarity in things has some sort of ability for leading things together. That is Democritus' view.[32]

In many ways this is a typical principle of the ancient world. It is important to recognise that there is no force of attraction working at a distance here, in the sense of a force operating between like entities. All there is, is a sorting principle. It requires the whirl for the process to take place.[33] Typically too for the ancient worlds, the examples are indiscriminate between physical and animate phenomena. Plato's *Timaeus* too advocates a like-to-like principle, though Plato is critical of cosmogony based on this alone, as for him the *kosmos* is a harmonious blend of opposites, something highly unlikely to be produced from a like-to-like principle alone. We might also compare this like-to-like sorting with Homer where 'god always leads like to like'.[34] Leucippus and Democritus, while they can quite reasonably be called materialists, can hardly be called mechanists, at least not in the seventeenth-century sense of the term. Here, instead of a mechanical analogue for physical and even biological processes, they use a biological analogue for physical processes.[35]

V. The fate of *kosmoi*

The third phase in atomist cosmogony is the creation of *kosmoi*. Hippolytus tells us that:

> Democritus holds the same view as Leucippus ... There are innumerable *kosmoi*, which differ in size. In some of these there is no sun or moon, in some they are larger than ours and in some more numerous. The spaces between *kosmoi* are not equal, in places there are more and in others less, some are growing, some are in their prime, some declining, some are coming to be and others failing. They are destroyed by falling into each other. There are *kosmoi* bereft of animals and vegetation and all moisture. In our *kosmos* the earth was generated prior to the stars, and the moon is the lowest, followed by the sun and then the stars. A *kosmos* is at its height until it can no longer accrete external material.[36]

It is interesting to learn here that the *kosmoi* have a life-cycle. This is confirmed by Simplicius, who says:

> The *kosmoi* of Democritus change into other *kosmoi* made of the same atoms. The *kosmoi* thus become the same in form, but not in individuality.[37]

Diogenes Laertius also tells us that:

> Just as the *kosmos* is generated, so too it grows, decays and perishes, according to some necessity, the nature of which he does not define.[38]

So *kosmoi* can be destroyed by collision or by general deterioration. It seems that the debris of these *kosmoi* is then reused for further *kosmoi*. Certainly there is a consistent process of the generation and destruction of *kosmoi*. As we might expect from an *ou mallon* distribution of atoms and an *ou mallon* view of where *kosmoi* begin to form, the existing *kosmoi* are unevenly spaced.

There is nothing in Leucippus and Democritus to indicate that the process of the formation of *kosmoi* has an end, or that it has a beginning. The universe, as an ensemble of *kosmoi*, has neither beginning nor end. In terms of zoogony, Leucippus and Democritus are rather less radical than Empedocles, or at least rather less forthcoming. We have nothing from them suggesting a process of accidents leading to the formation of species, only a brief comment that in some *kosmoi* there will be no plants and animals, presumably because there will be no water.

VI. Multiple universe theories

There are multiple universe theories in modern cosmogony. One of the problems that ancient and modern multiple universe/ *kosmoi* theories seek to answer can be expressed like this. It is not plausible that a unique universe/ *kosmos* could arrange itself as our universe/ *kosmos* is arranged without there being some aspect of design involved. If we reject the idea of design and a unique universe/ *kosmos*, then the alternative is that our universe/ *kosmos* may be one of an infinite array. Within that array, if all possible arrangements are instantiated, then there can be no surprise that our *kosmos* has, as it appears to us, a fortuitous or beneficial arrangement.

The modern version of this problem is concerned with the values of the fundamental constants, such as the gravitational constant. For our universe to come about, the degree of 'fine tuning' of these constants is something of the order of one in 10^{50}. We have no explanation of why the fundamental constants have their observed values. So an infinite number of universes are supposed, either all with different permutations of values for the fundamental constants, or with randomly allocated values.[39]

The ancient version of this problem is concerned with how our *kosmos* can come to be from the primordial state of matter. In modern cosmogony, effectively this problem has been solved as we now understand how gravity plays the key role in bringing matter together and then forming galaxies, stars and solar systems. With Leucippus and Democritus, the vortices produce different *kosmoi*. Perhaps the clearest statement of the general principles behind the atomist views comes from Lucretius:

> So many primordial particles, in a multitude of ways have been propelled by their own weight and impacts for an infinite amount of time, and have made trial of all things their union could produce; it is hardly surprising if they have come into arrangements and patterns of motion like those repeated by this world.[40]

Leucippus and Democritus are then the first to propose the idea of an unlimited number of co-existent *kosmoi*. The motives behind this are twofold. There is a rejection of any sort of organiser or intelligent organising principle, and the rejection of the idea that a unique *kosmos* could come about without any such principle. This marks them off from the steering principles of the Milesians. There is secondly their *ou mallon* reply to Parmenides, taken in conjunction with their atoms and the void solution to the problems of change and one and many. With no preferred times and no preferred places for cosmogony, in an unlimited space, it is hardly surprising that *kosmoi* come to be in many places and at many times. This marks them off in turn from Empedocles, who adopts a different solution to change and one and many, and while he has a preferred place for cosmogony he does not have a preferred time. So while Leucippus and Democritus have unlimited co-existent *kosmoi* and unlimited successive *kosmoi*, Empedocles has an unlimited succession of single *kosmoi*.

VII. Principles of cosmogony

The universe of Leucippus and Democritus conforms to some interesting cosmological principles. It is the first to accord to the Copernican principle. This can be stated as either

(1) There are no preferred points of observation in the universe, or
(2) The universe appears the same, on a large scale, from every viewpoint.

This is not quite the same as the distinction between open and closed universes, first made by Koyré and interestingly applied to the ancients by Furley.[41] Furley says that 'The Closed World, according to its adherents, is identical with the universe.' Certainly that is true enough for Plato and Aristotle. It is not true though that everyone who advocates one *kosmos* in antiquity also makes that *kosmos* identical with the universe. Their *kosmos* may be bounded while their universe is not. I have argued that both Anaximander and Anaximenes hold that there is one *kosmos*, and that *kosmos* is embedded in what remains of the primordial substance, the unlimited or air. Quite possibly that is the case for Thales as well, though we have too little evidence to be sure on this point. That Leucippus and Democritus have an uneven distribution of *kosmoi* is not of great significance here. The modern universe is considered to conform to the Copernican principle (indeed it is held to be a basic tenet of modern cosmology) without the modern universe having an absolutely uniform distribution of matter, as is clear from a look at the night sky. On a large scale, however, that difference in distribution is held to be negligible. When we are informed that Leucippus and Democritus did not have a uniform distribution of *kosmoi* (the comparison is allegedly with Anaximander, who is supposed to have a uniform distribution), I presume this unevenness is within reasonable limits or can be supposed to conform to the large scale assumptions of the Copernican principle.

There is another cosmological principle which Leucippus and Democritus conform to, and this is the perfect cosmological principle. Where the Copernican principle asserts the *kosmos* to be similar on a large scale from any point in space, the perfect cosmological principle asserts that the *kosmos* will also appear the same, on a large scale, from any point in time.

This principle was formulated by advocates of the 'steady state' (SS) theory of cosmogony/ cosmology in the middle part of the twentieth century. SS was a serious competitor to the big bang (BB) on both theoretical and observational grounds until a series of observations and experiments went decisively in favour of BB during the 1950s and 1960s. Indeed, the very term 'big bang' was coined by steady state theorists as a derogatory term for any theory of instantaneous creation. SS argued that BB was unscientific, on the grounds that we could not make repeated observations of cosmogonical processes, and that the cosmogonical processes halted once creation had occurred. SS argued for a continuous creation of matter.

Leucippus and Democritus have the first theory with a cosmogony which accords with the perfect cosmological principle. There is a permanent production and dissolution of *kosmoi*. The *kosmos* of Heraclitus accords to the perfect cosmological principle as well, though without either cosmogony or the death of the *kosmos* this is rather vacuous. Epicurus, as we shall see, is more explicit about the universe as a whole being unchanged through time. Leucippus and Democritus, while not having a continual (or indeed, any) creation of matter, do have a continual creation of *kosmoi*. The processes which bring about *kosmos* formation are not restricted to an early stage of the *kosmos*, but are an integral part of its nature, as with Empedocles.

VIII. The application of *ou mallon* arguments

I have argued that Leucippus and Democritus use the *ou mallon* principle quite widely, in epistemology, stoichogony (the shapes and sizes of atoms) and for an assortment of problems in cosmogony. There are some areas though where we might question why they do not apply the *ou mallon* principle as well. One is at the very start of the process of cosmogony. Why, when the atoms separate off and are crowded together, is a vortex formed? Unless there is some necessity which forces the production of a vortex, one might expect a multiplicity of products from this process, one of them being a vortex. It may be that the problem here is the transmission of Leucippus and Democritus' views. Perhaps they did believe some necessity generated vortices out of this jumble of atoms, perhaps they believed there were many products one of which was the vortex, and later commentators thought the vortex was the only product worth mentioning.

I have been careful so far to keep the terms *kosmos* and universe separate, as Leucippus and Democritus believe in many *kosmoi* within a single universe. They individuate their *kosmoi* quite strongly. Aetius tells us:

> Leucippus and Democritus envelop the *kosmos* in a circular 'garment' or 'membrane' which comes about through the hook-shaped atoms becoming entangled.[42]

A *kosmos* is then quite sharply differentiated from the surrounding, and any other *kosmoi*, by the membrane which surrounds it. With many *kosmoi* inhabiting one unlimited void, this is a wise move. Leucippus and Democritus do not seem to question the idea that there is one universe which contains the many *kosmoi*.

A further *ou mallon* consideration though is this. One might argue that if we treat the universe as a whole, it has a definite arrangement of *kosmoi* and matter within it. Is there any sufficient reason why it should have this specific arrangement of matter? If the answer to that is no, then we might have to say that no arrangement is preferred to any other arrangement, and given the structure of the argument about *kosmoi*, we might then suppose there to be an infinite number of universes as well. This is certainly a problem for modern multiple universe theories. An infinite regress can be generated in at least two ways. If the motivation for multiple universes is a version of the Copernican principle (nothing unique about the earth, solar system, galaxy, galaxy cluster, universe), then the argument is clearly regressive. If the motivation for multiple universes is that aspects of our universe can only be explained by considering it to be part of an infinite array of universes, then we must ask whether the resulting infinite array of universes also has aspects of order which need to be explained by a further infinite array. If so, the regress may be infinite. One way of halting this regress is to deny properties to the infinite array of universes, that is to deny that it constitutes a universe in itself. It is no longer then a trivial question whether we live in a universe, a single entity, or a multiverse.

I suspect that these sorts of considerations did not occur to Leucippus and Democritus. Radical though their application of the *ou mallon* principle is, they are also slightly inconsistent in their application of it.

IX. Conclusion

Leucippus and Democritus produce some radical and influential new ideas in cosmogony. While Empedocles had an unlimited number of successive *kosmoi*, Leucippus and Democritus had an unlimited number of co-existent *kosmoi*. They were the first to explore this important option. Their reaction to a range of problems for cosmogony set by Parmenides is also remarkable. In essence their reply to the 'why now' and 'why here' challenges to *kosmos* formation is simply that there are no preferred times and places. This principle they take further in relation to the shapes and sizes of atoms. The universe, and our *kosmos*, are entirely purposeless for the atomists. There is no principle guiding the organisation of the *kosmos*, neither the benevolent panpsychism of the Milesians nor the principles of love and strife of Empedocles. The views of Leucippus and Democritus are highly influential. With Empedocles, they establish the multiple accidents line of thought in cosmogony and zoogony.[43] Plato and Aristotle will be severe critics, though Plato never mentions Leucippus or Democritus by name.

8

Some Other Presocratics

In this chapter I deal with some of the rest of the Presocratics, Xenophanes (*c.* 570 – *c.* 475 BC), Pythagoras (*c.* 560 – *c.* 480 BC) and his followers, Archelaus (fl. 450 BC) and Diogenes of Apollonia (fl. 440-430 BC). None of these figures really merits a chapter to himself in a work on cosmogony, though all have something of interest to say, either in neighbouring disciplines, as influences on later figures, or as providing small variations on previous thinkers.

I. Xenophanes and theology

Xenophanes is not hugely important in the history of cosmogony, though his influence on religious thinking is important in itself, and as an influence on later theologians and cosmogonists. Xenophanes is critical of Homer and Hesiod:

> Homer and Hesiod have ascribed to the gods all those things which are shameful and reproachful among men: theft, adultery and deceiving each other ... Mortals believe that the gods are born, and that they have clothes, speech and bodies similar to their own... If cattle, horses and lions had hands, and could draw with those hands and accomplish the works of men, horses would draw the forms of gods as like horses, and cattle like cattle, and each would make their bodies as each had themselves ... The Ethiopians claim their gods are snub-nosed and black, while the Thracians claim theirs have blue eyes and red hair.[1]

So for Xenophanes, there is:

> One God, greatest among Gods and men, entirely dissimilar to mortals in *nous* and body.[2]

> [This god] always remains in one place and not moving, nor is it appropriate for him to go to other places at other times, but aloof from manual labour he agitates all these by the thoughts of his mind.[3]

Xenophanes asks important questions for the founding of a rational theology, questioning the number of the gods, whether our conception of them is anthropomorphic, and what the morality of the gods might be. The notion of a single god, with none of the human vices, who always acts for the best, will be important for Plato's cosmogony and much subsequent cosmogony.[4] We know little of what Xenophanes had to say about cosmogony, apart from two brief passages:

> All things that are generated and grow are earth and water.[5]
> We were all generated from earth and water.[6]

This is standard Presocratic material and unfortunately we know little of Xenophanes' reasoning here.

II. Xenophanes and the *kosmos*

Hippolytus tells us that for Xenophanes:

> The sun is generated each day from small pieces of fire, the earth is unlimited and is not surrounded either by air or the heavens. There are unlimited suns and moons, and all things come from earth.[7]

Given that the earth is unlimited, it is likely that these are successive rather than co-existent suns and moons. According to Aetius:

> Xenophanes says that there are many suns and moons according to the regions, divisions and belts of the earth.[8]

So while Xenophanes may have a rather odd and fragmented account of the nature of the earth, there are not an unlimited number of co-existent suns and moons. We also learn from Aetius that:

> Mankind is destroyed when earth is carried down into the sea and mud is generated, then another beginning is generated, and this foundation occurs for all of the *kosmoi*.[9]

It is unlikely that the earth as such is destroyed here, for according to Xenophanes:

> We see the upper limit of the earth by our feet, in contact with the air, but below its reach is unlimited.[10]

More likely there is a sufficient catastrophe for humans, and perhaps all life to be wiped out, when life is again generated from earth and water. So we should take *kosmoi* in the Aetius passage to refer to a human order which is destroyed in each deluge, especially as Aetius also tells us that:

> Xenophanes says that the *kosmos* is ungenerated, eternal and imperishable.[11]

It is likely then that Xenophanes held some form of cyclical theory concerning land becoming sea or mud, and this eventually becoming land again, perhaps with the destruction of humans in each cycle.

III. Pythagorean cosmogony

It is difficult to reconstruct Pythagorean cosmogony with any degree of certainty. Our evidence is fragmentary and unclear. We cannot say for sure that there was a Pythagorean cosmogony as opposed to several competing versions. We are unsure quite how religious thinking and speculation concerning the physical world are supposed to relate together in Pythagoreanism. Were there separate strands of Pythagoreanism with different emphases or was there a single strand combining these religious and physical elements? It is unclear whether the material we do have on cosmogony should be treated as a logical analysis of the nature of physical objects, or as an account of their physical genesis. It is also unwise to assume without further evidence that the Pythagoreans would have recognised the distinction between analysis and cosmogony. Aristotle, our main witness, treats the material as of physical origin, and although he asserts this clearly and seems aware of the alternatives, he is hardly an infallible guide on such matters, especially as he is extremely hostile to Pythagorean ideas. Aristotle recognises that the Pythagoreans use more abstruse principles than the *physiologoi*,[12] using mathematical rather than sensible principles, but is still adamant that their discussions do concern the nature of the universe and that they give a cosmogony which pays attention to events, actions, principles and reasons.[13]

It is possible to give a schematic of Pythagorean cosmogony which runs: the one – numbers – geometrical objects – physical objects, but whether one can make a coherent, detailed theory of cosmogony out of that on the current evidence is another matter. On the generation of numbers from the one Aristotle tells us:

> The elements of number are the even and the odd, and of these one is limited and the other unlimited. The one consists of both of these, since it is both even and odd, and number comes from the one, numbers constituting the whole of the heavens. [14]

But he is critical of the nature of this one:

> It is absurd, or rather impossible, to make a generation of eternal things. There is no question though that this is what the Pythagoreans do. They state clearly that the one has been constituted, whether from planes, surfaces, seeds or something which they are at a loss to explain, and at once the closest part of the unlimited began to be drawn in and limited by the limit. Here they mean to speak physically, of cosmogony.[15]

> The Pythagoreans believed in one kind of number, the mathematical. They

> hold that it is not separate, but sensible substances are constituted out of it. They construct the whole heaven out of numbers, not abstract units, but units which have size. However, on the subject of how the first extended one is constructed, it is likely that they are in difficulty.[16]

On the issue of the *kosmos* breathing, Stobaeus tells us:

> In the first book of 'On the Philosophy of the Pythagoreans', he (Aristotle) writes that the heavens are one, and that from the unlimited are drawn time, breath and the void. It is the void which distinguishes the nature of each thing.[17]

This may be a biological analogy; certainly it can be compared with the Pythagoreans on the birth of animals:

> Directly after birth an animal, whose body is mainly hot, will draw in a breath from the outside, which is cold, and then expels it as if of necessity.[18]

On a more specific piece of cosmogony, Philolaus tells us:

> The first thing to be harmonised, the one, is in the middle of the sphere and is called the hearth.[19]

He seems to follow the general Pythagorean view on the relation of one, number and *kosmos*:

> Nature in the *kosmos* was constituted out of the limited and the unlimited, as indeed was the whole *kosmos* and everything in it.[20]

He further tells us:

> All things which are known have number. Without this nothing can be understood or known.[21]

IV. Pythagorean cyclical cosmogony

According to Eudemus of Rhodes (*c.* 370 – *c.* 300 BC), some of the Pythagoreans held a theory of eternal recurrence:

> One might be concerned as to whether the same time recurs, as some say. Sameness is spoken of in a multitude of ways, and it appears that there is a recurrence of summer, winter and the other seasons and periods. So too motions of the same type recur, as the sun will complete solstices and equinoxes and other travels. However, if we believe the Pythagoreans, the same things in number recur, and I will tell this tale, staff in hand, to you sitting there and all other things will be similar as well, then it is reasonable to suppose that time will be the same as well. When motion is one and the same, and similarly all things are the same, then before and after are one

> and the same, as is their number. So all things will be the same, with the result that time is as well.[22]

I agree with Sorabji that although Eudemus does not spell this out, his argument here effectively demolishes the Pythagorean idea of exact recurrence, if we accept Aristotle's view that time is the number of motion with respect to prior and posterior.[23] If all the motions and the things participating in motion are the same, then it follows that time is the same as well. So there is no recurrence at a different time.

This is the first instance of successive *kosmoi* which are identical, a view which will be taken up again, with variations, by the Stoics. With the Stoics, I will argue that their motivation stems from a belief in a degenerating *kosmos* and a providential god. Aristotle, in his *On Philosophy*, offers three scenarios for successive *kosmoi*, all of which he rejects – that they are successively better, the same, or worse. The Stoics opt for successive *kosmoi* being exactly or largely the same, their disagreement with Aristotle being on whether the *kosmos* degenerates. I can offer nothing so specific to explain the Pythagorean idea of successive identical *kosmoi* as we know so little of their cosmology and theology. It is notable though that Empedocles' theory of successive *kosmoi* is substantially different from the Pythagoreans, and the Stoics as well. Where Empedocles does allow a significant role for chance, the Stoics most certainly do not, and as far as we know there is no role for chance in Pythagorean cosmogony. What importance does Pythagorean cosmogony have? I do not see that any of the specific theories of the Pythagoreans have any great short- or long-term influence on Greek cosmogony. Without doubt the most important thing they do is raise the issue of the role of number and mathematics in cosmogony. However, it is difficult to see that they give any precise answer to these questions. That may be a somewhat harsh judgement, as the development of the proper role of mathematics took a great deal of time and effort for science, and was not resolved until at least the seventeenth century. The Pythagoreans are however given to rather vague gestures towards the role of numbers rather than any precise analysis or concrete proposals. To what degree they influence Plato is an old and much debated question. Clearly there is some influence, though the extent and nature of it are rather more open questions.

V. Archelaus

Archelaus of Athens provides us with a small addendum to Anaxagoras. Simplicius tells us:

> Archelaus of Athens, a pupil of Anaxagoras who is said to have been associated with Socrates, tried to bring something new into cosmology and other disciplines, but gave the same first principles as Anaxagoras.[24]

Hippolytus, however, gives us some more information, and it is clear that Archelaus differed in some matters from Anaxagoras:

> Archelaus held that there was a material mixture similar to that of Anaxagoras, and his *archê* was the same. However, he held that from the beginning there was a certain mixture in *nous*. The beginning of motion was the separation from each other of the hot and the cold, of which the hot moves while the cold does not. When water liquefies it flows into the centre, where it is burnt to generate air and earth, the air being carried up and the earth being positioned below ... concerning animals, he says that when the earth first became warm in its lower parts, where the hot and cold were mixed, men and many animals appeared, all living in the same manner and nourished by mud. These lived only for a short time, but later others were born from one another.[25]

So Archelaus believes that from the outset *nous* was mixed, unlike the initial purity postulated by Anaxagoras.[26] The separation of the hot and the cold initiates motion, rather than *nous* itself.[27] There are differences in what the elements do, though this does not seem to be of any great significance. The zoogony abandons Anaxagoras' idea of seeds.[28] Archelaus then is a rather minor figure, essentially an Anaxagorean who tries, not altogether successfully, to integrate Anaxagoras' ideas with some Milesian ideas. There is no suggestion in Archelaus of any multiple *kosmoi*. Of course he may have differed from Anaxagoras on this issue, but one might expect that if Anaxagoras had held a multiple *kosmoi* view, we would find some trace or mention of it in Archelaus.

VI. Diogenes of Apollonia

Diogenes of Apollonia, who wrote around 440-423 BC,[29] was essentially a follower of Anaximenes. According to Simplicius:

> Diogenes of Apollonia, almost the youngest of those who were concerned with these matters, for the most part wrote eclectically, on some issues following Anaxagoras, on other Leucippus. He says that the nature of the universe is air, infinite and eternal, out of which the forms of all other things are generated by rarefaction, condensation and change of state.[30]

Air for Diogenes is intelligent and steers, as we have seen previously in relation to Anaximenes and the other Milesians. This intelligence is the basis of the world order, according to Diogenes:

> It would not be possible, he says, without *nous*, for it (the underlying substance) to be so divided that it has a share of everything, winter, summer, night, day, rain, winds and good weather. Other things as well, if one wishes to think of them, will be found to be in the best possible disposition.[31]

Diogenes, according to some sources, believes both in an organising principle for the *kosmos*, and in multiple *kosmoi*. Diogenes Laertius tells us that:

> His opinions were these. Air is the element, and there are unlimited *kosmoi* and an unlimited void. Air generates the *kosmoi* through condensation and rarefaction. Nothing is generated from what is not and nothing is destroyed into what is not. The earth is spherical, is supported in the middle and is constituted according to the revolution due to the hot and the coagulation due to the cold.[32]

Pseudo-Plutarch is of the same opinion:

> Diogenes of Apollonia supposes that air is the basic element. All things are in motion and there are an infinite number of *kosmoi*. Cosmogony works like this: everything was in motion, and became rare in some places and dense in others. Where the dense come together by chance it made the earth and other things for the same reason, while the lightest took the higher position and completed the sun.[33]

Simplicius also attributes multiple *kosmoi* to Diogenes of Apollonia,[34] though he is clear that Diogenes advocates successive rather than co-existent *kosmoi*. Guthrie argues in favour of co-existent multiple *kosmoi*, on the grounds that Diogenes wrote after Leucippus, that he believed in the existence of the void and in an unlimited amount of matter.[35] That is possible, though Pseudo-Plutarch agrees with Simplicius that only successive worlds ought to be attributed to Diogenes. That Diogenes Laertius says that the earth is supported in the middle would suggest successive *kosmoi* as well, though there could be many earths each supported in the middle of different *kosmoi*.

VII. Diogenes of Apollonia and multiple *kosmoi*

Did Diogenes believe in multiple *kosmoi* at all, though? According to the standard modern view of Heraclitus, the doxographical tradition is wrong to attribute successive *kosmoi* to him. According to what I have argued about the Milesians, Anaximander and Anaximenes in particular, they were also wrong to attribute multiple *kosmoi*, successive or co-existent, to them. A better case can be made for multiple *kosmoi* in Diogenes. Certainly he has the conceptual apparatus for a belief in multiple worlds. He believes there to be a void, he has the distinction between being and not being, and he believes there to be an unlimited amount of matter, even if he makes no use of the *ou mallon* principles of the early atomists. He also has to face the problems in cosmogony set by Parmenides. According to Pseudo-Plutarch, Diogenes also believes in a significant role for chance, if the dense comes together to form the earth and other things by chance. This is suppressed in some translations,[36] but the Greek here is *sunekurêsen*, and the primary meaning here is to come together by chance. We have seen this word used by Empedocles, and certainly this seems to be the sense in which Empedocles used it. One might also argue that the 'revolution due to the hot' implies some form of vortex, though we have no other supporting evidence for this, and a vortex need not imply chance of multiple *kosmoi*, as we have seen with Anaxagoras.

Against this, I am not convinced that Diogenes employs chance in cosmogony. Simplicius quotes a good deal of Diogenes, and there is no trace there of any use of chance. I would agree with Simplicius that:

> He demonstrates at great length that there is much intelligence in the *archê* he hypothesises.[37]

Indeed, Diogenes is keen to emphasise the role of intelligence in ordering the *kosmos* at all times. We might expect the use of chance in explaining the diversity of physical phenomena or, following Empedocles, in zoogony. However, in Fragment 5 Diogenes discusses precisely these issues and is adamant that the full diversity of physical and biological phenomena can be explained by intelligence/ air.[38]

If writers in the doxographical tradition are wrong in attributing multiple *kosmoi* to Diogenes, why do they go wrong? Aristotle, when discussing who held what view on substance, ties together Anaximenes and Diogenes as both believing the primary substance to be air, with no further comment.[39] Diogenes might then well be treated as similar to Anaximenes and given the successive *kosmoi* view, the doxographical tradition being wrong for the same reasons. Diogenes is also mentioned by Alexander along with Anaximander in relation to the theory that the earth is drying out in a terminal manner.[40] But I see nothing in any of the passages quoted by Simplicius which would support such a view. As with Anaximander, even if the earth begins in a relatively wet state and becomes, or is still in the process of becoming drier, this does not establish that it will eventually dry up, as at some point an evaporation/ rainfall cycle may be set up which is a stable state.

VIII. Diogenes the eclectic?

Diogenes had a reputation for being an eclectic thinker, as we saw with our first quotation from Simplicius. If Diogenes did have a steering principle and a role for chance and multiple *kosmoi*, here may be the reason. I have been arguing the view that the ancient Greeks either opted for a unique *kosmos* with some form of design principle (steering, *nous*, etc.) or invoked chance and opted for multiple *kosmoi* (atomists, Empedocles), and that none believed in a unique *kosmos* with no design principle. Diogenes, if he has been properly represented by the doxographers, presents an exception to my classification. As I commented in the introduction, though,

I am looking for a good, not necessarily a perfect fit for this classification. An eclectic thinker such as Diogenes, who may have drawn on several sources without realising the full implications of holding those ideas in unison, may represent an exception though not a particularly harmful one.

IX. The Derveni papyrus author

The author of the Derveni papyrus puts forward a cosmogony which is interesting in itself, though rather fragmentary. It is also of some significance in relation to Anaxagoras and Diogenes of Apollonia, and to a lesser extent to Anaximenes.[41] As usual, there is no absolute coming to be in the Derveni papyrus. Everything is formed from things which have previously existed.[42] The Derveni author supposes there to be a state in which fire dominates and mixes everything together, prior to the *kosmos*.[43] This fire mixes everything and prevents anything from separating out.[44] There can be no question of the *kosmos* forming here in any accidental fashion. Fire prevents any such thing from happening, so there must be some intervention by some other agency in order for cosmogony to begin. One might compare here Plato on the pre-cosmic chaos, as this chaos is non-progressive for Plato too. Plato does not have a process which prevents *kosmos* formation, but he asks deeper questions about what we might expect in a chaos, and whether we can assume that elements such as fire have their current properties in the pre-cosmic state.

According to the Derveni author, the cosmic *nous* removes fire to a safe distance, so that things can begin to separate out. The sun is generated by being 'separated and encircled' and the things above and below the sun then coagulate.[45] The stars, invisible during the day because they are dominated by the much larger sun, but visible at night, are of the same matter as the sun but are dispersed around the heavens. They 'float in necessity' so that they do not come together to form another sun.[46] The moon is composed of the 'whitest of all', but its matter is not hot, and seems to be distinguished from the matter of the sun and stars.[47] None of this is particularly surprising in the context of Presocratic cosmogony. However an interesting passage concerns the stars, of which the Derveni author tells us:

> Each of them is floating in necessity, so that they cannot come together with each other. If they did, those having the same powers as those from which the sun was put together would come together in a mass. If the god had not wished the things now existing to be, he would not have made the sun.[48]

What I take this to show is that multiple co-existent *kosmoi* are out of the question for the Derveni author.[49] There is one and only one sun because that is the choice of god. There would seem to be no role for chance in the Derveni author's cosmogony. None is mentioned, and we find the Derveni author saying:

> ... does not allow to obtain it by chance. Is it not due to these that the *kosmos* has order?[50]

The text here is damaged and the reconstruction rather hypothetical, but certainly one way of reading this is that the *kosmos* does not attain its order by chance, but from some other factors. The text then goes on to say:

> Similarly Heraclitus ... the common ... overturns the private.[51]

This may be an allusion to Heraclitus Fragment 89:

> For those who are awake the *kosmos* is one and common, but those who sleep turn away each into a private world. We should not speak and act like sleeping men.[52]

Heraclitus, I have argued, had no role for chance either. Not surprisingly we do not find any boundary or individuation criteria for *kosmoi* in the Derveni papyrus, even though we do find such criteria for the sun.[53]

How much we take that to be indicative of the views of other Presocratics is another matter. In relation to Anaxagoras, here we have a cosmic intelligence but a denial of multiple *kosmoi*. Anaxagoras may differ of course, but it is at least an interesting parallel.[54] If this sort of god/ cosmic intelligence makes a unique *kosmos*, one can see why Anaxagoras might say '*if* there were to be other *kosmoi* (but there are not), then they would be identical to our own'. In relation to Diogenes of Apollonia, one of the dominant themes in the Derveni papyrus is of the dominance of air and its relation to some form of cosmic intelligence.[55] Again, Diogenes may differ but the parallel is interesting. Does the Derveni papyrus suggest successive *kosmoi* at any point? I do not believe that it does. There is no role for chance, there are no criteria for differentiating successive *kosmoi*, there is no cause for the deterioration of the *kosmos* and the organising principle for the *kosmos* is immortal.[56] In relation to Anaximenes, again I would point out the dominance of air and the fact that the Derveni papyrus is clear on the issue of there only being one *kosmos*, so that many of the considerations relating to Diogenes also apply to Anaximenes. In relation to the question of the role of air in Anaximenes, it is interesting that the Derveni papyrus uses *epikratein* to describe what air does,[57] where Anaximenes arguably has *sungkratein*.

X. The Derveni papyrus and the cosmological constant

Betegh has made the very interesting suggestion that the way in which fire is held apart in the Derveni author can be compared to Einstein's use of a cosmological constant.[58] For the Derveni author, fire will coalesce unless it is prevented from doing so. There is a problem for any finite, static universe with gravitating matter. Over a long period, that matter will coalesce under the influence of gravity. As we do not see this happening, on a cosmic scale, there is a need to hypothesise another effect countering the influence of gravity. Einstein hypothesised a cosmological constant to counter the long-range effects of gravity within a static universe when developing relativity. While I agree that a comparison with Einstein is possible and interesting, a more precise parallel would perhaps be with the idea that, within a static, finite Newtonian universe god was thought to act to stop the stars moving together under their mutual gravitational influence.[59]

While the Derveni issue is clearly not a gravitational problem, it is a type of problem which occurs often in antiquity, where like-to-like principles, or principles of natural motion, if left unchecked, will simply bring like things together. Plato and Aristotle both consider forms of this difficulty, as we shall see, where the *kosmos* might ultimately sort out into earth, water, air and fire. There is a perennial problem in cosmogony with processes which, if left unchecked, will first produce and then destroy a *kosmos*, the prime example in antiquity being Empedocles.

Betegh also has concerns that what keeps fire apart is an *ad hoc* aspect of the Derveni author's theory.[60] This depends, in part, on whether mind/ air is doing something special in keeping fire apart.[61] One might argue that if everything is ordered, and kept in order by mind/ air, then fire being kept apart is no different from the formation and maintenance of the earth or anything else. It is mentioned as it is an important part of the process of cosmogony, given the nature of the pre-cosmic state, but actually requires nothing more than any other more mundane aspect of cosmogony.

Even if something extra is required to keep fire apart, that need not be considered to be *ad hoc*. Certainly it would not be *ad hoc* in the sense of an *ad hoc* modification to a theory, the classic case here being the scholastics who argued that the moon was surrounded by an invisible, perfectly spherical layer in order to counter Galileo's telescopic observation of hills and valleys on the moon. It is part of the theory from the start, not a reaction to new observations which go against a theory. Nor is it *ad hoc* in the sense of being arbitrary. One concern about Einstein's cosmological constant was that it was an interpretation of a constant of integration in the equations describing the universe. That constant had no determined value and could be set at zero, or given a value (either positive or negative), and its nature then hypothesised. It was arbitrarily given a value such that it opposed gravitational forces.[62] Betegh argues that fire being kept apart in the Derveni papyrus is *ad hoc* but defensibly so.[63] I am not convinced that it is *ad hoc* at all, especially if the author recognises from the outset that the tendency of fire to coalesce will need to be opposed.

XI. Conclusion

Xenophanes, for all that he is important in the history of theology, has nothing of great import to say of cosmogony. His theory of a cyclical destruction of land into the sea and the production of land out of the sea, again with the periodic destruction of mankind, is of interest, at least in that it establishes that Presocratic thinkers could hold such cyclical theories without resorting to successive *kosmoi*. Xenophanes may not have been disagreeing with Anaximander on this, as KRS suggest, but just giving a different sort of cyclical theory.[64]

It is difficult to know what to make of the Pythagoreans, either what they were proposing in terms of cosmogony in any concrete sense, or what their influence may have been. I do not see that any of their specific theories in cosmogony, where they are not so diffuse as to count as specific, have any great importance or influence. By far the most important issue the Pythagoreans raise is the question of the role of number in cosmogony and cosmology, and indeed in science in general. Their influence in relation to Plato is also difficult to assess.

Archelaus provides us with some small variations on Anaxagoras, but nothing of any great originality or lasting importance. Diogenes of Apollonia, in my view, gives us some small variations on Anaximenes. I doubt that he believes in multiple *kosmoi*, though his is the strongest case we have seen so far for a belief in both an organising principle and multiple *kosmoi*.

The Derveni papyrus, while somewhat fragmentary, does give us another interesting Presocratic cosmogony. Given the influences of Anaxagoras and Diogenes of Apollonia on the Derveni author, it may be possible to draw some weak inferences about the number of *kosmoi* in Anaxagoras and Diogenes, as the Derveni author seems clear that there is only one *kosmos*.

9

Plato

Plato (429-347 BC) gives us the first fully teleological cosmogony. There were of course precursors, arguably in the steering principles of the Milesians, and certainly in the activities of Anaxagoras' cosmic intelligence. Plato though gives us the first thorough teleology, fully encompassing cosmogony, zoogony and stoichogony along with criticisms of any previous multiple accidents theories in those areas. Plato argues that not only is there a unique *kosmos* but also that there are a small number of well defined types of ultimate particle[1] and well defined, permanent biological species, all designed with the best in mind.

I. Plato's teleological intent

The outlines of Plato's cosmogony are fairly clear. Prior to the intervention of the demiurge, Plato's craftsman god, there is not a *kosmos* but a chaos. There is complete disorder and in the absence of the earth, sun, moon and planets there is no measurable time. The demiurge acts upon this chaos and produces a *kosmos*. Plato gives us a clear statement of his teleological intent in cosmogony early on in the *Timaeus*:

> For god desired that all things should be good, and nothing paltry as far as was possible, he took over all that was visible, which was not at rest but in discordant and disorderly motion, and led it into order out of disorder, judging the former to be entirely better than the latter.[2]

One key difference with Anaxagoras here is that Plato makes it entirely clear that order is better than disorder, and that god will lead the universe into the maximum state of order. A key assumption in cosmogony for Plato is that the primordial chaos is non-progressive. Left to its own devices chaos will continue to be chaos and good order will not grow out of it. If some order should come about by chance, it is likely to dissolve immediately rather than breed further order as it does in the modern view of the evolution of the universe. Therefore something else is required to explain how the *kosmos* acquired that order.

II. Like-to-like principles

The *Timaeus* gives us a like-to-like principle, whereby similar sorts of entities, primarily the elements of earth, water, air and fire, congregate together.[3] So we are told that:

> Standing on the earth, and separating off types of earth, and sometimes earth itself, if we draw them into the unlike air by force and contrary to nature, both cling to their kind. The smaller is more readily forced into the dissimilar than the larger.[4]

This is repeated in a more generalised form:

> The thing to be recognised in all of these cases is that it is the motion of each towards its own kind that makes the moving thing heavy, and the place into which it moves 'down'.[5]

In addition to the like-to-like principle, there are two sources of motion. We are told that in the pre-*kosmos* chaos the receptacle, because it is filled with powers that are not balanced or evenly distributed, is itself shaken by these things and by this motion shakes them in turn.[6] It thus acts like a winnowing basket,[7] and just as the process of winnowing sorts the corn from the chaff, so the four elements are separated out, the most unlike furthest away and the most like closest together.[8] At *Timaeus* 57c and 88de we are told that this shaking continues in the *kosmos*. At *Timaeus* 58aff. we are told of the compressive effect which the rotation of the *kosmos* has on the elements, making them intermingle and denying the possibility of a void on a large scale. While the like-to-like principle and the sources of motion will be used to account for terrestrial gravitational effects, this is very far from anything like a modern theory of gravitation. There is no universal attraction of matter to all other matter. Earth will congregate with earth, but not the other elements. Nor is there any force which operates at a distance. Similarly at *Timaeus* 80bc electricity and magnetism are explained in a general manner as due to contact action and mutual replacement, and there is an outright denial that any attraction is involved.[9] Thus it is wise to refer to a like-to-like principle for Plato rather than to forces.[10]

III. Non-progressive chaos

A combination of the shaking of the receptacle and the like-to-like principle will not produce a coherent, well ordered *kosmos* but simply a sorting of the elements. The idea that some physical principle, on its own and undirected, could produce the sort of order that the *kosmos* currently displays is alien to Plato. That is quite reasonable given usual analogues for these processes. Winnowing corn does not produce any interesting order, but merely the separation and congregation of the entities involved. The following passage at *Laws* 889b seems critical to Plato's cosmogony and his criticism of his predecessors:

> Let me put it more clearly. Fire, water, earth and air all exist due to nature and chance they say, and none to skill, and the bodies which come after these, earth, sun, moon and stars, came into being because of these entirely soulless entities. Each being moved by chance, according to the power each has, they somehow fell together in a fitting and harmonious manner, hot with cold or dry with moist or hard with soft, all of the forced blendings happening by the mixing of opposites according to chance. In this way and by these means the heavens and all that pertains to them have come into being and all of the animals and plants, all of the seasons having been created from these things, not by intelligence, they say, nor by some god nor some skill, but as we say, through nature and chance.[11]

A 'fitting and harmonious' ordering will not be generated by like-to-like principles, as that ordering has a mixture of types, and indeed, as Plato emphasises here, of opposites. The only way that Plato's opponents here can explain how such an ordering came about is by chance, and with opposites that looks extremely unlikely. Here we see a problem generated by a lack of a conception of gravity. While gravity will bring together things that (for the Greeks) are unlike, principles such as like to like will not. This passage is also significant in that it links together three problems, stoichogony, cosmogony and zoogony, which Plato sees as requiring a teleological solution. It is also interesting to note that Plato has the demiurge bring together unlike things, where typically beforehand we find that, as in Homer, 'God always leads like to like.'[12]

IV. Versions of chaos

Plato's account of chaos has greater depth than I have suggested so far, and this is important for Plato's cosmogony. Timaeus tells us that:

> Before this, all things were lacking in reason and measure. When the task of ordering the universe was attended to, fire, water, earth and air all possessed traces of their own nature, but were so disposed as would be likely in the absence of god, and such being their nature at that time, god then firstly fashioned them by means of shapes and numbers. They were organised by god to be as fair and beautiful as possible, a state which they were not previously in.[13]

Again at *Timaeus* 69c he says:

> As was said at the beginning, all these things were in a state of disorder until God created in them commensurability with respect to themselves and each other, so far as it was possible for them to have proportion and commensurability. For at that time nothing partook of these things except by chance, nor was there anything deserving to be named as we now name things, such as fire and water and the others. But all these he first ordered, and then out of them he constructed the universe.[14]

It is possible to discern three levels of disorder here:

(1) Are the ultimate particles intrinsically ordered (do they have the required shapes)?
(2) Are the ultimate particles extrinsically ordered (do they form up as earth, water, air, fire)?[15]
(3) Are the elements extrinsically ordered into a *kosmos*?

This lack of intrinsic order has some important consequences for the application of a like-to-like principle to the primordial chaos. Such a principle will not work to produce any sort of ordering in the primordial chaos if we do not have elements. If we interpret *Timaeus* 63c-e more strongly as a general like-to-like principle, still this will fail to produce any sort of ordering unless there is some likeness in the initial chaos. Certainly Plato is aware of the idea of a chaos of complete unlikeness. The *Politicus* myth gives us a vision of the deteriorating *kosmos*, and at 273e a god who eventually intervenes because he is:

> Concerned that it should not be storm driven by confusion and broken up into an endless sea of unlikeness.[16]

One might argue though that this somewhat overstates Plato's position on elemental chaos. Just prior to *Timaeus* 53ab, we are told that the receptacle has a winnowing effect, such that the four kinds of element are separated out, the most dissimilar farthest apart, the most similar nearest. So at *Timaeus* 53ab, while the elements only have traces of their nature, perhaps they have sufficient traces for the winnowing effect to take place,[17] otherwise the passage at *Timaeus* 53aff. looks strange. However, *Timaeus* 53b is quite specific that it is the four kinds that are sorted:

> In the same way at that time the four kinds were shaken by the recipient, whose motion was like that of an instrument for shaking, and separated the most unlike kinds furthest from each other and forced together the most like kinds, so that the different kinds had different regions, even before the universe was organised out of them.[18]

However, if this winnowing effect does take place, there seems to be no limit on it, so eventually we would reach a situation where the four elements are separated out.[19] There is no mention of this winnowing effect or its results at *Timaeus* 30a or 69c, both passages seeming to indicate that there is ordinary chaotic motion.

I do not believe that Plato is being inconsistent here. Rather, he is asking deeper questions about the order of the *kosmos*, about what we can assume about a chaotic state and about the implausibility of the *kosmos* coming about by chance.[20] Even if the elements of earth, water, air and fire are fully formed, this is not enough, combined with the available sources of motion, for a *kosmos* to form. If the elements are well formed though, what of their genesis? Can something this well ordered simply exist by chance? So perhaps prior to the work of the demiurge, there are only traces of the elements, making the possibility of a *kosmos* emerging in the absence of action by the demiurge even more remote. One can run this type of argument again at a still more fundamental level. The elements are constituted from two basic types of triangle. Can something this well ordered simply exist by chance? Perhaps the demiurge imposes 'shape and number' on an utter chaos to give the building blocks which will ultimately be organised into the elements. If so, then the chaos is even further away from being able to form a *kosmos* by chance. So given Plato's analysis of the elements, there are three levels of implausibility in the *kosmos* coming to be by chance: the extrinsic order of the elements, the intrinsic order of the elements and the intrinsic order of the basic triangles.[21]

Origen, an early Christian thinker, puts this type of argument to use in arguing for the creation of matter *ex nihilo* by a Christian god. Origen argues that matter is well fitted for whatever god wishes to create out of it.[22] Could something so fitting for its purpose exist by chance? Origen's answer to this is no, god produced matter *ex nihilo* that was conducive to his purpose.[23]

One can read here an attack not only on Leucippus and Democritus, relating to their use of a like-to-like principle, but also an attack on Empedocles. Will love or strife, principles of association and dissociation of the elements, produce a *kosmos* on Plato's analysis? If Plato is correct in arguing that we need to hypothesise a substructure for the elements, then what account could either the atomists or Empedocles have of either the intrinsic order of the ultimate particles or their extrinsic ordering into the elements? Is the atomist *ou mallon* account of the shapes and sizes of atoms adequate anyway? Does Empedocles have any account of the intrinsic order of his elements?

V. The generation of the world soul

We can find similar considerations to the issue of the generation of the physical *kosmos* in Plato's discussion of the formation of the world soul. Plato describes how the demiurge mixes together 'indivisible existence' and the 'existence which is divisible and comes to be in bodies', and produces a third type which is intermediate between the two. He then produces a third, intermediate type of sameness, and a third, intermediate type of difference, from their indivisible and divisible forms.[24] We are then told that:

> He took these three and blended them into a single whole, forcing the different, which is of its nature difficult to mix, into union with the same.[25]

This is all very interesting in relation to *Laws* 889b, where it is hard to envisage unlike things coming into a 'fitting and harmonious' order. It is hard to see how any like-to-like principle will be able to explain the mixing of the indivisible and divisible types of being, sameness and difference into a third type intermediate between the extremes. Nor is it at all easy to see how that might come about by chance. As for the mixing of the three intermediates, again there can be no question of like to like, and indeed the demiurge has to force difference into this blend as difference is hard to mix (*dusmikton*) with anything else. So for Plato neither the physical *kosmos* nor the soul of the *kosmos* could be the product of chance or necessity. Plato has some more to say about how the world soul is divided up by the demiurge:

> He divided in this manner. First he took one part from the whole, then a part that was double this. The third was half as much again as the second, and three times as much as the first, the fourth twice as much as the second, the fifth three times the third, the sixth eight times the first and the seventh twenty seven times the first. [26]

This gives us a sequence of

$$1 - 2 - 3 - 4 - 8 - 9 - 27.$$

Timaeus treats these as two sequences, $1 - 2 - 4 - 8$ and $1 - 3 - 9 - 27$, and goes on to fill these intervals with the harmonic and arithmetic means,[27] giving a sequence of:

$$1 - 4/3 - 3/2 - 2 - 8/3 - 3 - 4 - 9/2 - 16/3 - 6 - 8 - 9 - 27/2 - 18 - 27$$

There is then some further filling of intervals where multiplying one of the numbers in this sequence by 9/8 does not exceed the next number in the sequence. Thus the first part of the sequence will run:

$$1 - 9/8 - 81/64 - 4/3 - 3/2 - 27/16 - 243/128 - 2$$

These divisions have a musical significance as 3/2 represents a musical fifth, 4/3 a fourth, 9/8 a tone and the remainder between the 9/8 multiplications and the next number is 256/243, close to a semitone. The world soul is then divided with great precision, and can be said to have been divided in an harmonious manner. It is hardly likely, from Plato's point of view, that these divisions could have come about by chance.

Does the nature of the world soul have consequence for physical cosmogony? Indeed it does. At *Timaeus* 36b we find that the demiurge splits the world soul into two lengths, and joins them in the middle in the shape of the Greek letter chi, X,[28] and then joins them at their ends as well so they form two great circles (see Figure 1). These are then made to revolve. This has an important physical consequence. One can see how anyone supposing there to be a vortex could generate the motions of the fixed stars from that vortex. Within that, one might have some explanation of the different speeds of the sun, moon and five naked eye planets, but one could have no explanation of the different paths of the planets. If, as is generally supposed, Plato's X shape is meant to represent, however roughly, the angle between the ecliptic and the equator, Plato can at least give some account of the paths of the sun, moon and five planets relative to the fixed stars. At *Timaeus* 36d the band of soul stuff revolving at the angle of the ecliptic is split into six parts, in accord with the first number sequence, 1 – 2 – 3 – 4 – 8 – 9 – 27, giving us some representation of differing orbit sizes, with three (presumably the orbits of the Sun, Mercury and Venus) moving at the same speed and the other four moving at different speeds. This division of the world soul allows the sun, moon and five planets combinations of circular motion with the axes of those motions offset, giving a far superior model of the heavens.

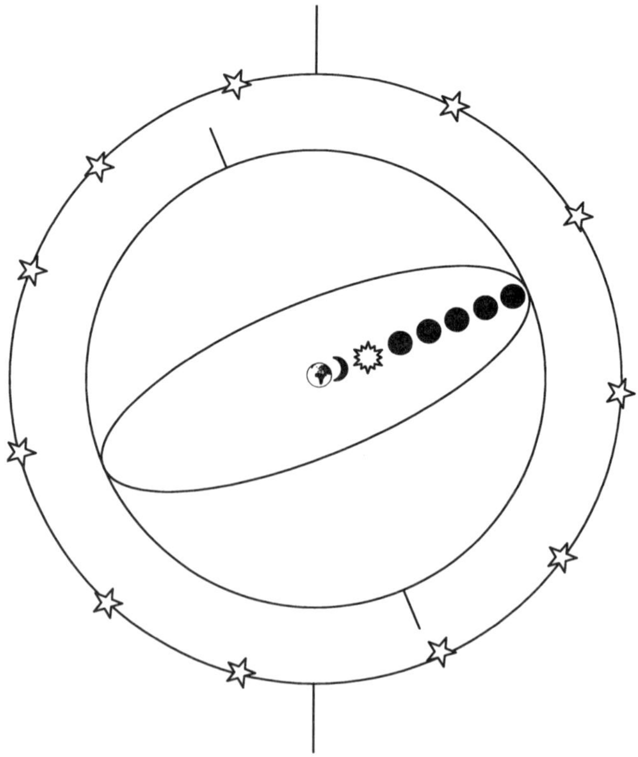

Figure 1. Plato's model of the heavens. The axes of rotation for the stars, and for the secondary motion of the sun, moon and planets are offset as in the Greek letter X. This not only gives a better account of the motion of the heavens, but also opens up the possibility of the superior models of Eudoxus and Callippus.

VI. Literal and metaphorical interpretations

There has been debate, both ancient and modern, on whether Plato meant the cosmogony of the *Timaeus* to be taken literally or not.[29] Aristotle was aware of this debate and asserted a literal reading, but reported that Xenocrates, the third head of the academy, opted for a metaphorical reading.[30] Those advocating a metaphorical interpretation argue that Plato's cosmogony was for the sake of elucidation, to show that an orderly *kosmos* was utterly dependent upon the existence of a benevolent deity, but not that the *kosmos* had actually been made. The description of chaos shows what the universe would be like in the absence of the demiurge, but of course the demiurge is always present. According to some versions of the metaphorical interpretation, Plato's description of chaos is incoherent or gives rise to contradictions. We are supposed to notice this, and draw the conclusion that chaos cannot have existed and so the talk of cosmogony from chaos must be metaphorical. These are the main arguments in favour of the metaphorical position:[31]

(1) Plato's cosmogony entails a beginning to time, and talk of a time before there was time. As Taylor puts it: 'No sane man could be meant to be understood literally in maintaining at once that time and the world began together, and also that there was a state of things, which he proceeds to describe, before there was any world.'[32]
(2) If Plato meant his cosmogony literally, then there are problems of consistency in relation to other works. The two main objections are that in the *Phaedrus* and in the *Laws* soul is apparently considered to be the only source of motion,[33] and that in the *Phaedrus* the soul is considered to be ungenerated.[34] If the soul is the only source of motion, then what of the chaotic motion prior to the formation of the *kosmos* in the *Timaeus*? Plutarch's suggestion that an evil soul must be in charge of this chaos has rightly been rejected.[35]
(3) Plato says that everything perceptible has a beginning, and so the *kosmos* must have a beginning.[36] However, chaos is also perceptible. So chaos too must have a beginning and something which precedes it, and the options here are all impossible. It cannot be the *kosmos* (the demiurge would not allow the *kosmos* to descend into chaos), it cannot be further chaos and surely Plato does not envisage creation out of nothing.
(4) According to Taran, the notion of a 'creating god' is absent from Greek cosmogonical thought.[37]
(5) The account that Timaeus gives is hedged around with qualifications that it is a likely or mythological account. This gives us licence to interpret the account metaphorically, either in part or whole.
(6) Some aspects of the account given in the *Timaeus* are clearly and uncontroversially metaphorical. Are there really independent clumps of

sameness, being and difference prior to their being blended together, and is there really a bowl in which soul is generated?
(7) The demiurge need not be interpreted as a real figure. He may be symbolic of Forms (either Forms in general, the Form of the Good, the Form of intelligence) or of the world soul (the soul itself or perhaps the intellect of the world soul).

These are the literalist replies:

(1a) Plato does not hold that time comes to be without qualification. Orderly time comes into existence with the *kosmos*, and can be measured by the motions of the heavens. Before that, there was temporal succession in the primordial chaos. It is then quite possible to talk of what happened prior to orderly time.[38]
(2a) There are many further issues of consistency between the *Timaeus* and other works where we do not and should not take such drastic action as relegating parts of the *Timaeus* to metaphor. Arguably, the *Timaeus* also develops or changes several earlier ideas without a clear retraction of those earlier ideas.[39] The *Phaedrus* appears to be on its own in asserting that soul is ungenerated. The *Laws* has the soul as being generated as does the *Timaeus*,[40] so suspicion here should fall on the *Phaedrus* rather than the *Timaeus*. The soul and motion are a problem, but it is surely better to read the *Laws* as saying that the soul is the source of all intelligent, orderly motion than to deny that the *Timaeus* has a cosmogony.[41]
(3a) There are several ways of halting this supposed regress. We should not suppose that Plato treated the *kosmos* and chaos as similar things. Indeed, as chaos lacks all order, structure and coherence, it may not even be an entity at all for Plato. So only individual chaotic events need to be explained, not chaos as a whole. The precedents of individual events in chaos are further individual events. So prior to the *kosmos* there are chaotic events, but there is nothing different prior to those events. Or one might take the view that what constitutes chaos are the receptacle and in some manner the forms (participation in which, perhaps accidentally, gives the 'traces'). Neither the receptacle nor the forms will require further explanation though.
(4a) The notion of a god who creates *ex nihilo* certainly is absent from Greek thought. Plato's demiurge though is an ordering god who orders the *kosmos* from a different prior state. Anaxagoras is a perfectly reasonable precursor here, and there are many mythological accounts of the generation of the world that Plato might draw upon.
(5a) We are also told that the *Timaeus* account is as likely as any other, and even in one place that it accords to the true account. There are places where there is greater local qualification of the account, but this dos not give us licence to treat the entire account of cosmogony metaphorically. Can the metaphorical view be both less likely than the given account

(being different from the likeliest account) and more likely than it in being a truer account?

(6a) Certain aspects of the account may be metaphorical, but as above this does not give us licence to take the entire account of cosmogony to be metaphorical.

(7a) Plato happily speaks of a demiurge in contexts which are clearly not metaphorical. There is no need to take the demiurge metaphorically in the *Timaeus*, and to do so would be a serious distortion of what the *Timaeus* has to say. The diversity of opinion on what the demiurge represents tells against the metaphorical view. Surely if Plato intended us to decipher metaphor here, he would have given us a better indication of how to do so, some determinate clue?

Cornford gives a slightly different view, that the *Timaeus* does not offer a cosmogony but all talk of the *kosmos* as coming into being should be interpreted as giving it metaphysical status (it is always becoming and never is) rather than giving a cosmogony.[42] However, this is almost impossible to reconcile with *Timaeus* 28bc where, (a) Timaeus asks if the *kosmos* has always existed or has come into existence from some beginning, and answers himself emphatically, in a single word, that it was generated, and (b) refers to the *kosmos* as something which undergoes generation and has been generated.[43]

VII. A qualified literal view

At a more strategic level, the argument for the metaphorical view is that the description of chaos is incoherent, so much so that Plato must have realised and intended this, and he intends us to draw the conclusion that he did not really believe in the pre-cosmic state. The literalist reply though is that this does not necessarily follow. Even if it is granted that the description of chaos is problematic and intentionally so, the challenge might be quite different. It may equally be that we are intended to draw some distinctions and so make the account coherent, or as I have suggested above, that we recognise the possible depths of chaos and how that might affect any account of cosmogony, and read the differing levels of chaos as problematic for the atomists and Empedocles.

The metaphorical interpretation was subsequently the majority view of later Platonists in antiquity. According to the modern metaphorical view, this is good evidence in favour of their interpretation.[44] In reply, the literalists say that Aristotle is clear that Plato's cosmogony is to be taken literally. At *On the Heavens* 279b12ff., he gives a brief survey of his predecessors' views on cosmogony. He says that while all thinkers believe the *kosmos* to have a beginning, some believe it to perish, some believe it to be everlasting once generated, and Heraclitus and Empedocles hold a cyclical view. Aristotle is critical of those who believe the *kosmos* to be generated but indestructible. Aristotle is about to launch into an argument, spanning several chapters of *On the Heavens*, that it is impossible for anything to be both generated and indestructible. According to him, only something which is ungenerated may be indestructible. It is difficult to believe that Aristotle would be mistaken on something so specific as whether Plato's cosmogony should be taken literally,[45] especially when he is aware of an alternative interpretation.[46] Aristotle tells us, at *On the Heavens* 279b33ff., that according to some Plato's expressed view on cosmogony was meant to be taken as something for the sale of instruction, as a mathematician might use a diagram, making something easier to understand.

There is significant opposition to the metaphorical view among ancient Platonists, notably from Plutarch, Atticus and Philoponus, and considerable debate on what 'generated' might mean in relation to the *kosmos*. The evidence of all later Platonists and commentators is suspect, though, as some have other agendas. Some wish to harmonise Plato and Aristotle, and so adopt a metaphorical view. In general, pro-Christian Platonists adopt a literal view, anti-Christian Platonists do not. There is also the issue of several highly influential anti-cosmogony arguments which may have persuaded Platonists, motivating them to interpret the *Timaeus* metaphorically. Questions around the issue of what god was doing prior to the generation of the *kosmos* (what was he doing? why did he delay doing good? why did he change his mind?) were much discussed. Another highly influential anti-cosmogony argument was Aristotle's that anything indestructible must also be ungenerated. One arguable implication of Aristotle's account in *On the Heavens* is that it is this argument, rather than any consideration about the coherence of the description of chaos, that provokes the metaphorical interpretation. If Xenocrates was convinced by this argument, and had to stand by the text of the *Timaeus* as well as he could, one can see why the metaphorical interpretation would be generated. Neither Aristotle nor Xenocrates make any mention of the coherence of the *Timaeus*' account of chaos. If it was Plato's intention to provoke his readers to reject this account, he failed badly with those who immediately succeeded him and knew him well. It would seem that the main reason in antiquity for the adoption of a metaphorical interpretation was an argument by Aristotle which postdates Plato and has no mention in the entire Platonic corpus. So in short, the metaphorical interpretation needs to provide a convincing motivation for Plato giving a metaphorical account of cosmogony, an account of why this supposed metaphorical ploy fails on Aristotle and an account of why the simpler explanation of the origins of the metaphorical view (Xenocrates cannot meet Aristotle's modal argument) fails. This final consideration throws another burden on the metaphorical view. It is one thing to show that some contradiction or inconsistency occurs in Plato's thought if the cosmogony is taken literally. One might well do much the same for many other ancient cosmogonies (where there is no doubt that they were intended to be taken literally) and even for some anti-cosmogony views such as Aristotle. It is one thing to find a difficulty in Plato's account of cosmogony, quite another thing to demonstrate that Plato was aware of the supposed difficulty, that he deliberately injected this into the *Timaeus* account and that the difficulties were great enough for him to reject cosmogony. If Plato himself had an objection to cosmogony, it is not to be found explicitly in Plato or in Aristotle. Plato indicates many specific points of disagreement with Presocratic cosmogony and zoogony. It is hard to see why Plato would do that if he believed the whole project of cosmogony to be fatally flawed. While he does make it clear why he favours a fully teleological account of cosmogony, stoichogony and zoogony, at no point does Plato differentiate himself from his predecessors by making it clear that he has no cosmogony at all. Nor does he give any indication that what he says about cosmogony is merely for the sake of education.

I favour the literal interpretation, then, but not without qualification. As we have seen, Plato's description of chaos is complex, with layers of concern about the extrinsic and intrinsic order of the elements, and the intrinsic order of the ultimate particles. So while I believe that Plato definitely had a literal beginning to the *kosmos*, we must take his description of chaos with some caution.

Perhaps he is looking for something more general here, which one might formulate as follows. If the current *kosmos* has come into being from a less ordered state, one that is to some extent chaotic, then that cannot be explained solely in physical terms, whatever the extent of the disorder. This may be the point of the unquantified comment on 'traces' of order. It covers many positions from no order onwards. If so, Plato has a rather stronger and more flexible position against the atomists and Empedocles. They would otherwise have the option of simply pointing to a disagreement over the nature of the primordial chaos, their view being less radical than Plato's and so perhaps being more likely to result in an ordered *kosmos*. So perhaps Plato is not stating a specific cosmogony, one which he would appear to be confused about, but is rather enunciating a general cosmogonical principle about the transition from any more chaotic state to the current *kosmos*. If so, the literal/ metaphorical question about his cosmogony may not be appropriate.

VIII. Zoogony

Plato's zoogony, as expressed at *Timaeus* 41bff. and 91dff., in many ways mirrors his cosmogony. Where there is good order, this is the product of design. That design produces singular, well designed species. With such design, there is no need for a multiplicity of accidents. So all life-forms are ultimately the products of the demiurge. Campbell notes four important points of difference between Plato's view and Presocratic theories:[47]

(1) Humans are created first.
(2) Species are formed by inter-species evolution.
(3) There are no extinctions.
(4) There is no spontaneous generation of life, either as a single event initiating life or as an ongoing process.

I would disagree with Campbell to some extent on his second point, while agreeing entirely with his other three. Humans, rather than animals, are created first and there is an important reversal in that animals for Plato are degenerate humans rather than humans being evolved animals. All creatures are designed and there is no need for spontaneous generation. There is no need for extinction, because design always generates viable species.

I am not convinced, for several reasons, that we should call the process that Plato has in mind evolution. Humans are generated first, and Plato clearly sees humans as the best species, superior to the birds, land animals and sea animals which are subsequently generated. There is no sense here of adaptation, survival of the fittest or extinction, or of a mechanism of evolution, be it the chance of Empedocles, or any form of adaptation or mutation. This is significant as it mirrors a belief that Plato holds deeply in relation to cosmogony. Chaos does not produce order of itself, so less ordered states do not become more ordered states. Zoogony cannot work for Plato by having small, less complex beginning and building towards something larger and more complex. Humans come first, and then there is deterioration.

Looking at *Timaeus* 91de, one might be led to believe there is some mechanism of evolution in Plato, as birds are said to derive from men by a process of transformation, harmless but light-minded men growing feathers instead of hairs, and men with no interest in philosophy dropping their heads closer to the ground and so going about on all fours. However, *Timaeus* 92aff. paints a slightly different picture. It is god who gives these creatures four feet, and generates bodies that are 'footless and wriggling' for the most stupid. It is also god who thrusts the unclean and wicked into water. So one might take the view that while each type of degeneration from humans has its basis in a psychological flaw, god designs and generates the bodies for each species within the groups of birds, land animals and marine animals. This is why Plato has no need for a mechanism of evolution, no need of extinctions and no need of any creature intermediate between the species which we know. Living creatures do not mutate into other living creatures, nor do they give birth to something other than their own species. *Timaeus* 42b tells us that human males who have not lived a sufficiently good and intellectual life are changed into women 'at the second birth' and so on down the scale in subsequent lives.

In defence of Plato's design view, there are issues of plausibility in the other theories of zoogony we have seen so far. Was Empedocles' account of body parts coming together by chance plausible? The *Timaeus* is quite satirical about this in its account of the human head:

> In order that it should not roll around on the ground, with its heights and depths of every kind, and be at a loss in scaling these things and climbing out of them, they gave it body as a means of support for ease of travel.[48]

The issue of plausibility was one that lasted a long time for materialist theories of zoogony. The project of mechanical biology, based on the principle of the mechanical philosophy of the seventeenth century, struggled badly. The basic problem for mechanical biology was that organisms appeared able to organise themselves beyond anything the mechanical biology could explain in terms of matter, motion and proximate causes.[49] Ancient material accounts of zoogony also struggled not least because they lacked the mechanical analogues that the later mechanical biology could draw upon.[50] Where material explanations of the origins of life were implausible, there was scope for accounts based on design.

IX. Stoichogony

Plato's account of stoichogony, the origin of the elements, also follows the same basic principles of Plato's cosmogony. If something has good order, then that order needs to be explained. For Plato, the shapes of the ultimate constituents of the *kosmos*, the stoicheic triangles, certainly do have good order. Plato's view here is again in stark contrast to that of Leucippus and Democritus. For Leucippus and Democritus, there are an infinite number of shapes and sizes of atoms, based on an *ou mallon* argument. Plato, on the other hand, restricts the shapes of the ultimate particles to two. Thus we find Timaeus saying:

> This we hypothesise as the principle of fire and of the other bodies ... but the principles of these which are higher are known only to God and whoever is friendly to him. It is necessary to give an account of the nature of the four best bodies, different to each other, with some able to be produced out of the others by dissolution ... We must be eager then to bring together the best four types of body, and to state that we have adequately grasped the nature of these bodies. Of the two triangles the isosceles has one nature, the scalene an unlimited number. Of this unlimited number we must select the best, if we intend to begin in the proper manner. If someone has singled out anything better for the construction of these bodies, his victory will be that of a friend rather than an enemy. We shall pass over the many and postulate the best triangles.[51]

This gives us:

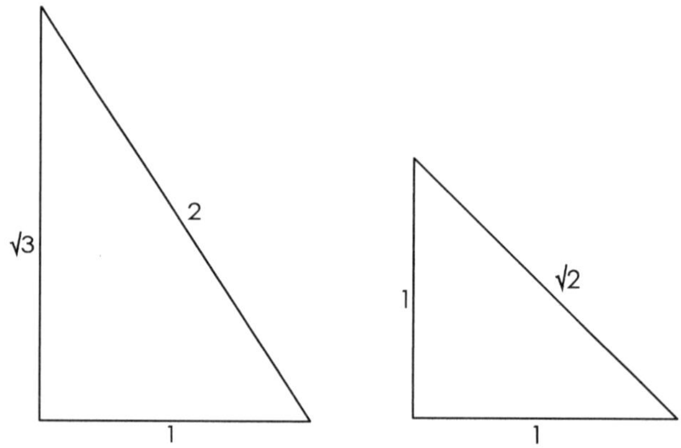

These then form up as:

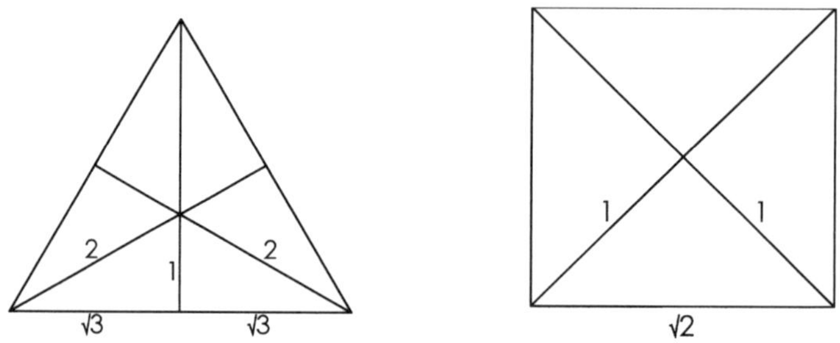

Which in turn form up as, for example, fire and earth:

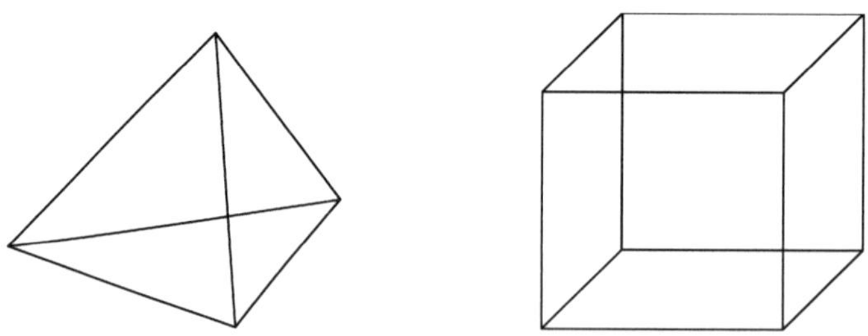

Plato's approach to the question of the shape and size of the ultimate particles is thus consistent with his approach to the questions of the number of *kosmoi* and the number of species. Plato uses teleological principles to minimise the number of entities here, and eschews the multiplicative approach. There is no question of a multiplicity of accidents in either cosmogony, zoogony or stoichogony for Plato. On the question of why Plato chooses these types of triangle I would refer the reader to Robert Lloyd's recent work on the symmetries of the three-dimensional objects they ultimately make up.[52]

X. A modern elements problem

Plato then has concerns about whether we can simply assume that the elements have been generated in a manner which is conducive to the formation of a *kosmos*. There is an interesting and important modern analogue to this. According to the Big Bang theory, in the first moments there are no elements as we would know them. As the universe expands and cools, electrons, protons and neutrons form, followed by hydrogen and helium. Elements heavier than hydrogen and helium are formed later by processes in stars. One of these processes is important for us carbon-based life-forms. Isotopes of hydrogen (deuterium and tritium) undergo fusion to generate more helium.[53] Helium nuclei then undergo fusion to generate beryllium:

$$2He^4 > Be^8$$

The beryllium is very short-lived, however, and very quickly another fusion reaction takes place, beryllium fusing with helium to form carbon:

$$Be^8 + He^4 > C^{12}$$

The energy levels mean that this reaction is a resonant one – it proceeds very quickly and very little beryllium is left. The next reaction is carbon and helium into oxygen:

$$He^4 + C^{12} > O^{16}$$

This reaction is not so efficient, which means that some carbon is left unburnt. Historically, when this theory was first proposed it seemed to be at variance with observations. On the then known energy levels of carbon, it seemed that very little carbon would be produced, far less than the observed levels and not enough to produce an abundance of carbon-based life-forms. A further, as yet unobserved energy level for carbon was hypothesised, sought for and found. If carbon did not have this energy level, it would not be possible for carbon-based life to have formed. Some saw this as evidence of a designer. Such a fortuitous energy level could not exist without having been designed in. Supporters of the anthropic principle hailed it as an anthropic prediction upheld (so we can make useful scientific predictions based on the idea that the conditions for our existence must have been fulfilled). The nature of the elements and their properties in relation to a universe which has the possibility of producing humans is not straightforward then for modern cosmogony. Small changes in the values of the fundamental constants, which would alter the electron orbits, could have serious consequences for the production and retention of carbon. We cannot assume, without further justification, that the elements have the features necessary for the formation of *kosmos*/ universe that could support life in the form that we know it.

XI. Unique *kosmos* or multiple *kosmoi*?

Some Presocratics postulate multiplicities, be they *kosmoi*, shapes and sizes of atoms or non-viable biological forms.[54] One might object to this on grounds of parsimony, but I would argue that Plato also objects to this approach on the grounds that the multiplication of entities in this manner is non-explanatory, and that the supposed *kosmoi* are not sufficiently ordered. Plato is not only worried about the arrangement of the constituents of the *kosmos* in relation to each other, he is also worried about the order that the constituents have themselves. The shapes and sizes of atoms are unexplained by Leucippus and Democritus who consider them to be infinite in variety and number, on the grounds that 'nothing is such rather than such'.[55] The following passage from the *Philebus*, while it is in the specific context of the analysis of limited and unlimited, might equally well apply to Leucippus and Democritus:

> The indefinite plurality of things and in things makes you in each case indefinite of thought and someone of neither status nor account, since you have never yet examined the number in anything.[56]

There are several examples of wordplay in the Greek here which associate allowing indefinite plurality in the world with being indefinite in thought and suggest that if you cannot give an account in either words or numbers then you are a person of no account. Similar sentiments, though perhaps less vigorously expressed, can be found in the *Timaeus* as well. The following passage comes directly after the *Timaeus'* description of how the two basic types of triangles combine to form the complex plane triangles and squares and these then form the three-dimensional elements:

> If in considering all this someone should raise the quite proper question of whether the number of *kosmoi* should be said to be unlimited or limited, he will suppose that the view that they are infinite is that of someone who is indefinite (*apeirou*) on a matter on which he should be definite, though to raise the question of whether in truth it is proper to speak of these as one or five is rather more reasonable. Our view declares it to be one, according to the likely account, but another, looking somehow to other things, might hold a different opinion.[57]

Again we have the play on indefinite in the world and of thought, though here it could be rendered 'inexperienced on a matter in which he should be experienced'.[58] The passage at *Timaeus* 55c has long been considered a rather puzzling one.[59] If we take the uses of *kosmos* here to refer simply to worlds, then it is hard to see why Plato should even begin to consider the possibility of five worlds.[60] However, Plato has just had Timaeus describe the organisation of the elements out of their component parts. Each element has *summetria* and *taxis* and so might be considered a *kosmos* in itself, and Plato is quite happy to use *kosmeô* and its cognates to describe the process of ordering the elements.[61] The passage can then be read in the first instance as an objection to the idea that there are unlimited shapes and sizes of atoms rather than unlimited worlds, something that Plato has had Timaeus object to previously at 31aff. Whether we speak of one or five *kosmoi* then depends on whether we are referring to the universe or to the elements from which it is organised. For Plato then neither postulating an indefinite number of worlds nor an indefinite number of shapes and sizes of atoms nor a multiplicity of accidental organisms in biology produces an adequate explanation of the nature of the world about us.

In biology Plato has ultimately been proved wrong, as we do now think in terms of a multiplicity of accidents driving evolution. However, although we have dropped the design connotations, physics thinks in terms of a small number of mathematically well defined ultimate building-blocks of matter, rather than the unlimited multiplicity of shapes and sizes of the atomists. On the issue of an indefinite number of *kosmoi*, there is still considerable debate. There are modern theories where our universe is one of many, perhaps infinitely many, which make up a multiverse. There are problems with these theories, however. If what generates many universes is the Copernican principle, that we do not live in a privileged or unique part of anything, then just as our galaxy is one of many, our universe may be one of many. But does that also mean our multiverse is one of many, and so on in regress? If what generates multiple universes is that there are determinate aspects of our universe for which we have no explanation (such as the values of the fundamental constants), then if there are inexplicable determinate aspects of the multiverse does that mean that there are a multiplicity of multiverses, and so on in regress?

If this is regressive, then I do not see that it is explanatory. If it is not, we ought to posit extra worlds only when we have concrete evidence of them, rather than as a means of explanation. One might try to deny that the multiverse has structure or could be treated as a single entity, but that would seem very odd, and I would certainly side with Plato on whether indeterminacy in this sense could ever be explanatory.

There are two important premises in play in some arguments in the *Timaeus*. The first is that what there is constitutes a single, whole entity and not an indefinite plurality. We can take that internally, in that what there is forms a coherent whole, or externally, in that our *kosmos* is all there is. The second is that there is a unique best. Put another way, there are no ties for first place when we ask what is best, but there is a unique best. There is one, and only one best arrangement for the *kosmos*, and the demiurge will instantiate that best arrangement as he always acts for the best, as is emphasised at the beginning of Timaeus' main speech at 29d7ff.

Plato has been accused of a fallacy of division at *Timaeus* 31a1ff.[62] The passage begins with Timaeus asking whether we ought to speak of one heaven, or many or an indefinite number. The critical point is *Timaeus* 31a3-4, where the answer is:

> One, if it is crafted according to a paradigm.[63]

Does Plato illegitimately assume that because the paradigm is unique, and the craftsman copies it, that the copy will be unique too? We have a perfectly plausible alternative. The copy is unique because the demiurge will make the best copy of it and there is one and only one best copy. The uniqueness of the copy is generated by the nature of the demiurge (acts for the best, no reason to do anything else) and not an attempt to copy the uniqueness of the model.

XII. Discovery and invention in cosmogony

Brisson and Meyerstein are the authors of a book entitled *Inventing the Universe: Plato's Timaeus, the Big Bang, and the Problem of Scientific Knowledge*,[64] so some comparison between their approach and my own is in order. On one issue we disagree from the outset. The first line of their work is '*Knowing* the universe is *inventing the universe.*'[65] If I were to condense my own view in a similar manner, it would be 'Knowing the universe is discovering the universe', reflecting a fundamental disagreement in the philosophy of science. Brisson and Meyerstein believe that for science there is a 'hiatus irrationalis'.[66] So between data, and any theory proposed to account for that data, there is a gap, unbridgeable by any rational means.[67] Here again I would disagree. Certainly there is a hiatus, but there are many theories as to how this gap may be bridged in a rational manner. Brisson and Meyerstein may reject all of these, but I do not. They go on to say:

> Philosophical investigation has wavered between two equally unpalatable solutions: posing the 'axiom which justifies all axioms', thus becoming meta-physical, non-rational or even mythical, or 2), after having lucidly established its inherently impassable limits, indefatigably driven by a desire to understand, by a nostalgia for the absolute, transgressing these same limits while unceasingly attempting to reinvent a simple and ordered universe.[68]

Their book addresses this problem.[69] My work does not depend on whether we interpret theories of cosmogony (either ancient or modern) realistically or instrumentally, nor does it seek to show that one of these philosophical views has dominated historically, or that one is more significant philosophically. It simply seeks to show that the Greeks formulated many question types in cosmogony as well as many solution types, and that many of these question and solution types are perennial for cosmogony. Brisson and Meyerstein say that:

> A careful examination of two fundamental cosmological models, Plato's *Timaeus* and the contemporary standard Big Bang model, shows the kind of knowledge called 'scientific' ultimately rests on a set of irreducible and indemonstrable formulas, pure inventions of the human mind, retained solely though recourse to this simple operative argument: 'it works'.[70]

I do not see anything intrinsic to the big bang theory which compels us to interpret it in this fashion. There are many perfectly viable realist interpretations. As for Plato, I have argued elsewhere that any instrumentalist interpretation of his astronomy (and that of Eudoxus as well) is demonstrably false.[71]

In short, Plato's astronomy fails the test of 'it works'. By combining regular circular motions Plato improves on previous models, but the model is very simple and fails to account for many phenomena which Plato was aware of.[72] It could 'work' (save the phenomena!) only if the regular circular motion restrictions (which are philosophically derived) were dropped. Plato does not take this route though, insisting on regular circular motion. I have argued above for a form of the literal interpretation of Plato's cosmogony.

Brisson and Meyerstein are interested in the similarities between modern views and those of the *Timaeus*, and in particular they are interested in axiomatising both views and comparing the axioms. So they say programmatically that:

> In many respects this list of axioms is similar to the list of premises that contemporary science has also to admit; the common character shared by these assumptions will be brought to light by our analysis of contemporary Big Bang cosmology.[73]

They find:

> A certain number of opposites, of which the pair being/becoming represents the fundamental antonymous couple, of the Timaeus, can be found as well in the Big Bang model: this is shown in this table:[74]

Timaeus		Big Bang	
Being	Becoming	Model	Reality
Original	Copy	Theory	Observation
Eternity	Change	Pattern	Complexity
Order	Disorder	Prediction	Complexity
Symmetry	Chaos	Symmetry	Complexity

After their analysis, they say:

> It is at the very least surprising that the list of axioms of the model proposed by Plato in the *Timaeus* in the fourth century BC, and the list of axioms underlying the contemporary Big Bang model, rely on the same abstract concepts. In effect, such notions as order, harmony, symmetry, measurement, stability, invariance, as well as the eternity and universality of the laws of nature, constitute the pivotal notions in both models.[75]

While Brisson and Meyerstein may find this concordance surprising, I do not. I have argued that philosophical cosmogony begins with the Greeks, and that notions of order, invariance and the eternity and universality of the laws of nature are key in differentiating philosophical cosmogony from mythology and creation tales. These ideas (in my view) are rationally defensible as the foundations of any philosophical cosmogony, and it is no surprise that scientific cosmogony builds on these notions. If these notions are the unconstrained inventions of the human mind, there should be surprise at the ancient/ modern concordance. As they are not, there is no surprise. Nor do I find it any surprise that there is a concordance of problem and solution types, as I believe there are perennial problem and solution types, which are not the free inventions of the human mind but are rather related to the inherent nature of philosophical, and subsequently scientific cosmogony.

XIII. The fate of the *kosmos*

What of the fate of the *kosmos* for Plato? The *Timaeus* is entirely clear that the *kosmos* will be everlasting. The demiurge says:

> I am the craftsman and father of these works, and they are indissoluble except by my will. Although all that is bound can be dissolved, to will to dissolve that which has been beautifully and well constructed would be bad.[76]

This view is well attested by Aristotle, not least because he believes it to be fundamentally flawed. Aristotle attempts to demonstrate that whatever is generated can also undergo destruction, and whatever is not generated cannot undergo destruction, and that there is nothing which is generated which is everlasting.

The *kosmos* of the *Timaeus* will not degenerate of itself. As I have argued elsewhere, the *Timaeus* seems to advocate a rather more stable view than some earlier works.[77] In the *Timaeus* the *kosmos* is 'ageless and unailing'.[78] We can find a weak indication of a degenerating *kosmos* at *Phaedo* 110a where the world is said to have been 'corrupted and eaten away'. More significantly, in the *Republic* we can find Socrates saying:

> Won't the true astronomer be similarly persuaded when he gazes upon the movements of the stars? He will hold that the architect of the heavens and

> what is in them has constructed these things as beautifully as is possible for such works. But concerning the proportions of night to day, of these to month, of month to year, and of the other stars to these and each other, wouldn't he consider it absurd to expect these things always to behave in the same manner and never to deviate (*paralattein)* in any way, as they are physical and visible, and absurd too to seek by every means to grasp the truth about them?[79]

I have translated *paralattein* at 530b2 as 'to deviate', but it is worth noting its resonances as given by LSJ of 'to change or alter a little, esp. for the worse ... to pass aside, turn from the path, to deviate, vary'. Interestingly the key word *paralattein* does not occur in *Timaeus*' account of the heavens, nor in other works (*Philebus, Laws, Epinomis*) which arguably agree with the *Timaeus.*[80] The position of the *Republic* seems clear on whether there is a constant relationship of the periods of the stars to each other, and hence whether there is regular celestial motion and whether the visible heavens are amenable to precise mathematical description. This has changed by the *Laws*, where the Athenian states categorically that:

> The usual opinion concerning the sun, moon, and other planets, that they at some time wander (*planatai*), is not the case; precisely the opposite is true. For each of these bodies always travel on one path, and not many, although this may not seem so.[81]

I have argued elsewhere that the motions of the heavenly bodies are entirely regular and amenable to precise mathematical description in the *Timaeus.*[82] The key passage is *Timaeus* 47aff:

> God devised and gave to us vision in order that we might observe the rational revolutions of the heavens and use them against the revolutions of thought that are in us, which are like them, though those are clear and ours confused, and by learning thoroughly and partaking in calculations correct according to nature (*logismôn kata phusin orthotêtos*), by imitation of the entirely unwandering (*pantôs aplaneis*) revolutions of God we might stabilise the wandering (*peplanêmenas*) revolutions in ourselves.[83]

It is also important to differentiate between the views of the *Timaeus* and the *Politicus*. In the *Politicus*, there certainly is degeneration of the *kosmos*:

> In the beginning it remembered the teaching of its creator and father most accurately, though this eventually dulled. The reason for this was the physical element in its constitution, which had been in it from the earliest stage, and partook of great disorder before the universe came to be in its present ordered condition ... When the world nurtures within itself living things under the guidance of the helmsman, it produces little evil and much good. However, when it becomes separated from him, it fares best during the time immediately after the release, but as time proceeds and it grows

> forgetful, the old condition of disorder gains sway more and more, and towards the conclusion of time little good and much of its opposite flourishes, and there is danger of the destruction of the world and those in it.[84]

That this change affects the heavenly bodies is confirmed at *Politicus* 269a. We might also infer this from Plato's general conviction that the stars have physical bodies and the myth's statement that all physical bodies degenerate. If the *kosmos* is continually degenerating towards chaos, and as *Politicus* 273b makes clear, this will be a radical and fundamental chaos (unless god intervenes, the *kosmos* will be 'storm driven by confusion and broken up into an endless sea of unlikeness', *Politicus* 273de), then clearly the periods of the planets will be subject to deviation.

One might also consider some more specific points of difference between the *Timaeus* and the *Politicus*. The *Timaeus* is adamant that the *kosmos*, once created, can function in a stable manner without any attention from its creator and can only be dissolved by an act of his will.[85] In the *Politicus* though either God must perpetually guide the *kosmos* or it degenerates of its own inherent nature, and is only saved from sinking into 'an endless sea of unlikeness' by the active intervention of God.[86] While the *Politicus* tells us that the *kosmos* is mounted on a pivot, the *Timaeus* is adamant that it is not.[87] The *kosmos* of the *Timaeus* is then everlasting and free from deterioration, and this may mark a change of position from earlier works.

XIV. Conclusion

Plato has reasons, beyond an overspill from his programme in ethics, for advocating teleology.[88] He is able to argue for a unique *kosmos*, for a small number of mathematically well defined types of ultimate particle, and for well defined species, all the products of a designer god.[89] He adopts a strong and consistent stance in adopting perennial solution types to perennial problems in cosmogony.

Plato rejects the multiplication of entities (*kosmoi*, types of particle, non-viable biological forms) as ontologically profligate and aetiologically inadequate. In its historical context, this is both a plausible and a viable programme. Plato argues strongly that it is simply not plausible that our *kosmos* could come about by accident. He does so in interesting depth. It is not merely a question of the organisation of the elements into a *kosmos*, but also of the nature of the ultimate particles and their organisation into the elements. Plato is adamant that mere chance, or even the action of a like-to-like principle or something similar will not produce a *kosmos*, with a fitting and harmonious mixture of opposites, from chaos. Plato's account of cosmogony was hugely influential and much discussed in later antiquity.

10

Aristotle

Aristotle (384-322 BC) expresses his views on cosmogony very simply and clearly in *On the Heavens*:

> The whole heaven is not generated nor can it be destroyed, as some have said, but is unique and eternal, not having beginning or end to its lifetime, having and embracing unlimited time in itself.[1]

So for Aristotle there is no cosmogony, there will be no end to the *kosmos* and the *kosmos* has existed for an unlimited amount of time. There are not multiple *kosmoi*, either in the co-existent sense as supposed by Leucippus and Democritus, or in the successive sense supposed by Empedocles, but there is a single, unique *kosmos*.

Aristotle's views were highly influential and were much discussed in antiquity. The development of Christianity brought a new challenge, as many theologians sought to argue first for the possibility of a created *kosmos*, then for its actuality. Aristotle's arguments for an eternal *kosmos* are variable in their quality. Some are based on his physical theories, and fall with the rejection of those theories. Others have a more enduring appeal. Also of significance here are Aristotle's discussions of the nature of infinity. Aristotle needs a theory of infinity in order to establish a coherent theory of change against the challenge of Zeno's paradoxes. However, this theory of infinity, which is widely accepted, is later used against the idea that the *kosmos* can have existed for an unlimited amount of time.

I. No origin of matter

Aristotle denies that matter can be created or destroyed. In the *Physics* he tells us that:

> If matter was generated, there must necessarily be something underlying first, from which it came and which would remain as a component. But this is the very nature of matter itself, so it would exist before it is generated. For I mean by matter what ultimately underlies something, from which a thing is generated and which remains with it as a non-accidental component. If it is destroyed, it is into this which is its ultimate destination, so it would be destroyed before it is destroyed.[2]

So Aristotle's definition of matter precludes the possibility of it being created out of something else or being destroyed into something else. Can matter be generated out of nothing? No, because, according to Aristotle, there is no such thing as a separate void. He says:

> The generation of any body is impossible, if no separate void is possible... One body can be generated from another, such as fire from air, but this is entirely impossible from something which does not exist beforehand. At most an actual body can be generated from a potential one, but if the potential body is not first an actual body, then there is a separate void.[3]

II. No origin of motion

Aristotle denies that there can be an absolute beginning to motion. So *Physics* VIII/1 opens with the question:

> Did motion come into existence, not having existed previously, and will it again, in a similar manner, cease to be so that there will be no motion?[4]

In his review of preceding opinions, Aristotle says that:

> All those who have spoken about nature believe in the existence of motion, as all theorise about the making of the *kosmos*, and about generation and destruction, which would be impossible if motion did not exist.[5]

He goes on to say that those who believe in unlimited *kosmoi* say that motion always exists, while those who believe in a single *kosmos* hold views appropriate to their theory.[6] There are only two possibilities for the absence of motion, according to Aristotle: the stasis of the universe prior to *nous* setting it in motion, as in Anaxagoras, and the stasis of the universe between the actions of love and strife, as in Empedocles.[7] Aristotle then offers two choices:

(1) If movable things are generated, not previously being in existence, then there must be some change or movement which generates the movable things.
(2) If movable things have always existed, but have not moved then there must be some change which makes them move.

The key assumption here is that each movement requires a previous movement to begin it.[8] This sets off an infinite regress of movement, which is Aristotle's goal. Movement has always existed and there is no first movement. We can see similar assumptions in Aristotle's criticisms of the early atomists:

> When Leucippus and Democritus speak of the primary bodies as always moving in the unlimited void, they ought to say what sort of motion this is

> and what their natural motion is. The elementary particles may be moved by force by others, but each must have a natural motion as well, in distinction to the enforced motion. Of necessity, the first motion cannot be instigated by force, but must be natural. We shall proceed to infinity, if there is nothing which instigated the first motion naturally, but there is always a prior motion moved by force.[9]

Aristotle believes that the same sorts of difficulties arise for Plato in his account in the *Timaeus*, when he speaks of the disorderly motion prior to the organisation of the *kosmos*.[10]

III. No origin of time

Aristotle treats time as intimately related to motion:

> Motion cannot be generated or destroyed, as it always exists. Nor can time, as there would be no before or after if time does not exist. Motion is continuous in the same way as time is. Time is identical with motion, or is some affectation of it.[11]

Time does not have a beginning, as to say 'before the existence of time' implies that there was a time before time came into existence. Similarly, one cannot say 'after the end of time', so time will stretch infinitely into the future as well as the past. Aristotle also takes solace in the fact that only Plato, of his predecessors, supposed an origin for time:

> Through this Democritus demonstrates that it is impossible for everything to have been generated, for time is ungenerated. Plato alone allows the generation of time, as he says it came into being at the same time as the heavens, and the heavens are generated.[12]

Aristotle's conception of 'now' also excludes a beginning for time. He treats the now as:

> The now is a mid-point, combining beginning and end. It is the beginning of future time and the end of past time. Time then is necessarily eternal.[13]

Any 'now' we care to mention is then the limit of a preceding time and so time will stretch infinitely into the past, and similarly any time we care to mention will be a limit of future time so time will extend infinitely into the future. Aristotle is naturally critical of Plato's view that time begins.[14]

IV. Proportionality and cosmogony

Aristotle does employ a 'why now?' argument against the origin of motion, and his reasons for doing so I believe are quite interesting. In relation to the views of Empedocles and Anaxagoras, he says that:

> Nothing natural, nothing according to nature is without order, as nature is the reason for the order in everything. The unlimited has no proportion to the unlimited, while all order has proportion. An unlimited time of rest, followed by motion at some point, there being nothing to distinguish a preference for now rather than earlier, excludes any sort of order, and is not a work of nature. The natural is absolute, and not sometimes thus and sometimes otherwise. Fire is always carried upwards and not sometimes like this and sometimes in another manner. When nature is not absolute, there is proportion.[15]

The motivation here is not a simple sufficient reason objection for picking out one point in time as opposed to any other, though that is part of the objection here.[16] Aristotle believes that key comparatives would lose their meaning if there were not some relation of proportionality between them. Consider the way in which pairs of comparatives such as heavier/ lighter, quicker/ slower, longer/ shorter (distance) and shorter/ longer (time) are related to each other for Aristotle. *Physics* VI/2 defines quicker as to cover a greater distance in the same time, or the same distance is a shorter time. Take any two speeds, such that b is quicker than a. They will map onto a pair of distances such that b' is further than a'. That must be so for all pairs of speeds and their related pairs of distances. If this were not so, then there would be at least one instance where being quicker does not entail going further in the same time. So the relation between speed and distance must be one that is order preserving, and that means that there must be some relationship of proportionality between speed and distance. That relationship does not necessarily have to be one of direct proportionality. As long as (in modern terms) we can write f (speed) α f (distance) (function of speed is proportional to function of distance), then order will be preserved and the meaning of quicker/ slower, etc. will be preserved.[17] Aristotle has no need to determine which relation of proportionality applies in each case, he merely needs to argue that in all cases there is such a relation. So we can find Aristotle saying:

> In conclusion then, the basic principle is clear. There is always a ratio between changes, for they are in time, and between two determinate time periods there is always a ratio, but there is no ratio between fullness and the void.[18]

Aristotle's criticism of other theories of change is often that key comparatives would lose their meaning (e.g. with motion in a void). Aristotle uses this analysis for all types of change, as part of his attempt to set up a comprehensive theory of change against Eleatic opposition. So Aristotle says:

> It is always the case that the greater power (*dunamis*) accomplishes the same act in a shorter time, whether that be heating, sweetening, throwing or any sort of change.[19]

> Suppose that A is heated, pushed, affected in any way or changed by B in a time C; let D be less than B, and let us assume that the lesser creates a smaller change in an equal given time. Suppose E to be that by which D is altered (*alloiomenon*). As D is to B, so will E be to some certain amount accomplished. We may assume that in the same time an equal force will effect an equal change (*ison alloioun*), a lesser a lesser in an equal time, and a greater a greater and so in proportion of the greater to the lesser.[20]

The English word proportion has very strong mathematical connotations, though even here we might say 'a punishment in due proportion to the crime'. The words that Aristotle typically uses (*analogos, logos, hosos*) perhaps do not have such strictly mathematical connotations, such that Aristotle may have something broader than a strict mathematical proportionality in mind. What Aristotle requires is that there is a *logos*, in the sense of account or explanation, for all sorts of change. In some situations, where quantification is possible, that *logos* can be a strict mathematical proportionality. Where qualitative judgements are required, the proportionality thesis still applies in order to generate a *logos* where the meaning of comparatives can be preserved.

The criticism of Anaxagoras and Empedocles goes deeper than the 'why now?' question then. If motion ceases, then for Aristotle time ceases as well. If that is so, then there will be no before and no after. In such a situation, there can be no orderly, comprehensible change.

V. Empedocles and Anaxagoras

It is interesting that Aristotle then thinks that Empedocles has the advantage over Anaxagoras, in that Empedocles gives an ordered succession of stages rather than a singular, arbitrary beginning.[21] Aristotle does not think that Empedocles does enough to justify the actions of love and strife, however. He believes that Empedocles needs to show that love and strife alternate in the required manner, and needs to give some justification of why the periods of love and strife are equal in length.[22] Aristotle also voices the following opinion in relation to Empedocles:

> The alternation of combination and dissolution does nothing other than make it eternal but changing in form, just as if one considered the generation of a man from a child and a child from a man as involving at one time destruction and at another existence. It is clear that when the elements come together, no random order or formation occurs, but the same one, especially according to the accounts given of this view, since they explain each arrangement in terms of an opposite. So if the whole body, being continuous, is arranged and ordered at one time in one way and at another time in another, then the *kosmos* would not be generated and perish, but its arrangements would.[23]

It is clear from this that Aristotle operates with a different notion of *kosmos* from his predecessors. A *kosmos* is not simply an orderly arrangement of matter, but may be an orderly process or succession of states, where there is no necessity for at least some of those states to be orderly, as long as there is order in their succession.

VI. Physical objections to cosmogony

Aristotle also has a set of considerations derived from his theory of the elements and his theory of motion in favour of a unique, eternal *kosmos*. *On the Heavens* I/2-4 argue that the heavens are composed of aether, that aether is neither heavy nor light, that it does not undergo generation or destruction, that aether cannot grow or diminish, that aether is not subject to any qualitative change, that aether's natural motion is circular and that no alteration has ever been observed in the heavens. The natural conclusion here is that the heavens have no origin and are unchanging. In I/8, Aristotle argues for a unique *kosmos* based on his views concerning natural motion and place. There is a centre to the universe, and earth, water, air and fire move towards or away from this point.[24] The idea of other *kosmoi*, each with a point for these natural motions is a nonsense. Aristotle also says that:

> All the *kosmoi* must be comprised of the same bodies, if they have a similar nature. Each of these bodies must have the same capabilities, that is to say fire and earth and those between. If we give the same name to these bodies as we do to ours without an identity of form, then this whole can only be called a *kosmos* in name.[25]

In support of the idea of a finite *kosmos* are the usual considerations concerning the finite nature of change and the point that in an unlimited *kosmos*, a body could attain an unlimited speed approaching the centre, which Aristotle holds to be impossible.[26] Ultimately, a key conclusion is that:

> It is clear that there is neither place, void nor time outside of the *kosmos*.[27]

So on a multiplicity of physical grounds, for Aristotle there can be only one *kosmos* and that *kosmos* is eternal.

VII. Zoogony and stoichogony

Aristotle's views on zoogony are well known. For Aristotle there is fixity of species, and the species of animals have existed for as long as the *kosmos*. So just as there is no cosmogony for Aristotle, there is no zoogony either. Naturally Aristotle is critical of others on this topic, and Empedocles is selected for special treatment. We saw in the chapter on Empedocles that

Aristotle was critical of how air might flow up at some times and in other directions at other times. Aristotle then continues:

> He says that the parts of animals were generated by chance for the most part.[28]

Aristotle, not surprisingly, is critical of Empedocles' scheme of explanation for the generation of living things: With reference to Empedocles, he says that:

> Generation is for the sake of being, rather than being for the sake of generation. So Empedocles was not correct to say that animals have many features on account of chance events during their generation, such as the backbone being as it is due to getting broken in bending.[29]

Nor will Aristotle allow that living things have been designed by a cosmic intelligence and have come into being at a certain time, as in Plato. Just as there is no cosmogony or zoogony for Aristotle, there is no stoichogony either. The terrestrial elements may of course change into each other for Aristotle, but there is no stage when matter is unformed or chaotic. The aether never undergoes any change whatsoever. Aristotle is highly critical of the idea that matter could be composed of, and indeed in any way generated from the basic Platonic triangles.[30] Matter itself, as we have seen, cannot be generated. Aristotle does not believe that there will be any end to the *kosmos* either.

VIII. Logical objections to cosmogony

In *On the Heavens* I/12, Aristotle argues for the following conclusions:

> (i) Everything which always exists is absolutely ungenerable and indestructible.[31]
> (ii) Everything which is absolutely ungenerable and indestructible always exists. [32]

Along with this goes the assumption that:

> Nothing can be indestructible or ungenerable through chance. Chance and luck are contrary to that which always or mostly happens. That which exists for an unlimited time, absolutely or from a certain time, exists either absolutely or for the most part.[33]

So a *kosmos* which exists for ever is incapable of not existing. Indeed, anything which exists forever exists of necessity. Such a view is rather odd to the modern eye and these passages have been much discussed. Perhaps Aristotle has gone astray either in argument or methodology,[34] perhaps there are unspoken metaphysical assumptions which help to produce these conclusions.[35] Whatever Aristotle's reason for holding this view, his target is clear enough. He wishes to oppose the Platonic view that the *kosmos* came into existence and will subsequently exist for all time. *On the Heavens* I/12 concludes:

> It is impossible for what was previously eternal to perish later, or what was previously not eternal to be eternal later. All generable and destructible things are subject to change. They change by means of contraries, and that from which natural things are constructed is the same as that by which they are destroyed.[36]

IX. *On Philosophy*: theological objections to cosmogony

Aristotle also argues against the destruction of the *kosmos* in his early work *On Philosophy*. Here many of the arguments are to do with god's attitude to his creation. Fragment 16 argues that god is changeless, on the basis that where there is a better there is a best, and god is clearly the best. Change may be initiated by something else or by god. God would not change himself into something worse, and there is nothing better for him to change to. The better cannot be affected by the worse, and so god cannot be affected by anything else. Therefore god is changeless. Fragment 18 argues that to say the *kosmos* is generated or can be destroyed is to say it is no better than the work of men's hands. It also relates that Aristotle said that:

> Of old I was concerned with a fear that my house would be destroyed by strong winds or exceptional winters or time or slipshod maintenance, but now there is a greater fear looming, that the *kosmos* will be demolished by argument.[37]

Fragment 19a argues that either the *kosmos* might be destroyed from within, or from without. It cannot be destroyed from without as there is nothing external to the *kosmos*. It cannot be destroyed from within because a part would then be greater than the whole. The *kosmos* has unsurpassable power, and directs all its parts and is directed by none of them. Fragment 19b argues that:

> The reason for the destruction of other animals is having an order contrary to nature, but in the *kosmos* each of the parts is ordered according to nature and has a home assigned to it, so the *kosmos* may be justly called indestructible.[38]

Fragment 19c begins by asking:

> Why should god destroy the *kosmos*?[39]

This cannot be so that god no longer has to engage in *kosmos* creation, or in order to create another *kosmos*. It is fitting for god to turn disorder into order, not vice versa. Aristotle then gives us three possibilities:

(1) If the new *kosmos* is worse, it can only be because god is worse. Why would god instantiate something worse when something better is available?
(2) If the new *kosmos* is the same as the old, the creator will have laboured in vain, and it would have been better not to destroy the original *kosmos* but to leave it in place.
(3) If the new *kosmos* is better, it can only be because god is better, and must have been imperfect when he made the original *kosmos*.

Aristotle further rules out (1) and (3) by asserting that:

> God is equal and identical to himself and does not slip towards the worse or stretch towards the better.[40]

Fragment 20 argues that the *kosmos* could not have been generated as there was no new design from which such a work could be given a beginning.[41] Effe has argued that there is some further material, perhaps even a further fragment which bears on Fragment 20.[42] The argument questions why and when god would produce anything new, such as the new design of Fragment 20, and concentrates on the time before the innovation. Was god asleep? Yet eternal sleep is death. As god does not suffer from fatigue, that could not be a concern. If god was unhappy with his environment, why did he not effect the change earlier?

The arguments of *On Philosophy* are variable in their merits, to say the least. Fragment 16 gives a suspect ontological argument for god and his nature, and I see little reason why we should accept that the better cannot be affected by the worse. Fragment 19a is correct in asserting that the *kosmos* cannot be destroyed by anything external if there is nothing external, but one would hardly be reassured by the idea that the *kosmos* cannot be destroyed from within because then a part would be greater than the whole. 19b gives us perhaps the worst argument, because it is clear that for Aristotle the parts of the *kosmos* are not all in their natural positions. If they were, there would be concentric spheres of earth, water, air and fire with no mixing of the elements. The oddest part of this argument is that Aristotle recognises that the parts of animals are not in their natural places, but he does not then recognise that animals are part of the *kosmos* and so not all the parts of the *kosmos* are in the natural place. Fragments 19c, 20 and the material suggested by Effe produce the best, and most influential arguments against a beginning for the *kosmos*. These types of issues are much debated by Christian and Neoplatonist thinkers in later antiquity.

X. Aristotle and cosmogony

Some of Aristotle's arguments for a unique, eternal *kosmos* are based on his long defunct theories of natural place and natural motion, and so have little relevance to modern discussions of cosmogony. However, some other worries that Aristotle raises are of a rather more enduring nature. Aristotle is quite correct to point out that there are problems in relation to proportionality and comprehensibility in some accounts of cosmogony. Aristotle would doubtless be opposed to the existence of 'naked singularities' in modern physics. The notion of matter in zero volume, at infinite temperature, pressure and density would be an anathema to him, and there are many modern philosophers and scientists who have significant doubts about the existence of such singularities.

Aristotle, I suspect, would also be an opponent of big bang cosmogony. Certainly he opposes the beginning of time supposed by BB, and I believe he would also be in opposition to the generation of space. Aristotle would oppose the idea of BB on the grounds that it is inexplicable. In his own terms BB would have no material, formal, efficient or final reason. In modern terms, our theories of causation/ explanation also fail with BB, rendering it difficult to give any reason why BB happened. In his criticism of Anaxagoras and Empedocles, Aristotle is more critical of Anaxagoras for supposing a single start for the *kosmos*, allowing that Empedocles at least has some system to his cycle of stops and starts. So I believe Aristotle would also object to the unique nature of BB. I think he would also wonder, quite rightly, why we get a determinate universe out of nothing. Where we get specific quantities of matter or motion (i.e. the expansion rate of the universe), those quantities are in due proportion in relation to the antecedent conditions. With a generation of the *kosmos* out of nothing, though, there can be no such proportions.

XI. Conclusion

Aristotle's anti-cosmogony stance was highly influential. There are trenchant criticisms of his predecessors, in particular Anaxagoras and Empedocles, as well as a significant challenge to Plato's account of cosmogony. Aristotle will of course have his followers, but he is also important for later thinkers to react against. Both Christians and Neoplatonists see Aristotle's arguments as a major obstacle to establishing the sort of cosmogony they would like to see. Accepting some aspects of Aristotle's account of the natural world makes that task more difficult, though as we shall see there are some ingenious ways around Aristotle's arguments, and some even use Aristotle's views on topics such as infinity to argue against Aristotle's anti-cosmogony stance. Aristotle's anti-cosmogony is not merely based on his theory of natural motion and natural place. He also has deeply held concerns about how comprehensible any cosmogony might be. As we have seen, some of those concerns also trouble modern cosmogony.

11

Epicurus and His Followers

Epicurus (342/1-271/0 BC) begins the second strand of atomist thought in antiquity. We have relatively little of his own work, but the general outlines of his cosmogony and his philosophical system in general can be gleaned from the summaries he gave of them in his letters. Lucretius (94-55 BC) gives us an extensive Epicurean account in his *On the Nature of Things*, which we have in its entirety. It is important that Epicurus follows Plato and Aristotle, as he is to some extent able to take on board their criticisms of Leucippus and Democritus.

There are considerable similarities between the views of Epicurus and those of Leucippus and Democritus. Epicurus believes there to be an unlimited number of *kosmoi*, with *kosmoi* coming into existence and others passing out of existence all the time. Time is unlimited both in the past and in the future, there is a limitless void and there are an unlimited number of atoms. The *kosmoi* are generated by chance, without purpose and without any intervention from the gods. There are though some interesting and important differences. There is no use of the *ou mallon* principle which was so important to Leucippus and Democritus. Atoms, while being unlimited in number and unlimited in shape, are limited in size to lower than the threshold of human perception. Epicurus also tries to distance himself from Leucippus and Democritus by rejecting what he perceives as their rigid determinism in favour of an undetermined swerve of the atoms which he hopes will preserve free will, but has important implications for cosmogony as well.

I. The Epicurean *kosmos*

Epicurus, and indeed Lucretius,[1] follow Parmenides on the question of whether something can come to be out of nothing:

> Nothing is generated out of that which does not exist. If it were, everything could be generated out of everything else with no need of seeds. And if that which disappears perishes into the non existent, all things would have been utterly destroyed, since that into which they are dissolved does not exist.[2]

Epicurus also follows Leucippus and Democritus in believing there to be an infinite number of atoms, an infinite amount of space,[3] an infinite amount of time,[4] and an infinite number of *kosmoi*:

> There are an infinite number of *kosmoi* both like and unlike this one. As has already been demonstrated, there are an infinity of atoms, and they are carried far away. The atoms which are such that a *kosmos* could be generated or made from them, are not exhausted in the completion of one *kosmos* or many, nor all those that are alike or those that are unlike. Thus there is no impediment to there being an infinite number of *kosmoi*.[5]

It is an interesting principle here that the atoms are 'carried far away' throughout the void, presumably because there is nothing to interrupt their motion.[6] Lucretius argues that just as we do not find single animals, but everything belongs to a group of animals, so we should not suppose there to be a single sky, earth sun or moon.[7] These are innumerable, and there is a limit to their lifespan just as for any animal on earth.[8] For Epicurus a *kosmos*:

> is a bounded part of the heavens, containing stars and earth and all the phenomena, whose dissolution will cause all within to be seized with confusion. It is cut off from the infinite, and concludes in a boundary, either rare or dense, either revolving or stationary. In outline it may be spherical or triangular or some other shape.[9]

It is unlikely that Epicurus thought that there were all shapes of *kosmoi*. Rather, he just wanted to contest the view that all *kosmoi* have the same shape.[10] It is interesting to note that those who suppose there to be co-existent multiple *kosmoi* give those *kosmoi* physical boundaries and have strong individuation criteria for those *kosmoi*. We saw that with the membrane that surrounds the *kosmoi* of Leucippus and Democritus. Neither the Milesians, Diogenes nor Anaxagoras, sometimes supposed to have multiple *kosmoi*, speak of physical boundaries or individuation conditions. Epicurus also disagrees with Leucippus and Democritus on the number of different shapes of atoms:

> In addition to this, the full atomic bodies, out of which all compounds are generated and into which they are all dissolved, have an incomprehensible variety of shapes. It is not possible for so much variety in things to have come about from a comprehensible number of the same shapes. For each shape, there is an absolute infinity of like atoms, but the varieties of shape are only incomprehensible in number, not absolutely infinite.[11]

Further, he disagrees on the number of sizes of atoms:

> The existence of all sizes of atoms is not required to explain differences in respect of qualities. Visible atoms would be known to us, but we do not see this, nor is it conceivable that a visible atom would occur.[12]

What is noticeable here, in relation to the positions adopted by Leucippus and Democritus, is that there is no use of the *ou mallon* principle. Indeed there is no use of this principle at all in relation to issues of cosmology and cosmogony in either Epicurus or Lucretius.

II. Epicurean cosmogony

Epicurus also differs from Leucippus and Democritus on where *kosmoi* can form:

> That there are an infinite number of such worlds is well established, also that such a *kosmos* is capable of coming to be inside a *kosmos* or intercosmically, by which we mean in the intervals between *kosmoi*. This will occur in a place with much void, but not in a place of complete void, as some have said.[13]

The final sentence here would seem to be an allusion to Leucippus and Democritus where the bodies that have been cut off from the infinite move into a great void, form the vortex and create *kosmoi*. It would seem that *kosmoi* can begin to form in any place according to Epicurus. However, they do need seeds:

> *Kosmos* formation occurs when the appropriate seeds flow in from one *kosmos* or an *interkosmos* or from many. Gradually, by additions and joinings and migrations to another place, as may occur, and appropriate irrigations of this matter, a state of completion and permanence is reached, which lasts while the underlying foundations are capable of being added to.[14]

This is important in relation to Parmenides' challenge and the principle of sufficient reason. If Epicurus and Lucretius reject the *ou mallon* principle, then they must have some other means of answering the 'why here' and 'why now' questions for the formation of *kosmoi*. Their reply is that *kosmoi* form wherever there is a seed for them to form. As Lucretius emphasises, there is a temporal aspect to this as well. With the seeds of plants, we see that they begin to grow only at certain times and in certain conditions.[15] Lucretius says nothing specific about the seeds of *kosmoi* here, but neither does he distinguish between the seeds of plants/ animals and the seeds of *kosmoi*.[16] Indeed, Lucretius employs a biological analogy for the growth and subsequent decline of the earth,[17] and likens the earth to a fertile woman.[18] Lucretius fills us in on the details of the formation of the earth and the other elements of the *kosmos*:

> Firstly, all of the pieces of earth, because of their weight and entanglement, gathered in the centre and took the lowest place. The more tightly they became united and entangled, the more they expelled that which would form the sea, stars, sun, moon and the walls of the great world. The constituents of these bodies are smoother, rounder and of significantly smaller size than the elements of the earth. Firstly, fiery aether immediately passed through the loose interstices of earth and being light, elevated itself carrying with it

> much fire ... light and aether then formed a coherent body curved in all places which spread widely in all directions to enclose all things in a greedy embrace.[19]

> The sun and moon and the other stars were not generated by themselves and subsequently taken in by the *kosmos*, but were formed and grew from the first by the accretions and whirlings of certain natures composed of small parts, either windy or fiery or some combination of these. For this is what sensation suggests.[20]

> The different parts began to separate, and like began to unite with like, beginning the organisation of the world, and the placing of the huge components. That is, the division of heaven from earth, a distinct place given to the sea and its expanse of water, in isolation from the heavenly fires such that the two should not mix. [21]

Epicurus is quite specific in opposing the idea of a vortex for the creation of *kosmoi*:

> More is required than a gathering together or the formation of a vortex in a void which will permit a *kosmos* to come to be, such as is imagined from necessity. Nor can a *kosmos* grow until it collides with another, as one of the so-called natural philosophers says. For this is in conflict with the phenomena.[22]

That the targets here are Leucippus and Democritus is evident from the combination of vortex and necessity. So more is required than a simple gathering together of atoms, and more is required than a vortex.

III. Against teleology and providence

Lucretius gives a selection of arguments against the ideas of providence and teleology. Only rarely are these arguments aimed at a specific doctrine or philosopher. Much more commonly they are aimed at a general viewpoint, and apply to Plato, Aristotle and the Stoics. His most straightforward anti-teleology statement is:

> Now I will explain the order by which the collection of matter founded the earth, heavens and deep oceans, and set the sun and moon in their orbits. Certainly the primary things did not intentionally and intelligently place themselves in their positions.[23]

There is an array of further argument in *On the Nature of Things*, and these arguments are often further discussed in Cicero's *On the Nature of the Gods*. Lucretius takes two things for granted, having already been argued for, that there are infinities of time, space, atoms and *kosmoi*, and that the best life is a tranquil one free from worry. One of the main motivations behind Epicurean philosophy is the relief from worry.[24]

(1) If there are unlimited *kosmoi*, could god control the universe?

> Who is strong enough to rule the sum of the measureless, who can hold in his hands and control the mighty reins of the depths? Who at one instant can turn all the heavens ? Who can warm the plentiful worlds with aethereal fire?[25]

(2) Why should the gods create a *kosmos* for humans? Why should beings living in tranquillity suddenly decide to do something different and create a *kosmos*? Surely not for the thanks they would receive from us! What fault did they find in the old that would motivate them to something new?[26]
(3) What harm would be done had humans never been born?[27]
(4) Where did the gods acquire the plan for the *kosmos* and for man? How did they acquire a knowledge of the primary particles and what they might be capable of in other arrangements?[28]
(5) There is no need for teleology. If we suppose there to be an unlimited number of particles which have interacted for an unlimited amount of time, and have created an unlimited number of *kosmoi*, then it is no surprise that we live in a *kosmos* like ours. Lucretius says:

> So many primordial particles, in a multitude of ways, have been propelled by their own weight and impacts for an infinite amount of time, and have made trial of all things their union could produce, it is hardly surprising if they have come into arrangements and patterns of motion like those repeated by this world.[29]

(6) The *kosmos* is flawed, and so can hardly be considered to be a divine gift. Large parts of it (rocky areas, swamps, seas) are uninhabitable. Two thirds of it are uninhabitable due to high or low temperatures. If humans did not clear land for farming, it would soon be overrun, there are many wild beasts, many diseases and early death is commonplace.[30] Lucretius also believes that the heavens are flawed, in defiance of those who see them as the highest work of god.[31]
(7) Why, in a world providentially orientated for humans, do human young need great care and nursing, while those of animals grow without the need for toys, tutoring or clothes? They do not need weapons or walls to defend themselves, as the earth supplies all of their needs.[32]

Cicero gives us some variations on these themes, through his Epicurean speaker, Velleius:

> (2a) The Epicurean god is blessed while the Stoic god is extremely overworked. If the *kosmos* is a god, what could be less tranquil than rotating without cessation at a high speed? Maintaining the succession of the seasons, variations and regularities, watching over the sea and guarding men's lives is a vexatious and tedious task. The Stoic god is not tranquil, and nothing is blessed if it is not tranquil.[33] Why should the gods suddenly decide to build

> a *kosmos*, after centuries of inactivity ? Time continues despite there being no *kosmos*, it being unthinkable that there should be a time without time.[34]
>
> (5a) The *kosmos* is the product of nature and was not manufactured. This is an easy process, and nature has made, is making and will make infinitely many *kosmoi*. You who do not see how this can be done are like the tragic poets who cannot bring a plot to its conclusion without the intervention of god. This you would not need if you recognised infinite space and an infinite number of atoms.[35]

Argument (1) depends on a conception of god and what can be done with the infinite. If god cannot know or control the infinite, if there are an infinite number of *kosmoi* he cannot know what to do in them for the best nor can he control them.

Argument (2) is a cluster of arguments around the Epicurean view that the best life is a tranquil one free from worry, and naturally the gods lead the best sort of life.[36] If the gods have this tranquillity though, why would they alter it? One might also read into this a 'why now' sufficient reason objection, and the 'new design' objection raised by Aristotle.

Argument (3) is phrased in terms of individuals, who if they had never existed would know nothing else, and so there would be no harm in them not coming into existence. A more modern line on this might ask if any harm would be done by humanity never coming into existence, pointing out the harm that humanity has done to the earth and other animals.

Argument (4) can be taken in two ways. The objection may be about how the gods could know about the infinite number of primary particles and the infinite number of ways in which they may be arranged. If the gods cannot know either, they cannot make a selection of the best *kosmos*. Alternatively, this may be an objection about how the gods come to know. Do the gods learn these things at some stage? If so, which? What prompts them into learning these things, when presumably they were quite tranquil with the knowledge they had beforehand.

Argument (5) is interesting for being specifically anti-teleological and giving a clear answer in terms of seeing our *kosmos* as one accident among an infinite array of other accidents.

Argument (6) gives a classic reply to the idea of benevolent creation/providence, that the *kosmos* is in many ways flawed and cannot be seen as the best possible creation. There are of course similar modern replies. To pick out a few of the more interesting and entertaining, I would agree with Bertrand Russell that:

> The believers in Cosmic Purpose make much of our supposed intelligence but their writings make one doubt. If I were granted omnipotence, and millions of years to experiment in, I should not think man much to boast of as the final result of all my efforts.[37]

Ambrose Bierce gave the following definition of 'ocean':

> Ocean, n. A body of water occupying two thirds of a world made for man – who has no gills.[38]

Monty Python offered the following ditty:

> All things dull and ug-ly,
> All creatures, short and squat,
> All things rude and na-sty,
> The Lord God made the lot.[39]

In the following chapters we will see the Stoic, Neoplatonist and Christian replies to these types of argument.

IV. Zoogony

Epicurean zoogony is anti-teleological. Lucretius argues that we ought not to suppose that eyes were made in order for us to see, or legs jointed so that we can take long strides. All such explanations are back to front.[40] More specifically, he says that:

> Nothing is generated in us in order that we might be able to use it, rather it is the fact that it has been generated which enables the use.[41]

Epicurus himself gives us relatively little on the origins of living beings.

> It cannot be proved that in one sort of *kosmos* these seeds from which animals and plants and all the rest of the things we behold have been formed, were not taken in, and that in another sort of *kosmos* this was not possible.[42]

Lucretius, on the other hand, waxes lyrical at some length on the generation of living beings. There are several influences on his writings, most likely the now lost Epicurus *Concerning Nature* 11-12 and the works of Empedocles on the positive side,[43] and of course he will be opposing Plato and Aristotle. Lucretius says:

> Firstly the earth gave forth types of bright green grass, about the hills and all over the plains, and the flowering meadows shone with greenness. The various trees began a race of growing in an unrestrained fashion through the air. As feathers and hairs and bristles first grow on the limbs of four-footed animals and the bodies of strong winged birds, at that time the new earth produced grasses and young trees, then animals, many types being produced in many ways. Animals cannot have dropped out of the sky, nor can land animals have emerged from salt pools. It remains then that the earth deserves the name of mother, as all things have been generated from the earth. Even now many animals are generated from the earth, produced by the rains and the warm heat of the sun. It is less wonderful then if at that

> time more and larger animals were generated, raised by the new earth and the new air.[44]

He also says:

> Again and again the earth deserves the name of mother, since she generated humans, and almost at a fixed time produced every type of animal that ranges wild and wide over the great mountains, as well as all the various types of birds of the air.[45]

In the beginning, the earth generates many oddities. There are creatures belonging to neither sex, others that are footless or handless, or without eyes or mouths, or having limbs inseparable from their bodies. For a species to survive, it must be able to get sustenance, there must be a means for life-giving seeds to flow out of the body, and male and female must have suitable organs for reproduction. Those species that could not regenerate or find food for themselves became extinct.[46] Although both Empedocles and Epicurus employ only chance and necessity, there is the significant difference that whereas Empedocles has parts of bodies properly formed and then conjoins them, Epicurus has his creatures constructed from scratch. There are two phases of extinctions for Epicurus. In the first place, creatures which are simply not viable, even if they have the earth to themselves, die out because they cannot feed themselves or they cannot reproduce. At the second stage there is competition between the viable species, which the most suited to the environment will win. So the brave (lions), the cunning (foxes) and the swift (deer) will win out over other species, and some species which mankind finds useful will win out by coming under their protection. The species are fixed, and there is no evolution of species, merely extinctions of those not well suited to their environment, and as other earths will be different from our own we might well expect other animals to survive in the different environment there. This attitude to zoogony can be contrasted with those of Plato, Aristotle, and the Stoics. The Stoics see the fit between a creature's structure and abilities and its environment as proof of the providence of god.[47]

V. The Epicurean swerve

Notoriously, the atoms of Epicurus do not simply drop in straight lines, but may swerve as well. There is no mention of the swerve in any of the extant writings of Epicurus but it is very clearly in Lucretius and it is attributed to Epicurus by many ancient sources.[48] Lucretius tells us:

> When bodies fall through the void due to their weight, at quite uncertain times and places they swerve a little from their course, just enough for this to be called a change of motion. If they were not accustomed to swerving, they would all fall down, like drops of rain through the deep void, and there would

> be neither collisions nor blows between the primary bodies. Thus nature could never have produced anything.[49]

Is this an argument for the swerve as the first cause of collisions? That is, does the series of collisions have a first member whose cause is the swerve? This is the view of Long and Sedley: 'The swerve theory is introduced to solve a problem of Epicurus' own making: given the equal speed and downward motion of atoms, how can collisions between them ever have started?'[50]

If Epicurus is worried about how collisions start, should he not also be worried about how the downward motions of atoms start? If he believes that motion requires an instigator, then surely both types of motion, fall and swerve, require something to start them ?[51] Lucretius' view though seems to be that both types of atomic motion have an infinite past, as he tells us that the atoms:

> have been propelled by their own weight and impacts for an infinite amount of time.[52]

This also seems to be Epicurus' view:

> The atoms are permanently in motion and some separate far from each other, while some are restrained vibrating, when they happen to become trapped by their entanglement or confined by interweaving. This is because the nature of the void separates each of them by itself, and is unable to provide them with any support. The fact of their solidity means that they rebound when they collide, as far as the entanglement permits them to return from the collision. There is no beginning for these, as the atoms and the void are eternal.[53]

One way of trying to escape the difficulty here is to suggest that the 'these' of the final phrase refers only to the falling of the atoms. The Greek simply has 'these' without specifying what 'these' are. However, it is fairly clear from the way that the two results of collisions are discussed immediately before that 'these' is intended to cover both types of collision, or at least all instances of the second type of collision.[54]

I agree with O'Keefe that the best solution here is to regard the swerve as an explanation, and as part of a reaction to Aristotle's criticism of Democritus. Aristotle's concern was that Democritus' atoms had no natural motion, and the atomist universe, having no centre, had no preferred direction to move in. With Epicurus though, it is the nature of the atoms to fall downwards, and to continue to do so, and also in their nature to swerve, and to continue to do so. So there is an explanation of why atoms fall, but not a cause of their initial motion, and an explanation of why atoms collide, but not a cause of their initial collisions for as time extends back infinitely for Epicurus, there is no first motion and no first collision.

VI. Some perennial problems

Epicurus and Lucretius face a problem that we have seen previously, of how a *kosmos* can form if we assume relatively homogenous prior conditions, in this case the parallel fall of atoms. As with Leucippus and Democritus, and as with modern cosmogony, their answer is to introduce some indeterminacy to generate the necessary inhomegeneity. This is not to suggest that Epicurus in any way 'anticipates' quantum mechanics, not least because I believe the relations between the Epicurean swerve and QM are often overdone.[55] The structure of the proposal though is intrinsically similar. Where we have initial homogeneity, we must have some means of generating inhomegeneity and escaping the principle of sufficient reason.[56]

It is interesting that Epicurus comes quite close to a formulation of the perfect cosmological principle (that the universe, on a large scale, looks the same from any point in space and from any point in time). This was so for Leucippus and Democritus as well. Epicurus tells us that:

> The universe always was as it is now, and always will be. There is nothing into which it changes, and nothing outside which could enter it and effect a change.[57]

> Nothing new happens in the universe if you consider the infinite time which has occurred.[58]

As it is a fundamental tenet of Epicureanism that there is change, contrary to the Eleatics, Epicurus must here be talking about the universe as a whole. Clearly, things within the universe change (there is the generation and destruction of *kosmoi*) but the universe as a whole remains constant over time. It is also interesting in this context that Lucretius says:

> There can be no centre when the universe is infinite.[59]

Epicurus and Lucretius are consistent in their application of their physical principles, and clearly take the view that there should be no difference between the processes which form *kosmoi* and the processes which are active in the *kosmos* today. That is evident from the fact that they take *kosmos* formation to be an ongoing process. The swerve, the ultimate explanation of the collisions, can also happen at any place and any time. So the processes which lead to *kosmos* formation are permanent features of Epicurean physics.

VII. Epicurus on rare and dense *kosmoi*

Epicurus also expressed an interesting problem in cosmogony, which then lay dormant for many centuries before becoming of the utmost importance for modern cosmogony. This question is the density of the *kosmos* or, put another way, the amount of matter in the *kosmos* in relation to its size. In modern cosmology there is the Friedman equation, which balances the factors which would cause the universe to expand against those which would cause it to contract. There are some other variables, but essentially the Friedman equation balances the expansion of the initial 'big bang' explosion against the contraction of gravity.[60] There are three possibilities here.

(1) If the total gravitation of the *kosmos* is stronger than the expansive effects, then the *kosmos* will collapse. It will cease to expand and implode at some stage.
(2) If the total gravitation of the *kosmos* is weaker than the expansive effects, the *kosmos* will continue expanding, but will do so ever more rapidly.
(3) If the total gravitation of the *kosmos* is equal to the expansive effects, then the universe will continue to expand but the rate of expansion will gradually slow down and approach zero.

Cosmologists have good reasons to believe that (3), or something very close to (3), is the case. If we know the rate of expansion of our universe, then it is possible to calculate the amount of matter required to produce (3). This is known as the critical density. The actual density of the *kosmos* divided by the critical density is known as omega, Ω. Ω is then supposed to be 1, or very close to 1. If there is too little matter in the universe (Ω greater than 1), then the universe expands too quickly for the processes critical to galaxy and star formation to occur, and without those processes there will be no sun or earth. If there is too much matter in the universe (Ω less than 1), then the universe undergoes gravitational collapse before the processes critical to galaxy and star formation can occur, so again there will be no sun or earth. According to orthodox modern cosmology, Ω has to be very, very close to 1 in order that we get something like our current *kosmos*. Very small deviations from 1, in either direction, will stop the processes required for galaxy and star formation. Cosmologists are so strongly committed to $\Omega = 1$ that even when we can only account for around 5% of the matter required, they search for the missing matter in the form of dark matter or dark energy rather than revise the $\Omega = 1$ assumption. What Epicurus has to say is this:

> The universe is boundless. That which is bounded has a furthest point, and that is always seen in relation to something else. As it has no furthest point it has no boundary. As it has no boundary it must be limited and not unbounded. The universe is infinite both in the number of bodies and the extent of the void. If the void were boundless, and the bodies limited in number, they could in no way stay still, but would be carried and scattered through the infinite void, not having collisions to support and arrange them. If the void were limited an infinite number of bodies could not be placed in it.[61]

As the modern concern is with the dynamics of an expanding *kosmos*, we can hardly expect Epicurus to have the density problem in its modern form. However, he does have the difficulty that if there is insufficient matter in the *kosmos*, the atoms will be 'scattered through the infinite void' and that the processes required for the formation of *kosmoi* will not occur because the atoms will be too widely spaced. Modern cosmology supposes a finite amount of matter, but where there is sufficient matter for gravitational collapse to take place there is the concern that if that process continues *ad infinitum* then there will be a finite amount of matter in no space at all, producing an infinite density (this is the black hole/singularity problem).

VIII. The destruction of *kosmoi*

According to Epicurus, there is a limit beyond which the 'underlying foundations' can no longer be added to, so the notion of the 'super-*kosmos*' of Leucippus and Democritus, which continues to accrete matter until it collides with another *kosmos*, is impossible. Epicurus, like Leucippus and Democritus, believes that each of the *kosmoi* has a limited life-span:

> We must believe that *kosmoi*, and indeed all limited aggregations having a close similarity to the things which we see were generated from the infinite, and that all things greater or smaller were separated off from distinct clumps. All are dissolved back, some quicker, some slower, some suffering this for one reason, some for another.[62]

Similar views can be found in Lucretius,[63] along with the view that just as you should not be concerned about your non-existence before your birth, you should not be concerned with your non-existence after your death.[64] Lucretius supplies us with some further arguments on why the *kosmos* is not eternal.

(1) If the parts are perishable, so is the whole, and it is evident that the parts of the *kosmos* are perishable.[65]
(2) If the *kosmos*, and mankind, have existed without a beginning, why does known history go back only to the Theban and Trojan wars?[66]
(3) What is everlasting needs to be solid, so it cannot be prised apart by blows from matter incident upon it, or like the void in being intangible, or not surrounded by space so that there is nowhere for the matter to disperse. The *kosmos* though is not like this, as the void is a part of it, it is not intangible like the void and there is space around it for the matter to be dispersed into.[67]
(4) Given the competition between the elements, a balance between them cannot be maintained eternally. Indeed, in the occasional dominance of

fire and water we see that the balance of the elements can be seriously disturbed.[68]

(5) The earth is like an animal, requiring sustenance and repair. Nature cannot provide that sustenance in full, and already we see that the earth is in decline. The fields are less fertile, and while in the youth of the earth all kinds of animals sprang from it, even huge wild beasts, now the earth has trouble producing even small animals.[69]

Argument (1) is an example of the fallacy of composition. We would not conclude that because the universe is constituted from small particles that the universe too is small, so we would not conclude that because the components of the *kosmos* are perishable so is the *kosmos*.

Argument (2) is interesting in that any reply that language, writing or record keeping are recent will be well met by the argument that that if the earth and humankind have existed for all eternity, why are they only recently invented? A more reasonable reply from an eternalist would be that there are cyclical natural disasters which the Epicureans admit (see (4)), which while they do not wipe out humankind entirely do wipe out civilisations and records.

Argument (3) is based on late atomist assumptions that anything which is not entirely solid can, and ultimately will be prised apart by blows from atoms.

Argument (4) shows little, as it does not demonstrate anything beyond occasional natural disasters occurring, and they might do so on a cyclical basis without detriment to the continued existence of the earth.

Argument (5) presents a case for the terminal decline of the earth, though the lack of fertility might be explained by over farming and a lack of crop rotation, and the argument about what animals the earth produces is dependent on the late atomist zoogony.

IX. Conclusion

I disagree with Bailey's assessment that 'The cosmogony of Epicurus does not show any very great originality.'[70] Certainly Epicurus leans on some of the basic principles of Leucippus' and Democritus' cosmogony, but it would appear that he also has had the good sense to take on board many of the criticisms of Leucippus and Democritus. His answers to those criticisms are original. We have an intervention of god-free cosmogony without a vortex, and the idea that *kosmoi* may be created at any time and place. We have an interesting consideration about the density of the *kosmos* and how that may affect cosmogonical processes. Epicurus adopts ideas about the universe as a whole which are close to the modern perfect cosmological principle. He makes an attempt at the consistent application of physical principles. We have the rejection of the *ou mallon* principle favoured by Leucippus and Democritus for the notion of seeds instead. The processes of cosmogony for Epicurus are not unique or arbitrary, but are ordinary, ongoing physical processes. Finally, we have the use of the atomic swerve. Whatever the merits of this move, and whatever relation it may have to modern quantum mechanics, this is a highly original move.

12

The Stoics

We are not quite so fortunate with Stoic material on cosmogony as we have been with Plato, Aristotle, Epicurus and Lucretius. Although the Stoics wrote a good deal on cosmology and cosmogony, relatively little of this has come down to us. As with many of the Presocratics, we depend on fragments preserved by the doxographers. The Stoic school, beginning with Zeno of Citium (333-262 BC), lasted long into the Roman era. Some of the most important Stoics were Cleanthes of Assos (d. 232 BC), Chrysippus of Soli (c. 280 – c. 206 BC), Boethus of Sidon (fl. second century BC), Panaetius of Rhodes (c. 185 – c. 110 BC), Posidonius of Apamea (c. 135 – c. 55 BC), Seneca (1-65 AD), Epictetus (c. 55 – c. 135 AD), Hierocles (fl. second century AD) and Marcus Aurelius (121-180 AD). According to Philo of Alexandria (c. 20 BC – c. 50 AD), a virulent opponent of the Stoics:

> The third sense of *kosmos*, which is the view of the Stoics, is something which survives *ekpurôsis*, a substance which has a certain ordering or disorder, and time is the measure of its motion.[1]

We have already seen something of Stoic cosmogony in the chapter on Heraclitus. The Stoics, or at least all of the early Stoics and most of the late ones, believed in *ekpurôsis*, the current *kosmos* being generated from the fiery end of the previous *kosmos*, and then itself ending in fire to provide the beginning for the next *kosmos*. This cycle of *kosmoi* was unlimited in the past and in the future. There was debate among the Stoics concerning whether one cycle was identical to all the other cycles. Some said yes, exactly the same people and events would recur indefinitely, though they then had to think carefully about issues of the nature of time, immortality and personal identity. Others said that there would be minimal differences between each cycle.

I. Stoic cosmogony

The actual process of cosmogony in Stoicism, of the formation of an ordered world from fire, is relatively straightforward. Let us begin from the closing point of one cycle, where everything has been converted into fire.[2] Chrysippus tells us that then:

> Fire changes in this manner. It changes through air into water. From water,

after earth has precipitated, air is evaporated. When air has rarefied, aether is spread around in a circle.[3]

The same material then persists from one cosmic cycle to the next. Fire changes to air, and then to water. As Seneca puts it:

> It is fire which grasps the *kosmos* and converts all things into itself. This, slackening, then groups together and nothing else is left in nature, after fire had been extinguished, except water.[4]

The other standard three elements (earth, air, fire) are then produced from this water. Diogenes Laertius tells us that for the Stoics:

> The *kosmos* is generated when the substance is converted from fire through air into moisture. The thicker parts of this then condense to become earth, while the finer are thoroughly rarefied, and when they have been thinned a great deal, they become fire. After this, from mixture come plants, animals and other sorts of things. [5]

There is a slight difference of opinion between Zeno and Chrysippus on the actual process of the formation of the elements. For Zeno the processes are settling out, combination, evaporation and ignition. For Chrysippus, the processes are composition and dissolution/ dispersal. Although the general outline of elemental change is common to all of the Stoics, there is less agreement on the exact nature of the processes involved.[6] Alexander of Aphrodisias tells us that:

> Only matter and god survive *ekpurôsis*.[7]

Cosmogony happens in a limited part of the universe. Surrounding the *kosmos*, for the Stoics, is an unlimited void. Sextus Empiricus says:

> The Stoic philosophers believe there to be a difference between 'the whole' and 'the all'. The whole they say is the *kosmos*, the all is the *kosmos* plus the external void. Due to this distinction, they say that the whole is limited, as the *kosmos* is limited, but the all is unlimited, since the void outside the *kosmos* is thus.[8]

The Stoics also made the important distinction between the order of the *kosmos* and the ordering of the *kosmos*. Diogenes Laertius says:

> The Stoics use the term *kosmos* in three ways. Of god, the idiosyncratic nature composed of all substance, who is indestructible and ungenerable, fashioner of the cosmic order, subsuming all substance into himself and generating it again from himself according to periods of time. They also say the cosmic order, of the stars, is a *kosmos*. Thirdly, they use it of what is composed of both.[9]

The *kosmos*, for the Stoics, is most definitely designed by god and is the best possible *kosmos*. According to Aetius:

> The Stoics considered god to be an intelligent, designing fire, treading the path[10] to the generation of the *kosmos*, encompassing all the seminal principles according to which all things are generated according to fate. A breath pervades the entire *kosmos*, which takes names according to the changes of matter it passes through.[11]

Diogenes Laertius also tells us that:

> God, intelligence, fate and Zeus are one, and many other names are given besides these. In the beginning, on his own god converted the entire substance through air into water. As sperm is enclosed in seminal fluid, god is the seminal principle of the *kosmos*, and remains in the moisture, making matter usable for himself in generation. Firstly, he generates the four elements, fire, water, air and earth. Zeno describes these things in his *On the Whole*, as does Chrysippus in the first book of his *Physics*.[12]

The Stoic *kosmos* is strongly deterministic. Aetius tells us:

> The Stoics consider fate to be a sequence of reasons, an unchangeable order and concatenation.[13]

Gellius says that:

> In book 4 of his *On Providence*, Chrysippus says that fate is a certain eternal natural order of the whole. One state of affairs closely follows and succeeds another, the interweaving of these being unchangeable.[14]

The model that the Stoics use for cosmogony is undoubtedly a biological one. Water is the vehicle for what, in various sources, can be called seed, breath or soul.[15] This parallels the Stoic model for human reproduction, where human semen is *pneuma* in water.[16] At the other end of the cosmic cycle, the *kosmos* undergoes *ekpurôsis*. Philo tells us that:

> Of necessity the *kosmos* must change into flame or into light, into flame according to Cleanthes, into light according to Chrsysippus.[17]

Chrysippus says:

> Death is the separation of the soul from the body, and the soul of the *kosmos* is not separated, but grows continuously, up to the point when it has subsumed all of the substance into itself. Thus it should not be said that the *kosmos* dies.[18]

That is interesting in relation to *On the Heavens* I/10, 280a11-280a24, where Aristotle argues that where successive *kosmoi* are supposed, as with Empedocles, the *kosmos* does not perish, but arrangements do. For the Stoics there can be no external causes of destruction for the *kosmos*. Beyond the *kosmos* there is only void, and as only body can be a cause for the Stoics, the void cannot affect the *kosmos*. [19]

II. The Stoics against chance

An important Stoic argument against any non-teleological, non-providential *kosmos* is given by Cicero, through his speaker Balbus in *On the Nature of the Gods*:

> Should I not be amazed here that anyone might persuade themselves that certain solid and indivisible pieces of matter are carried by their own weight and from the fortunate combination of these bodies a world of the greatest splendour and beauty is generated? I fail to see why someone who thinks this can occur does not also believe that if innumerable copies of the twenty-one letters, made of gold or anything else, were thrown together in some vessel then shaken out onto the ground, it is possible that they would form a readable Annals of Ennius. I doubt whether fortune would give us even a single verse![20]

This we might compare with Lucretius *On the Nature of Things* V,186-94, where Lucretius effectively comes to exactly the opposite conclusion. It should be no surprise at all, according to Lucretius, that a *kosmos* like ours should emerge from an infinite amount of time and an infinite number of atoms. Cicero also says that, according to those who believe in atomism:

> From bodies without any heat, or with any quality or any sense the world has emerged complete by chance, or rather innumerable worlds, some being born and some perishing at each moment. But if the coming together of atoms can produce a world, why can it not produce a colonnade, a temple, a dwelling, which are much less difficult things to produce?[21]

The key assumption of Stoic zoogony, stated at the outset of Cicero's discussion of living things, is that all animals are the subject of intelligent design.[22] All creatures have been generated with appropriate coverings (skin, fur, feathers, etc.) and the limbs and structure appropriate to their way of life. Providence provides ample food for each type of animal.[23] Cicero runs through many wonders of the animal world, on the basis that these things demonstrate intelligent design and providence, and could not have been produced by chance alone.

III. The expansion and contraction of the *kosmos*

It is likely that the material which is to form the *kosmos* undergoes expansion and contraction during the process of cosmogony for the Stoics. This comes across most clearly in the cosmogony of Cleanthes. As there is a surrounding void, there would be nothing to prevent this happening. Aetius tells us that the *kosmos* expands and contracts for the Stoics.[24] Plutarch and Philo give us the interesting information that the seed of the *kosmos*, unlike the seed of anything else, is larger than the resultant product.[25] The reason for this is the transformation of the initial elements. If fire contracts to become air, and in the cosmogonic process at one point all the matter is fire, then the air will take up less space than the fire.[26] Similarly, when air transforms into water we would expect this to take up less space. As the *kosmos* proper then forms out of water, we would expect some expansion.

This seems a natural consequence of the physical processes of cosmogony. I doubt though that we can properly say that the *kosmos* expands and contracts for the Stoics. These elemental transformations are really pre-cosmic, and a world that consists of only fire, water or air, even if it contains seminal principles, could hardly be called a properly organised *kosmos*. One might argue that as the *kosmos* is gradually turning to fire, then the *kosmos* is gradually expanding, though I see no recognition of this in any of our sources. I see nothing here which relates to any modern idea of an expanding *kosmos*, especially since the modern expansion generates further space while the Stoic elements/ *kosmos* would expand into a pre-existent void. Nor is there anything hugely original here. It is very likely that the *kosmoi* of the Milesians expanded as they formed, and certainly the *kosmoi* of the early atomists would increase in size as well. A subsequent contraction of the *kosmos* is more original, though the idea of a cycle of *kosmoi* is still a relatively novel one, the only precursor having been Empedocles.

IV. Successive cycles

An important question for the Stoics is whether successive cycles have exactly the same contents and events or not. Nemesius tells us that for the Stoics:

> The *kosmos* is re-established in its original condition. The stars move again in a similar fashion, and each thing which happened in the first period is completed without any change. There will be again a Socrates and Plato and each of humanity with the same friends and citizens. They will suffer and meet the same things, and deal with the same things, and every city, village and field will similarly be reconstituted. This reconstitution of everything happens not once but many times, or rather, this reconstitution is accomplished an unlimited number of times. Those gods who are not subject to destruction, due to their knowledge of this one period, know from it all that will happen in future periods. There will be nothing foreign to what has happened previously, but all will be utterly without change of even the smallest detail.[27]

Simplicius discerns a problem, however:

> In saying that the same I is generated in the regeneration the Stoics rightly enquire whether the I now and the I at some other time are one in number, through being the same in substance, or are torn apart through being placed in different cosmogonies.[28]

Origen tells us that:

> Those who were ashamed of this doctrine said there were few and very insignificant changes of things occurring in one period when compared to the preceding period.[29]

Alexander, in his commentary on Aristotle's *Prior Analytics*, gives us some more detail on what sort of changes we might expect from one cycle to another:

> The Stoics believe that after *ekpurôsis*, all things happen again in the *kosmos* according to number, such that the same idiosyncrasies in nature that existed before will occur again in this *kosmos*, as Chrysippus says in On the *Kosmos* ... They say that the only discernible changes between later and prior idiosyncrasies of nature are due to certain external accidents, and these changes, for the same Dion continuing and living, do not alter him. He does not become someone else if formerly he had spots on his face, but subsequently does not. Such changes, they say, do happen between the idiosyncrasies of nature in one *kosmos* and another.[30]

There are two important issues for cosmogony in relation to the Stoic attitude to time and successive *kosmoi*. According to Simplicius:

> Aristotle said that time is the number of motion. Of the Stoics, Zeno said that time is simply the dimension of motion, while Chrysippus said that it is the dimension of the motion of the *kosmos*.[31]

Aristotle actually said:

> Time is the number of motion with respect to before and after.[32]

Do those Stoics who take one *kosmos* to be identical to the next one, 'utterly without change of even the smallest detail', consider the time to be the same for these two *kosmoi* as well? As we saw with Eudemus' criticism of the Pythagoreans on this issue, making the time identical would actually destroy the notion of successive *kosmoi*.[33] There would only be one *kosmos*. More likely the Stoics conceived of a set of identical events occurring in time, a cycle of events but without cyclical time. Perhaps the best analogy for the Stoics would be a spiral. The events form a circle in two dimensions; time pulls the circle into a spiral. Each event, represented by a point on the spiral, then has a direct correlate in all previous and future turns of the spiral, but they are separated by time. It is interesting that Lucretius also considers the possibility of our atoms being gathered together again in the same disposition so as to produce the same human again. As the atoms combine in all ways over all eternity, this is certainly a possibility for Lucretius. However, he is uninterested in this possibility on the grounds that any memory of a previous life would have been broken.[34]

The other issue is what happens during *ekpurôsis*, if the *kosmos* comes to a halt, as it might be supposed to do when it is all fire, all water or all air. I doubt, given Aristotle's critique of the difficulties involved in having a *kosmos* stop moving and then start moving again, that the Stoics ever envisaged the *kosmos* coming to a halt. If so, time would come to a halt as well. There would also be a problem of what god was doing during any period of stasis in the *kosmos*.

V. *Ekpurôsis* and providence

It is clear then that many of the Stoics held a theory of *ekpurôsis* and a perpetual destruction and construction of the *kosmos*. Some held that succeeding *kosmoi* would be identical, others held that there would be insignificant changes between *kosmoi*. It is also clear, though, that there is no role for chance in Stoic cosmogony, given their rigid determinism and strong sense of god's providence. Why did they hold this theory of eternal recurrence? In relation to this, it is necessary to consider how providence can work when the *kosmos* undergoes *ekpurôsis* and reconstitution. As Mansfeld puts it: 'In what way is the exercising of providence to be reconciled with such a grim event as the destruction of heaven and earth and all living beings found therein?'[35]

There are two strategies which the Stoics employ here. The first is to decouple providence and fate. So rather than everything being both providential and fated, everything which is providential is fated, but not everything which is fated is providential.[36] Philo tells us:

> The Stoics suppose there to be one *kosmos*. God is the reason for its generation, but not for its destruction. This is due to the action of the tireless fire in existing things, which has the ability, over great periods of time, to dissolve everything into itself, and out of this to reconstruct the *kosmos* again to the plan of the designer.[37]

Seneca too considers natural disasters to have their own causes, rather than being directly attributable to god.[38] Mansfeld has detected a second strategy with the Stoics, which is for everything to be the result of providence but to deny that *ekpurôsis* is a bad thing.[39] Certainly the Stoics do not describe *ekpurôsis* in negative terms. Chrysippus, as we have seen, says that death is the separation of the soul from the body but the soul of the *kosmos* is not separated from it, but continues to grow with it and so the *kosmos* does not die.[40] One might then consider the state of the universe at *ekpurôsis* to be the best state the universe can achieve, where god alone is present in the substance of the universe. As Mansfeld recognises, this then leads to the question of why the *kosmos* should then regenerate given that the cosmic state is an inferior one to *ekpurôsis*. It may be that total fire can only be sustained while there is fuel, and so this state has a limited life before the next stage of the cycle.

A third approach for the Stoics would simply be to deny that any stage of the cycle is superior to any other stage. This is not to deny change, but one might balance say the degree of organisation of the *kosmos* against the development of god and argue all stages are equally good. I do not see evidence that the Stoics adopted this approach though. Rather, they treated the cycle as the best possible cycle. There were better and worse stages in that cycle, but that would be inevitable and god could do nothing about it. What is would always be in the best possible state for it at that moment, as part of the best cycle.

I do not believe that the Stoics needed to work enormously hard to produce a plausible position relative to the arguments of Plato and Aristotle. Plato, in the *Timaeus*, has the *kosmos* as eternal after it has been generated by the demiurge. As we saw with Plato, though, that may well be a change of opinion on Plato's part. There is plenty of material in Plato's other works which suggests a degenerating *kosmos* and even, in the *Politicus*, a cyclical *kosmos*.[41] There is material in the *Republic* suggesting a degenerating *kosmos* even though the *Republic* argues strongly for an unchanging god.[42] Plato has no problem in these works in accommodating a degenerating *kosmos*, even a cyclical *kosmos* with an unchanging god.

There is a basic difference of opinion with Aristotle on the constitution of the *kosmos*. In the *Meteorology*, Aristotle says that:

> If the intervals were full of fire and the bodies composed of fire, then each of the other elements would already have vanished.[43]

The Stoics of course simply say that the heavenly bodies are made of fire, and this supposed conflagration does in fact happen. A second critical difference is that it is not the case that the Stoics suppose there to be a stable *kosmos* which god then for some reason leads into conflagration; the *kosmos* for the Stoics contains the seeds of its own destruction.

The Stoics need not be hugely concerned by the arguments of Aristotle's *On Philosophy*. In relation to Fragment 16, they can maintain that god is changeless, certainly in intent and from cycle to cycle. In relation to Fragment 18, the *kosmos* clearly is better than the works of men, even if it undergoes a cycle of changes. That those changes are eternally recurrent would also distinguish the *kosmos* from any work of men. In relation to

Fragment 19a, the Stoics agree that the *kosmos* cannot be destroyed by any external cause. That the *kosmos* turns into fire does not mean a part becomes greater than the whole. In relation to Fragment 19b, Aristotle's argument that the elements are in their natural place we saw was a poor one anyway, but the simple reply for the Stoics would be to argue that in their *kosmos* the elements undergo a natural cycle. In relation to Fragment 19c, the critical question perhaps is whether successive *kosmoi* are better, the same or worse. Clearly the Stoic reply has to be that they are the same, but with an important qualification. Given a stable *kosmos*, like that assumed by Aristotle, it would indeed be futile for god to destroy that *kosmos* and create another one exactly the same. The Stoic *kosmos* though is not stable. In relation to Fragment 20, the Stoics do not need to claim that there is any 'new design' in their cosmic cycle. If the cycle extends infinitely into the past, the design has always been there and is simply reinstantiated with every cycle. God does not have to come up with anything original, and there is no time before the *kosmos* to be concerned about.

VI. The Stoic conception of fire

Why does the Stoic *kosmos* undergo the process of *ekpurôsis* and reconstitution, with subsequent *kosmoi* identical to or insignificantly different from their predecessors? There is no consensus among modern commentators as to why this is so, most listing a number of possible factors without settling on any one driving reason.[44] The usual factors cited are the great year (the period between identical positions of the sun, moon and planets), the role of fire, the nature of Stoic determinism and the notion of the *kosmos* as a living entity. Firstly, it is important to distinguish between the Stoics and Empedocles, the early atomists and Epicurus and his followers. In each of these multiple *kosmoi* theories, the *kosmoi* are significantly different. The nature of our *kosmos* can then be explained as one accident in an unlimited array of other accidents, with chance playing a major role in cosmogony. This is not so for the Stoics though, as they specifically exclude chance with a rigid determinism, have god as the designing influence for the *kosmos* and have a succession of identical or insignificantly different *kosmoi*.

In my view, the key question for the Stoics is how to couple together a degenerating *kosmos* with the idea of a providential god. If there is only one best *kosmos*, then god must keep recreating that *kosmos* each time the current *kosmos* degenerates. It may be that god is blameless in that degeneration. He creates the best possible *kosmos*, but due to factors beyond him, that *kosmos* degenerates. The key factor in the degeneration is played by fire, and it is interesting that the Stoics have a markedly different attitude to fire to many other ancients.[45]

If one treats fire as an element, rather than a process, and as an element that can transform into each of the other elements (e.g. earth, water and fire, as for Aristotle), then it is relatively easy to treat the *kosmos* as being thermodynamically stable.[46] Burning may transform the other elements into fire, but there are other processes which will transform fire back to the other elements again, leaving an overall balance between the elements.

There is also the consideration that for Aristotle the heavens are composed of aether, and aether is not subject to the qualifications hot or cold, and of course is entirely changeless. So the heavens can happily be eternal. What we see as the heavens are in fact patches of the upper atmosphere which have been ignited by the motions of the heavenly bodies. As long as the heavenly bodies keep moving, and as their motion is entirely natural there is no reason why they will not, then we will continue to see them in the same way. There is no sense in Aristotle that the continued motion of the heavens, and the ignition of small parts of the upper atmosphere, will lead to a gradual warming of the terrestrial region. This ought to be a problem for Aristotle, because presumably the terrestrial realm can neither conduct, convect or radiate energy to the celestial realm, as this would be to heat the celestial realm and generate a qualification of aether and a change in the heavens.

Other ancients, with simpler views of the heavens, see thermodynamic stability there too. Plato treats fire as an element, and one that can transform into air and water, if not earth. Plato considers the stars to be mostly fire, but as he treats fire as an element rather than a process the stars are eternal. Even if we go to the idea that the stars are hot stones, the heat caused by their motion, as long as the motion is inexhaustible and the stones do not suffer attrition during their orbits, there is no reason why the stars should not be eternal. The Stoic view, however, is that everything will be converted into fire. We are told by Alexander of Lycopolis that:

> Everything which burns and has something to burn will burn it entirely. The sun is fire, and so will it not burn whatever it has? Zeno concluded from this that the all will be entirely destroyed by conflagration.[47]

Plutarch tells us that:

> Cleanthes, defending the idea of *ekpurôsis*, says that the sun, as leader, assimilates the moon and the other stars to itself, and changes them into itself.[48]

Stobaeus also informs us that:

> Zeno says that the sun, moon and other stars are intelligent and wise, and comprise of designing fire. There are two types of fire, one undesigning and changing fuel into itself, the other designing, producing growth and protec-

> tion, as with plants and animals, where it is constitution and mind. This is the fire which is the substance of the stars.[49]

Cleanthes has two interesting things to say about the sun:

> The sun is composed of fire and feeds upon the vapours of the ocean, as no fire could remain in existence without a source of nourishment.[50]

> The sun returns from the furthest extent of its orbit in order that it should not be too far from its source of nourishment.[51]

There is a slight difficulty here in that if the sun is designing fire, will it not 'produce growth and nourishment' rather than burn everything? Some have seen this as a serious defect in the Stoic account,[52] and there has been some debate on this issue.[53] Mansfeld's solution is that during cosmogony fire has an entirely benevolent effect, but this changes as the *kosmos* ages,[54] and as Mansfeld points out, the Stoic theory may have a third type of fire, the 'flame' of Cleanthes or the 'light' of Chrysippus, at *ekpurôsis*. I would add two points in favour of Mansfeld's proposal. An example, which we have seen before, of the sun's beneficial action in cosmogony might be the setting up of the weather cycle, the sun evaporating off moisture which then falls as rain, which would promote growth, etc. If the sun consumes moisture, however, the sun might then dry out the *kosmos* with the ultimate fate of *ekpurôsis*. Secondly, it would not be unprecedented for a consistent effect to have both a cosmogonic and a destructive role. In Empedocles, arguably love and strife have both roles, creating and then destroying a *kosmos* in the process of creating one out of many or many out of one. In relation to this, it is perhaps also significant that Empedocles also held a theory of a cosmic cycle. A further consideration here is Heraclitus' attitude to fire, given the acknowledged Stoic debt to Heraclitean ideas. Heraclitus too treated fire as a process rather than an element, as we saw in Chapter 3. Given that the Stoics have fire transform into water and into air, we might have the same concerns about how fire somehow occupies a privileged role in the Stoics as we did with Heraclitus. The answer can be similar. The Stoics need not give fire an ontological or a temporal priority, but they can give it a causal priority. As time extends back infinitely for the Stoics, I take it that their account does not start with primordial fire in the same way that the Milesians begin with one element. Rather, there is no starting point to the unlimited cycles of the *kosmos*. It may be that the Stoics were convinced that the *kosmos* was degenerating, and settled upon fire as the central aspect of that. There are many examples of degenerating cosmologies prior to the Stoics. It may be that the Stoics' conception of fire led them to a belief in a degenerating *kosmos*.

I do not wish to argue that if one conceives of fire as an element, then one cannot be aware that fire needs fuel or that one would immediately adopt an eternal *kosmos*. However, if one conceives of fire as an element it is easier to have a sun, and stars which shine permanently. Such was the Christian idea of cosmology for many centuries. The sun that we see is no different, in thermodynamic terms, to that seen by Adam and Eve, and it will continue shining until god decrees otherwise, and the rest of the *kosmos*, most notably the solar system, is also entirely stable. It was not until the nineteenth century and the rise of thermodynamics that this view was seriously challenged. The idea that fire or heat was a material substance also had a long lifespan, the Aristotelian notion of fire as one of the four elements being superseded by the idea of phlogiston, a heat-carrying substance. When experiments indicated that phlogiston would have to have, in differing reaction, positive weight, negative weight and no weight at all, the idea of caloric, a weightless heat-carrying substance was developed, the weight changes being explained by other means.

If *ekpurôsis* does begin from the sun, then the Stoics hit upon the real ultimate fate of the *kosmos*. In due course, our sun will change in nature to become a red giant expanding in size to greater than the current orbit of the earth, so the earth, and much of the solar system will indeed meet a fiery end.[55]

VII. Reasons for successive *kosmoi*

Ekpurôsis is also linked with the theory of the great year. According to Nemesius:

> The Stoics say that when the planets return to the same sign, in length and breadth, as they were when the *kosmos* was first constituted, after specific periods of time they produce the *ekpurôsis* and destruction of existing things. The *kosmos* is re-established in its original condition.[56]

The notion of the great year, the period when the sun, moon and five planets return to the same relative positions, can be found in Plato.[57] Various estimates of the length of the great year were made in antiquity. The Stoic link between the great year and *ekpurôsis* is hardly surprising, as one great year might be considered to be the key cosmic cycle. Although the idea of the great year sits reasonably well with a recurring *kosmos*, it is hard to see this as the prime motivation for such an idea. Barnes has suggested the following argument.[58] Suppose we accept that:

(1) The universe has only a finite number of states.
(2) Time is infinite.
(3) There cannot be time without change.
(4) The universe is deterministic.

Once the universe has been through its finite number of states, it must repeat a state. As there is a deterministic sequence between the first occurrence of that state and the second occurrence, the universe must then repeat that sequence ad infinitum. Did the Stoics believe these premises? (2)-(4) are relatively uncontroversial, as we have already seen. In relation to the number of states in a *kosmos*, it is interesting that Chrysippus says:

> That which is not true nor will be true may yet be possible.[59]

Chrysippus says this in reaction to Aristotle's view on possibility, which we looked at in relation to whether it was necessary that the *kosmos* exists, and does so eternally. If there are possible states of the *kosmos* which are not actualised, that would at least suggest that for Chrysippus there are a finite number of states of the *kosmos*. Wallis points out that for the Stoics, god can know what happens in each of an unlimited number of cycles as he knows all that happens in one cycle,[60] citing the Stoic view as reported by Nemesius.[61] If there is an objection to god knowing the infinite, as with an infinite number of cycles, then one cycle must be finite in the required manner for god to have knowledge of that, which would suggest a finite number of states in one *kosmos*. It is also significant that the Stoics recognise the great year in astronomy, and link this to *ekpurôsis*, as there would seem to be a limited number of patterns for the heavenly bodies and the initial pattern does eventually recur. I agree with Sorabji, though, that we cannot find the determinism argument in the Stoics, although it may be found in later thinkers such as Origen and Plotinus, whom we will come to in the next chapter.[62] While the Stoic account of how god may know the contents of many cycles is interesting, it is not used as an argument that there are many cycles, rather it explains how, if there are many cycles, god can have knowledge of all of them.[63]

Another important idea which sits well with the idea of recurrence is the idea of the *kosmos* as a living being. One might well take the view that if the *kosmos* has a birth, and can be reasonably compared to a living being, then it also has a death. That will not quite work for the Stoics, for as we have seen they deny that the *kosmos* has a death, but a similar cycle of generation/ degeneration would suit them well. Again though, it is hard to see this as the prime motivation for Stoic recurrence. Many people before the Stoics had modelled the *kosmos* on a living being, but had not felt the need for eternal recurrence. Nor would this explain why the Stoics were so keen on the idea of a succession of identical or insignificantly different *kosmoi*. I hold to the view that the main driving force behind the Stoic theory of recurrence was their idea of a degenerating *kosmos* with fire a the critical factor, allied to their conception of a providential god.

VIII. Philo on Boethus' arguments

Philo of Alexandria, in his *On the Eternity of the Kosmos*, launches a concerted attack on the Stoic ideas of *ekpurôsis* and reconstitution. He tells us that:

> Boethus of Sidon and Panaetius abandoned the idea of *ekpurôsis* and regeneration in favour of the more godlike notion of the indestructibility of the *kosmos*. It is said that Diogenes, when he was young, was a follower of the doctrine of *ekpurôsis*, but came to doubt this later in life and withheld judgement.[64]

He then gives the arguments 'Of Boethus and his school', which can be summarised as follows.[65]

(1) There is nothing inside or outside the *kosmos* which could cause its destruction. Therefore the cause of its destruction would have to come to be *ex nihilo*, which is impossible.[66]
(2) There are three ways in which something may be destroyed, dissolution into parts, annihilation of a quality or mixture. The *kosmos* is not ordered in such a way that it can be dissolved into parts, the Stoics themselves claim that the order of the *kosmos* survives *ekpurôsis* and if all the elements are destroyed together there is nothing for them to change into other than not being, and that, like creation *ex nihilo*, is impossible.[67]
(3) What does god do during *ekpurôsis*? Nothing at all is the natural answer, but that is impossible. If god is the soul of the *kosmos*, and all motion in the *kosmos* stops, then god will die, and that is unthinkable.[68]
(4) Philo gives us this argument:

> In what manner will there be regeneration, if all things are resolved into fire? If substance is consumed by fire, of necessity fire will be extinguished, having nothing to feed it. If fire remains as the seminal principle the cosmic ordering would be saved, but if it is destroyed it is not. This is doubly horrible and impious, not only supposing the destruction of the *kosmos* but also denying its regeneration, as though god delighted in chaos, inaction and total disharmony.[69]

> At *ekpurôsis*, if everything has been converted to fire, then all the fuel for fire has been used. Fire will then be extinguished and no new *kosmos* will come about. Some Stoics say that fire is not entirely extinguished after *ekpurôsis*. Fire cannot be partially extinguished in this manner, especially if there is complete conversion of the fuel. Neither fire nor the seminal principles can survive the total conversion of *ekpurôsis*, as neither will have anything for their nourishment.[70]

(5) In pairs of opposites, it is impossible for one opposite to exist but not the other. At *ekpurôsis* though, hot, rare and light would exist but not their opposites, which is impossible.[71]

(6) What reason is there to pick fire rather than one of the other elements?[72]

Some of these arguments are relatively easy for the Stoics to answer. In answer to (1) and (2) the Stoics would simply assert that there is a reason for the destruction of the *kosmos* within the *kosmos*, and that the destruction consists of all the other elements being converted into fire. To the modern eye, (4) presents the most serious challenge. If there is a complete conversion to fire, and all the fuel has been used, how can there be a new *kosmos*? The modern consideration will be a thermodynamic one. If all the fuel for fire has been used, how can there be any fire in any future cycle? Philo's objection is rather that if all the fuel has been used, all fire must be extinguished and so there will be no seminal principle for the new *kosmos* to grow from, even if the Stoics alter their view slightly to try to allow some fire to survive. This will force the Stoics to rethink the idea of total conversion, if it is necessary that some fire survives. If one is committed to an Aristotelian account of the elements, then (5) is a good criticism. (6) is somewhat unfair to the Stoics as the cycle runs fire into air into water into the terrestrial elements again.

IX. Philo on Theophrastus' arguments

Philo also records that Theophrastus collated four Stoic arguments for a cosmic cycle.[73] These are:

(1) The earth cannot have existed for an unlimited time, or even for a great amount of time, otherwise rainfall would have levelled the earth's surface.[74]
(2) Of the elements, water is beginning to disappear as evidenced by the lowering of the sea level at Delos.[75]
(3) As each of the four elements is destructible, the *kosmos* is as well.[76]
(4) Humanity cannot have existed for an unlimited time, or even for a great amount of time, as certain skills have only recently been invented. If man has not existed for an unlimited time, neither have the animals.[77]

Philo replies to each of these arguments in turn:

(1) Mountains, like trees, grow slowly. The attrition of rainfall is offset by the growth of the mountains.[78]
(2) Sea levels are lower in some places, but are higher in others. There is then no reason to suppose a systematic drying out.[79]
(3) Philo's reply harks back to Boethus' second argument. The elements are either destroyed into each other, when we still have a *kosmos*, or are destroyed into not being, which is impossible.[80] As we have seen with Lucretius, the simplest reply here is that this is a fallacy of composition.
(4) If the arts have only recently been invented, within a countable number of years, does that mean that the *kosmos* and humans have only recently

been generated? The reason for the recent generation of arts is that humanity suffers from natural disasters, either fire or flood, which wipe out civilisation and cause humans to have to begin again. Thus the recent invention of arts is part of a natural cycle which does not involve the destruction of the *kosmos*.[81]

While some Stoics gave up the idea of *ekpurôsis* in favour of an eternal *kosmos*, others defended *ekpurôsis* well into the Roman era. As Sedley has argued,[82] Lucretius also replies to at least the third and fourth of the arguments mentioned in Theophrastus here in *On the Nature of Things*.[83] He argues against the eternity of the *kosmos*, as we would expect, and there is evidence here of an ongoing debate concerning whether the earth is eternal or has a limited life-span.

X. Conclusion

The Stoics produce a novel cyclical account of the *kosmos*, even if the actual processes they use in cosmogony are by now commonly used. There is nothing arbitrary about Stoic cosmogony, given their rigid determinism and the nature of their god. God always does the best thing in any given situation, and all events are part of a providential cycle. Cosmogony for the Stoics is never anomalous. There is no creation *ex nihilo*, and there are no sufficient reason questions about when cosmogony occurs. Stoic determinism dictates that cosmogony occurs as part of an entirely regular cycle and that we can give a causal account of that cosmogony.

As we have seen previously, there are modern cyclical theories of cosmogony. Like Empedocles' theory, though, they produce a different *kosmos* with each cycle and seek to explain aspects of our *kosmos* by seeing it as part of an infinite array of *kosmoi*, all with accidental properties. There is nothing accidental in the Stoic *kosmos*, and successive cycles in the classic Stoic theory are identical, or in the later minority view, insignificantly different. The Stoic problem, I have argued, is not how our *kosmos* gains its character, as that is given to it by a providential god. Rather, it is how to marry the ideas of a degenerating *kosmos* and a providential god together. Thus god, as part of the natural cycle, regenerates the *kosmos* in exactly the same form each time.

It would be nice to credit the Stoics with the first thermodynamic considerations in cosmology, but this would be a step too far. Their treatment of fire, as consuming fuel, and eventually consuming everything in the *kosmos*, is interesting relative to previous accounts in which cosmological processes, such as the burning of the sun and stars, had been treated as open ended. Whether their treatment of fire convinces them that the *kosmos* must degenerate, or whether their belief in degeneration leads them to their view on fire, I think must remain an open question.

13

Early Christianity and Creation

With the rise of Christianity, we find the idea of creation *ex nihilo* being taken seriously and adopted for the first time. I do not find creation *ex nihilo* in any thinker prior to the rise of Christianity. Sorabji lists some possible exceptions,[1] but I would argue in each case that the balance of evidence is against anyone holding a creation *ex nihilo* view.

I. Creation *ex nihilo* prior to Christianity

Hesiod is criticised by Epicharmus and Epicurus for saying that Chaos (chasm)[2] came into existence, yet being unable to say what it came from.[3] That, of course, does not attribute a creation *ex nihilo* theory to Hesiod, but merely points to a deficiency in his account, if we take it as philosophical rather than poetical. If Hesiod had taught creation *ex nihilo*, one would expect Epicharmus, Epicurus and others, not least Aristotle, to have a good deal to say in explicit criticism of it.[4] I doubt that creation *ex nihilo* was even an issue prior to Parmenides, in that Hesiod could be loose in his account of the origins of Chaos without meaning to suggest creation *ex nihilo*, or being taken to suggest it by his contemporaries.

The author of *Melissus, Xenophanes and Gorgias* criticises Xenophanes' assertion that god could not have come into existence from either like or unlike, suggesting the further possibility that he could have come into existence from what is not.[5] However, this is much more likely to be pointing at a logical deficiency in the argument (Xenophanes ought explicitly to rule this out) than to be an advocacy of creation *ex nihilo*.

Plato may have been interpreted by a small number as advocating creation *ex nihilo* in the *Timaeus*, but that does not mean that these interpreters advocated creation *ex nihilo* themselves. Neoplatonists such as Porphyry and Hierocles can speak of the creation of matter, and the idea that god needs no matter for his creation, without this implying any creation *ex nihilo*. Their view, following on from Plotinus, was that god can generate 'out of himself'. Aristotle, as we have seen, does not allow the creation of matter *ex nihilo* but does allow forms to come into existence out of nothing.[6]

Nor is there a definite position of creation *ex nihilo* outside the Greek and the early Christian traditions. In the Jewish tradition there are some inconclusive passages in the Syriac *Apocalypse of Baruch*:

> You who at the beginning of the world called forth what previously was not.[7]
> By a word you called into life what was not there.[8]

One could interpret these passages as creation *ex nihilo*, but equally for the first passage it might be the organisation of the world which is called forth into existence from chaotic pre-existing matter, and in the second passage life is called into existence from non-animate pre-existing matter.

II. The biblical account

The early Christians originate serious contemplation of creation *ex nihilo*. This is not to argue that they are unanimous in accepting this view. There was a considerable divergence of opinion, in response to a battery of arguments concerning creation in general and Christian creation in particular. I have chosen my words carefully when I say that the idea of creation *ex nihilo* comes with the rise of Christianity. I do not see any evidence for this idea in the Judaic tradition. Nor do I find any clear-cut evidence for this view in the Bible. That is the view of many scholars, though I would add, in symmetry, that I do not find clear-cut evidence in favour of creation from pre-existing matter either. So:

(1) There are passages which refer to god's act of creation without referring to pre-existing matter, but none which state there was no pre-existing matter.
(2) There are passages which refer to god ordering a *kosmos* out of pre-existing matter, but none which state that this matter has always existed and has not been earlier created *ex nihilo*.

What we do not find is the kind of explicit formulation that Augustine gives:

> All formed things were generated from matter, and this matter itself was made from absolutely nothing.[9]

For those who wrote the Bible, the bifurcation of these two views was not a live issue and the Bible itself is ambiguous. I would agree with Winston who states: 'The first explicit formulation of *creatio ex nihilo* appeared in second-century Christian literature.'[10] This became an issue only around the end of the first century AD in the debates between Christian theologians and Neoplatonist philosophers. On balance, most passages lean towards the pre-existing matter view without excluding creation *ex nihilo*.

III. The Christian Bible

Here I should say that I do not believe the Bible to represent the word of god, or the authors of the Bible to be divinely inspired.[11] As the authors were many and the Bible is large, we cannot expect perfect consistency between books. Indeed, we may not find that within single books where there is multi-authorship, or where that book draws on a multiplicity of tales or traditions. I take the editing into the 'true' Bible and apocrypha to be arbitrary in that it was a succession of editorial decisions on the part of humans with an assortment of motives and not a divine or divinely inspired decision. So while I am happy to apply the principles of generosity in interpretation that we would apply to other ancient texts, I also take it as quite possible that, as with other ancient texts, some passages may be ambiguous, unclear or contradictory. I do not take the passages relating to creation we shall be concerned with to be primarily philosophical texts, or even, in many cases, to be primarily about creation. Often their emphasis is on there being one and only one god, on whom everything depends rather than the philosophical aspects of creation.

IV. The Bible on creation

Clearly the key text in Christian cosmogony is the opening two verses of Genesis. These are usually rendered:[12]

> In the beginning God created the heavens and the earth. The earth was without form and void, and darkness was on the face of the deep; and the Spirit of God was moving over the face of the waters.

How ought we to read Genesis 1:1-2? Even if we settle on the above rendering, which I believe is the most favourable for the creation *ex nihilo* view, the text still appears to be ambiguous. We could take 1:1 to refer to the creation of a formless and void earth and what follows in the six days of creation to be how the formless earth and the heavens become a *kosmos*. However, we need to take 1:1 more broadly or perhaps more metaphorically than that, otherwise there is a problem with the deep and the waters of 1:2, which are not specifically mentioned in 1:1 and might otherwise be thought to pre-exist.

Alternatively, we could take 1:1 as a summary of what god will do over the six days of creation, which is then given greater detail. On this reading, 1:2 is the initial state of affairs with at least some form of matter existing prior to the ordering activities of the six days. Other renderings of the Hebrew are also relevant here, as they tend to favour the idea of ordering out of pre-existing matter: [13]

> When God began to create the heavens and the earth, the earth was without form and void.

In favour of this view one might argue that all of the acts of creation in Genesis are introduced by 'God said', so we get:[14]

> And God said 'Let there be light'; and there was light. (Genesis 1:3)

So too for the other five days of creation, but not for Genesis 1:1. One might also cite here John 1:1, where the idea of the word is paramount:[15]

> In the beginning was the Word, and the Word was with God, and the Word

> was God. He was in the beginning with God; all things were made through him, and without him was not anything made that was made.

In reply one might suggest that Genesis 1:1 is a different sort of activity, creation *ex nihilo* as opposed to ordering. Genesis 1:2 is also capable of a somewhat different rendering, arguably more favourable to the pre-existing matter interpretation:[16]

> When God set about to create heaven and earth. (Genesis 1:1)
> The earth being then a formless waste, with darkness over the seas and only an awesome wind sweeping over the water. (Genesis 1:2)

This is the rendering which is most favourable to pre-existing matter. Here we have a description of the pre-existing chaos. However, there is no denial that god created the pre-existing chaos. Some see here an early intimation of baptism. Jerome says:

> Now for its mystical meaning – 'The Spirit was stirring above the waters' – already at that time the baptism was being foreshadowed.[17]

Of other relevant texts, Job 38-9 has god speaking to Job out of a whirlwind, contrasting god's creative powers and experience with Job's knowledge, beginning with the following question from god to Job and continuing in much the same vein:[18]

> Where were you when I laid the foundation of the earth? (Job 38:4)

It is interesting, and perhaps significant though that in this extensive list of what god has done or can do and what Job has not done or cannot do there is no mention of creation *ex nihilo*, though there is no explicit denial of it either. Job 28 too has no mention of creation *ex nihilo*, but does speak of:[19]

> When he gave the wind its weight, and meted out the waters by measure. (Job 28:25)

That looks like the ordering of the elements out of a pre-existing chaos. Job probably predates Genesis, though I do not see any pattern of an increasing influence of either *ex nihilo* or pre-existent matter over time. The New Testament has no mention of creation *ex nihilo*. Isaiah has been taken as good evidence for creation *ex nihilo*. There we are told that:[20]

> For thus says the Lord, who created the heavens (he is God!), who formed the earth and made it (he established it; he did not create it a chaos, he formed it to be inhabited). (Isaiah 45:18)

One commentator asks:

> Could the author tell us more plainly that matter was not pre-existent?[21]

I have to reply, well, yes he could actually. He could state clearly that everything was created by god *ex nihilo*, and not from pre-existing matter. 2 Maccabees gives perhaps the clearest *prima facie* statement of creation *ex nihilo*:[22]

> I beseech you, my child, to look at the heaven and the earth and see everything that is in them, and recognise that God did not make them out of existing things. Thus also mankind comes into being. (2 Maccabees 7:28)

However, this too is ambiguous. God may have made everything out of non-existent things, or out of things which are not now existing (i.e. unformed matter).[23] Clearly god did not make mankind out of the non-existent. There are parallels in non-philosophical usage for something coming to be from nothing (usually children being generated out of not being) where no one would want to suggest that a creation *ex nihilo* theory is in play.[24] Revelation 22:13 is quoted in favour of *ex nihilo* creation:

> I am the Alpha and the Omega, the first and the last, the beginning and the end.[25]

I do not see that this demands *ex nihilo* creation, however, and if we are to take god as the literal instigator of creation here, then he is also the literal ender of creation, which is harder to defend. An interesting parallel here is from Plato in the *Laws*, where the Athenian Stranger says:

> God, as in the ancient tale, holds the beginning and end and middle of all things having existence.[26]

This is not taken though as in any way suggesting a creation *ex nihilo*. Of the apocrypha, the Book of Wisdom refers quite bluntly to the

> all powerful hand which created the world out of shapeless matter.[27]

There is no mention at all of any creation *ex nihilo*. As the author is keen to praise god's miraculous abilities, if creation *ex nihilo* had been an option here, it is likely they would have taken it.[28] Jubilees, while retelling most of Genesis and some of Exodus, sheds no new light on Genesis 1:1, simply saying that on the first day god created the heavens and the earth,[29] and while 2 Enoch does speak of bringing the non-existent into being, that can be interpreted along the same lines as 2 Maccabees, and it is likely that 2 Enoch is later than the first century AD.[30]

V. Word and deed in Genesis

There is also a question in Genesis whether god creates by word or by action. So we might contrast Genesis 1:3, where god says 'Let there be light', and there is light without any further action from god, with passages such as Genesis 1:6 where god has to perform a creative act as well:[31]

> And God said 'Let there be a firmament in the midst of the waters, and let it separate the waters from the waters.' And God made the firmament and separated the waters which were under the firmament from the waters which were above the firmament. And it was so.

It has been suggested that the final Hebrew version of Genesis may be an amalgamation of two prior accounts, one based on words of creation and another based on deeds.[32] Alternatively, the 'Let there be' phrase might express god's intent, which is followed by the deed of creation, though then one must render the phrase as 'and it was so' (e.g. god went on to do this) rather than 'so it was done' (e.g. god has done this simply by his word). There are difficulties in all directions here as the Hebrew does not maintain a standard formulation for each of the days of creation (so light has word but no deed) and there are significant differences between the Hebrew and the Greek texts.

VI. Biblical creation and other creation tales

The creation myth of Genesis displays some interesting similarities to other creation myths we have examined, in particular the Babylonian Atrahasis epic, the Enuma Elish creation story, the Gilgamesh epic and the Sumerian creation-flood story (the Eridu Genesis). This is so not only for the types of events but also for the order of events, which makes it difficult to rule out some form of influence.[33] Heidel gives the following table:[34]

Divine spirit and matter are coexistent and coeternal	Divine spirit creates cosmic matter and exists independently of it
Primeval chaos; Ti'amat enveloped in darkness	The earth is a desolate waste, with darkness covering the deep
Light emanating from the gods	Light created
The creation of the firmament	The creation of the firmament
The creation of dry land	The creation of dry land
The creation of luminaries	The creation of luminaries
The creation of man	The creation of man
Gods rest and celebrate	God rests and sanctifies the seventh day

If we take the view that creation is out of pre-existing matter in Genesis, and that is in a chaotic state before god creates the *kosmos*, then the parallels are even more striking.

This is not to say that the Genesis myth introduces nothing new, that the influences from previous mythologies were exclusively Babylonian,[35] nor that the whole of Genesis can be shown to display such tight parallels,[36] although there are other interesting parallels later in Genesis.[37] However, I would agree with Gunkel when he says, 'Genesis 1 is not the free composition of an author but is rather the deposit of a tradition.'[38]

VII. The Christians on creation

The early Christians adopt a wide variety of opinions on the nature of creation. All consider god to be uncreated, and all are agreed that at least the *kosmos* began. Beyond that basic agreement though, there is a considerable diversity of opinion.

(1) God created the *kosmos ex nihilo*. In the first half of the second century AD we can find this view ossifying from opinion into dogma. Hermas gives us his commandment 1:

> First, believe there is one god who generated and shaped everything, and generated everything out of non-existence into existence.

This was maintained either as a matter of faith or a matter of reason. Those defending *ex nihilo* creation as a matter of faith could look to:

> By faith we understand that the world was created by God, so that what is seen was made out of things which do not appear.[39] (Hebrews 11:3)

Frank comments: 'The Christian philosophers were fully aware of the fact that their idea of creation, just as their idea of god, was based on their religious faith, on revelation, not on intellectual reasoning.'[40] However, there were Christians who argued for creation *ex nihilo* and against other views without simply relying on creation *ex nihilo* as a matter of faith. As examples of argument for creation *ex nihilo*, we might look to Tatian, who says:

> Matter does not lack an origin as god does, and it is not equal in power to god by lacking an origin. It is generated, though it was not generated by anything other than the creator of all.[41]

Tatian considers that the universe begins with the word of god, and that god is in no way diminished by his act of creation, being like a torch from which many other flames are kindled.[42] Chalcidius argues that:

> They demonstrate that matter was generated, and generated in such a way that it now exists where it did not do so previously, as material readied for mortal craftsmen is supplied by other craftsmen, supplied for them by nature, and supplied for nature by god. No one prepares it for god however as there is nothing older than god.[43]

In the next chapter we shall look at Origen in more detail, and he has perhaps the most sophisticated and interesting argument for creation *ex nihilo*.

(1a) God creates matter and then the *kosmos*. A variation on simple *ex nihilo* creation was the view that god created matter and then created the *kosmos*. So Basil says:

> After casting about in his mind and determining to bring into being time which had no being, he imagined the world and created matter in harmony with the form that he wished to give it.[44]

While Ambrose says:

> Things were first created and afterward put in order lest it be supposed that they were not actually created and that they had no beginning.[45]

According to Augustine:

> The matter is first called by the name of the universe, that is, of heaven and earth, for the sake of which it was made from absolutely nothing. Second, its formlessness is conveyed by the mention of the unformed earth and the abyss, because among all the elements earth is more formless and less bright than the rest. Third, by the name *water*, there is signified matter that is subject to the work of the Maker, for water can be moved more easily than earth. And thus on account of the easiness by which it can be moved, the matter subject to the Maker should be called water rather than earth.[46]

> God is correctly believed to have generated everything out of nothing, for even if everything was generated out of matter, that matter itself was made out of nothing. We should not be like those who do not believe an omnipotent god could not generate something out of nothing, as when they think of carpenters and other craftsmen who cannot generate anything without material.[47]

(1b) A more nuanced way of defending *ex nihilo* creation, and in particular god's creation of matter, was held by Gregory of Nyssa (*c.* 331-396 AD). Sorabji has argued that Gregory espoused a form of idealism.[48] According to him, what we perceive as objects were not constituted from matter and form, or matter and qualities, but were simply collections of qualities. So Gregory gives a list of qualities and then says:

> All these are thoughts and ideas, none of them is matter on its own. When these are combined, then they become matter.[49]

The motivation for this theory is a concern over whether an immaterial god can create matter.[50] If matter, as Gregory believes, is a collection of ideas, then god can collect these ideas together, and god can generate ideas out of nothing.[51] Thus there can be creation *ex nihilo*, god generating ideas and collecting these together into what we call matter.

(2) Some held that matter was eternal, and so god generated the *kosmos* out of pre-existing matter. Tertullian wrote a treatise against Hermogenes on this issue, which we shall look at in some more detail in the next chapter, and Origen also attacked the idea of pre-existent matter.[52] The Gnostics and the Manicheans both held that matter was eternal.

(3) Some Christians argued that god created the *kosmos* from himself, and not *ex nihilo* or from matter. Three biblical passages in support of this view might be:

> There is for us God, the Father, from whom are all things and for whom we exist.[53] (1 Corinthians 8:6).

> For as woman was made from man, so man is now born of woman. And all things are from god.[54] (1 Corinthians 11:12).
> For from him and to him are all things.[55] (Romans 11:36)

Quite possibly this is not an idea original to the Christians, but is influenced in particular by Plotinus and Neoplatonism with the idea of all things flowing from the one.

(4) Philo of Alexandria (a Jew) argued that god creates the intelligible first, and this is what is referred to in Genesis 1:1-5.[56] Whether Philo believed that matter was created *ex nihilo* is a vexed issue.

VIII. The Christians on pre-creation

A question frequently raised in relation to god's supposed creation of the *kosmos* was: Why did he not create the *kosmos* sooner? This question provoked a further diversification of Christian views. Ultimately this sort of question goes back to Parmenides, though it is wise to note some differences. Parmenides is interested in questions about the generation of anything at all, not just the generation of a *kosmos*. Parmenides' formulation is also indifferent to the qualities of the creator or the creation. Parmenides requires a sufficient reason for anything, good, bad or indifferent, to come into existence, and so asks why sooner rather than later.

With a supposedly good god and a good *kosmos*, the question naturally becomes why not sooner.

(1) The simplest reply, that there was no time before the creation of the *kosmos* and so there is no question 'why not sooner', was given by Augustine (354-430 AD). There were others who adopted this sort of position prior to Augustine, but Augustine gives us a clear formulation in relation to the why not sooner problem.[57] He says that:

> Both the world and time had one beginning, and one was not prior to the other.[58]

Augustine goes on to say:

> The world was generated with time, but not in time. What is generated in time is generated both before and after other times. Before is past time, after is future time. No time passed though before the world. [59]

The perceived basis for this answer went back to Plato, who was read by some as saying there was no time before the generation of the *kosmos*, rather than saying there was no orderly time. Augustine also recognises that in some circumstances there will be the 'why here' question as well, when there is an infinite space. As we have seen before, with the strongly geocentric nature of Greek cosmology, with a limited space that has a centre, or an infinite space where there is believed to be a centre, the 'why here' problem is easily solved. When there is an infinite space without significant points though, there is a serious sufficient reason problem for why there should be only one *kosmos* and why it should be sited where it is.[60]

(2) The Christian god was thought by some to have generated intelligible entities prior to generating the *kosmos*. Basil tells us:

> Therefore, if he makes the world appear in the beginning, it is not proof that its birth preceded all other things that were made. He only wishes to tell us that, after the invisible and intellectual world, the visible world, the world of the sense, began to exist.[61]

Ambrose tells us:

> The angels, dominions and powers, although they began to exist at some time, were already in existence when the [visible] world was created.[62]

I am not convinced this solves the problem, as one might ask what god was doing before he was generating the intelligibles. Augustine believes that god generated intelligibles but did so not in time,[63] in accord with his view that there was no time prior to the generation of the *kosmos*.

(3) There were two types of successive *kosmoi* views:

(3a) Clement believed that matter was eternal, and that there were many *kosmoi* before that of Adam.

(3b) Origen believed there to be a limited number of *kosmoi* prior to this one (and a limited number after), and I discuss this view in more detail in the next chapter. Origen also believed that god generated timeless intelligibles, and holds this to be a solution to the question why god did not do the good act of creation when he had done nothing good before.[64] He says:

> The creation was always present in form and plan. There was never a time when this precursor of things to be did not exist in Wisdom.[65]

He quotes Psalms on wisdom:

> In wisdom have you made all things.[66] (Psalm 104)

Origen identifies Christ with wisdom,[67] and believes wisdom to be without a beginning.[68] So the form and plan of the *kosmos* have always existed.

(4) Irenaeus held the view that god generates the matter out of which he generates the *kosmos*, and also that we should not enquire into what god was doing before the generation of matter and the *kosmos*, as scripture will not give us the answer.[69] Augustine, somewhat tongue in cheek, says that before the creation god was preparing hell for those who ask questions such as 'What was god doing prior to creation?'[70]

The modern physical answer to the pre-creation issue is relatively straightforward, though some of the other problems it raises are not. There is neither time, space nor matter prior to the big bang, so there is no question 'why not sooner'. There is no question 'why here', because there is no space either. The problem this raises is the one of causation. If we base our account of causation on spatial or temporal factors, then it would seem that we cannot have a causal account of the origins of the *kosmos*.

IX. Creation *ex nihilo*

Christian creation *ex nihilo* differs in several important aspects from the sort of creation *ex nihilo* envisaged by Parmenides who, in asking the 'why now' question, seems to suppose the existence of time. The Christians can sidestep this issue by having time generated along with matter. The 'why here' question is also sidestepped in a similar fashion, by having space generated along with time and matter. The issue of the distribution of matter within space once creation has occurred is relatively simple to solve given the existence of god and a limited space. The *kosmos* can be placed in the centre, and matter distributed as god thinks best.

One of the great difficulties with Parmenidean creation *ex nihilo* is what one gets from absolutely nothing. It would appear that there is nothing to determine how much matter would be produced from a genuine creation *ex nihilo*. The Christian reply would presumably be that god creates as much matter as he sees fit. Whether that is adequate or not is another matter. Sorabji comments: 'Creation out of nothing cannot be excluded as involving any logical impossibility.'[71] While I agree on this point, both secular and theological versions of creation *ex nihilo* appear to involve some physical incomprehensibilities. For both there is the question what one gets from nothing, and why. Our universe has a determinate, calculable amount of matter/ energy. What determines that we get this from nothing? To make Aristotle's point here, nothing cannot enter into any proportional relationships, so even if we can determine what it is that brings matter/ energy into existence we cannot say how much matter/ energy. For the theological versions there is the problem of how an incorporeal god can act on nothing to produce matter/ energy. That is quite beyond me and I have never seen anything remotely adequate to suggest this as a possibility. Supposing god to be omnipotent does not help here, as it brings us no nearer to understanding how a god might act on nothing, and brings with it further puzzles on the lines of whether god can perform the logically impossible.[72] For the secular version of creation *ex nihilo*, there is the question why this should happen once and once only. The theological version can reply that god only needs to create the universe once. The secular version though has the problem of why the creation of matter/ energy does not recur. Simplicius puts a good case against creation *ex nihilo*, which is more than just a 'why now' objection. He says:

> If something exists at a later time rather than an earlier, it cannot have come from nothing. What comes from something comes to be according to a chain of events preceding it and following on it. The Trojan war, for example, was able to happen at a specific time because of the chain preceding it, at the same time as it and following on it. If something is generated from nothing though, why should it be generated at one time rather than another?[73]

Simplicius seems to have recognised that there can be no chain of causality where there is nothing, a point that is problematic in modern explanations of cosmogony. If there is no chain of causality, there can be no determination of when something comes into existence and no explanation of why it should do so at a specific point of time. According to Simplicius, we can explain why today is happening now by reference to the prior, current and future motions of the heavens.[74] Simplicius adds:

> Where something did not exist but does exist later, if there was no prior capability, why did it come into existence then and not at some other time?[75]

If something did not have the ability to exist (which, as it did not exist, it could not have) or there was no outside capability for it to exist (which with true creation *ex nihilo* must be so as well) then why should anything come into existence at a specific time? Typically for an ancient, Simplicius also sees the problem of destruction into not being as a mirror problem:

> If something was produced from not being and was destroyed into not being, I believe no one could give a reason why it did so now and not earlier. That which always is, however, as it is not generated by means of some transformation, does not require the question of why now and not earlier.[76]

Once one accepts that god is omnipotent, and can produce the *kosmos* out of nothing, and indeed that god is omniscient as well, a number of other theological problems relating to creation have relatively straightforward solutions. Does god willing a change, that is willing the *kosmos* to come into existence, involve a change of god's will?[77] Not if the change has always been part of god's plan. Nor does always willing the existence of something imply that there is a will that that thing should always exist.[78] If god does will the existence of the *kosmos*, and is a sufficient cause of the *kosmos*, what can delay god generating the *kosmos*? An omnipotent god can no doubt delay creation until he considers the time to be right.[79] Does a new change require something to begin it, which in turn requires a further change, as Aristotle argued? An omnipotent god though could well be able to create a new change without the prior change required for ordinary physical change.[80] The problem with these replies is that to say, for any problem with theological cosmogony, that god is such as this is not a problem, or god is omnipotent (either physically or logically), looks all too easy.

X. The originality of Christian cosmogony

The creation of matter *ex nihilo* is not the only original aspect of Christian thought.[81] The creation of space is something not really discussed by Greek philosophy. The Milesians seem to take it for granted that there is space for their primordial elements and *kosmoi* to exist in. Parmenides with his bifurcation of what is and what is not bypasses any serious discussion of the nature or origin of space. The atomists assert the existence of space, but for them it exists eternally along with matter. Plato, assuming that the demiurge does engage in an act of organising the *kosmos*, has disorderly matter and motion prior to the *kosmos*, and space for that matter to be in. Aristotle naturally has space co-existing with everything else that he believes to be eternal. That the Stoics have a *kosmos* which expands and contracts is interesting, but it does so on an eternal basis in an eternal infinite void. So the notion of the creation of space is new with the Christians as well. I doubt very much that this is a biblical notion. Certainly nowhere is it spelt out that space is generated by god, and of course the pre-existing matter view would require pre-existing space as well. With the adoption of a clear notion of creation *ex nihilo*, however, we begin to find the idea of the creation of space. Indeed, one might argue that the complete lack of any discussion of the origin of space is indicative that there was no serious reflection on the creation *ex nihilo* theory in the *Bible*. We can find Philo saying:

> God generated space and place at the same time as bodies.[82]

So too we can find Augustine saying:

> How, God, did you generate heaven and earth? Clearly it was not in heaven or on earth, nor in air or water, as these are part of heaven and earth. Nor did you make the universe in the universe, as there was no place for it to be generated until it was made.[83]

There are also some aspects of Christianity, which while they are not perhaps entirely original, are worth noting as significant changes of emphasis. There is the development of the notion that god is omnipotent, at least in a physical sense.

This is driven from two sources. Firstly, in a situation where religions were competing for followers, it was important for the Christians to have a god who was not only the only god, but also the most powerful god imaginable. Intellectually, if this god had to generate matter, space and time out of nothing, then he had to be able to do virtually anything. There is also a move, again perhaps driven by a need to separate Christianity from other religions, towards the worship of god rather than the worship of nature. So we find Augustine saying:

> We know God through his creation and understand the invisible functioning of his power due to that which he has made. I am not willing to worship what he has generated for us.[84]

Philo attacks those who believe the world to be eternal and without a beginning (i.e. Aristotle and his followers) as having greater admiration for the *kosmos* than the *kosmos* maker.[85] God for the Christians is also transcendent. That is nothing original, as the followers of Plato were well aware, but it did mark a significant difference from the views of Aristotle and the Stoics.

While Christianity is associated with a one-off, indeed miraculous, origin for the *kosmos*, there are views of the role of god post-creation which make the initial act of creation less anomalous. The most extreme of these views, which can be found in Islamic and seventeenth-century Christian thought, is that god must re-create the *kosmos* every instant. I do not think we find this view among the ancient Greeks or the early Christians, but there are other views which make the continued existence of the *kosmos* dependent on god. This stems perhaps from Plato's reaction to Anaxagoras. In Anaxagoras, the cosmic mind retires from the *kosmos* quite early, while Plato is insistent that a cosmic intelligence has an ongoing role in the *kosmos*. The dissolution of Plato's *kosmos* would take a positive act from the demiurge, rather than indifference or a lapse of concentration. The demiurge does delegate tasks through the lesser gods, so the *kosmos* could perhaps continue in the face of indifference from the demiurge alone. If the lesser gods lost interest, then the *kosmos* would degenerate, but it would not immediately cease to exist nor would it degenerate into chaos instantaneously. Later Platonists, quite likely under Christian influence, took a more radical view, and Zacharias claims to have heard a Platonist offer the view that if god's attention lapsed, then the *kosmos* would be destroyed utterly.[86] On the Christian side of things, Augustine tells us that if god's power should be withdrawn at any stage, the *kosmos* would be dissolved. The issue here for the Christians is what god does after creation. Clearly, as with the question what god does before creation, god cannot be idle. While this is not the same issue as whether creation is a singular occurrence, or something that is part of the ongoing structure of physics, it is related.

XI. Conclusion

Scholarly opinion on how to translate Genesis 1:1-1:2 is divided.[87] There are textual issues (how the written Hebrew should be vocalised), and issues of grammar, syntax and parallels with similar constructions in other biblical passages. I doubt that a definitive solution is possible, on current evidence. I suggest the reason that the text appears to be ambiguous is simply because it is ambiguous. At the time when it was written, the distinction between creation *ex nihilo* and out of pre-existent matter was not an important issue.

There was a considerable divergence of Christian views on creation. Clearly a Christian god generated the *kosmos*, but whether that was from pre-existent matter, from nothing, from god or from ideas was another question. So too there was a diversity of opinion on what a Christian god would have been doing prior to the generation of the *kosmos*.

14

Later Platonism and the Debate with Christianity

This chapter deals with the later development of Platonic views on cosmogony and with the debate about cosmogony in the later part of antiquity. This is quite a broad debate, with contributions from Christians, other religious groups, assorted later Platonists, Peripatetics, Stoics and Epicureans. I cannot deal with the debate exhaustively here. Rather, I hope to bring out the main themes in relation to the role of god in generating the *kosmos* and what god has to work with in that generation. This naturally includes the question whether matter is eternal or generated by god, though there are other possibilities here too.

I. Later Platonists and the *Timaeus*

As we have seen, there has been debate over Plato's views on the creation of time and the *kosmos* in the *Timaeus*. There was also a diversity of views on this issue in later antiquity. The idea that the *Timaeus* should be interpreted metaphorically is first mentioned by Aristotle,[1] and was adopted by the first heads of the academy, Speusippus and Xenocrates.[2] This view was also adopted by Eudorus of Alexandria and most of the Middle Platonists,[3] and was the virtually unanimous view of the Neoplatonists. Taurus recognised that a key passage against the metaphorical interpretation was *Timaeus* 28b, where we are told that the *kosmos* was created (*gegonen*). According to Taurus, this passage had misled many into believing that Plato envisaged a beginning in time for the *kosmos*. Taurus goes on to distinguish four senses of *gegonen* which do not entail a beginning in time. Something may be said to be generated (but not generated in time) when:

(1) It is of the same type as other things which have been created. Something may be visible without ever having been seen.[4]
(2) It is theoretically composite, even if it has not been combined. The *kosmos* is a combination of factors and can be analysed into its constituents.[5]
(3) It is always in the process of generation. The *kosmos* is always changing and is in a constant state of becoming.[6]
(4) It is dependent for its existence on something external. The *kosmos* is dependent on god for its existence.[7]

While Taurus may not have been entirely original in these formulations, he was influential. Of these definitions, (4) was popular and seems to have been Taurus' own view,[8] (3) was also popular and was the view of Alcinous.[9]

As we saw, Aristotle criticised Plato for holding that there was a beginning to time and that there was disorderly motion prior to the organisation of the *kosmos*.[10] This line is later taken up by Eudemus (*c.* 350 – *c.* 290) and Diogenes Laertius (fl. 230).[11] In defence of Plato on this issue, Atticus takes the view that Plato believed there to be disorderly motion and a related conception of disorderly time prior to the organisation of the *kosmos*, orderly motion and time being generated with the *kosmos*, while Plutarch takes the view that there is only a formless matter of time prior to the *kosmos*, and denies that this is time.[12] Philo avoids the conflict in another way, by interpreting Plato as having both matter and time come into existence such that there is no preceding time and no preceding disorderly motion.[13]

Proclus argues against the beginning of time employing Aristotle's argument that to refer to a time before there was time is incoherent. Philoponus disagrees, believing that one can give harmless, non-temporal senses to words such as 'before' and so one can quite reasonably talk about a state before there was time.[14] We shall see more of the debate between Proclus and Philoponus on the eternity of the *kosmos* below. Platonists believing in a beginning to the *kosmos* have criticisms of Aristotle's view here and more generally, they criticise Aristotle for removing his first mover, thinking god from any ongoing interaction with the *kosmos* and for failing, on this basis, to give an adequate account of providence.[15]

II. Neoplatonism

The later Platonists then either opt for a *kosmos* that has been ordered by the demiurge from pre-existing disorderly matter, or believe that the *kosmos* has always existed. Unlike with the Christians, there is no question of the creation of the *kosmos*, or of matter *ex nihilo*. There is a further Neoplatonist view though, stemming from Plotinus, that god generates the *kosmos* neither from pre-existing matter nor *ex nihilo*. Instead, the Neoplatonic one generates everything else that exists from itself. So we can find Plotinus saying that:

> Anything which exists after the first must of necessity be generated from the first.[16]

The one is compared to a spring which has no other source but itself which supplies water for its rivers but does not undergo any change itself. The one is also compared to the life-force of a tree, which spread through the tree while still remaining dependent on the roots.[17] For Plotinus, the generation of matter comes at the end of the sequence of what exists after the one.

It is not entirely clear whether Plotinus believes in successive *kosmoi*. According to Plotinus, there is a seminal principle for all humans. If there is not a succession of *kosmoi*, then there will be an infinite number of these seminal principles, and this Plotinus holds to be impossible. So there is a limited number of seminal principles, and some passages suggest a succession of *kosmoi*.[18] It is not clear, if this is his view, whether succeeding *kosmoi* will repeat earlier *kosmoi* exactly. Slightly later in *Enneads* V,7,3, however, he appears to take a strongly deterministic line with every subsequent development of the *kosmos* being latent in the seminal principles. There is a beginning to this succession of *kosmoi* though, as for Plotinus time comes into existence when the first process of the generation of the sensible world begins.[19] Against the successive *kosmoi* view in Plotinus is *Enneads* II,1, which argues strongly for an everlasting *kosmos*.[20] One would then interpret the passages as proposing cycles within a single *kosmos*, rather than separate successive *kosmoi*. The questions when and why the sensible world is generated are asked, according to Plotinus, only by those who wrongly believe that there is time prior to the formation of the *kosmos*.

An important difference, at least at first sight, between Neoplatonists and Christians is whether creation is willed by god or not. The classic passage for Neoplatonism is in Plotinus, where he hypothesises what nature itself might say if it were asked why it produced its works. The material world 'falls from its contemplation', and nature itself was born from higher things 'not by their act but by their being'.[21] Proclus says:

> Thought is actualised in thinking, this being the same as being. Existing is in making (for that which makes without motion always has its existence in making). Thus thought is making.[22]

> God makes because he thinks, but he does not think in order to make.[23]

There are differences between the Neoplatonists on precisely what it is that does the thinking and making, and to varying degrees they stratify what to Plotinus is very much a smooth emanation from the one. Typical analogues for god's creation are the casting of a shadow,[24] the moon's reflection of the sun's light and a foot which makes an imprint.[25] In each case, there would be no change in god. The Christian criticism is that creation should be a conscious, willed act of god. So we can find Ambrose saying:

> The *kosmos* was made according to those who hold that the *kosmos* is co-eternal with god like it is a shadow of the divine power, and they also hold that it exists of itself. While they hold that the cause is god, they do not

> believe that the cause comes from his own will and power. Instead, they believe it to be like the relation of body and shadow.[26]

Whether the later Platonists in general and Plotinus in particular completely exclude will from creation is another matter. I agree with Sorabji that while for Plotinus the One does not will creation,[27] the lower levels may will the creation of levels below them, and in this he is followed by later Neoplatonists. The difference is in how the will is used. For Plotinus there is no forethought or plan of creation – the One simply generates by being what it is. Porphyry tells us:

> The initiator of the *kosmos* eternally rests within himself and has generated all things from himself yet remains within himself.[28]

Porphyry also has two more interpretations of what 'generation' may mean to add to those of Taurus:

(1) Words and syllables may be said to be generated as they can be decomposed into letters.[29]
(2) Things which depend on a process of coming into existence, such as a house or a ship, may be said to be generated but things which have no process of coming to be, such as a flash of lightning, may not.[30]

Porphyry also addresses the question of how matter can be generated from the immaterial. He says that human semen produces body parts as qualitatively diverse as bones (solid and hollow), nail and hair, lung and liver, blood and phlegm, marrow and fat, sinews and finally membranes. These things are generated from the principles present in the semen, not the matter. The demiurgic principle, on the same lines, is able to organise the *kosmos* using principles and not initially requiring matter.[31] Augustine suggests that some Platonists argue for a *kosmos* without a beginning on the basis that a creation would require a new idea on the part of god; and that new idea would need to be explained as a chance occurrence (why does god change his thought now rather than at another moment?) and a change in the nature of god.[32] Augustine also suggests that some Platonists adopted the idea of successive *kosmoi* so that god is never idle and as each cycle is finite, god can know one cycle and repeat that eternally, perhaps with minor variations.[33]

III. Sallustius

Sallustius (fl. 350 AD), whose work *On the Gods and the Kosmos* has survived intact, gives us a good example of later Platonist thinking on cosmogony. He believes that the *kosmos* must be both ungenerated and imperishable.[34] If it perishes, god must make a better, worse or similar *kosmos*, or allow disorder. This is of course very reminiscent of Aristotle *On Philosophy* Fragment 19c. Sallustius has similar possibilities to offer. God will not generate a worse *kosmos*; if he could make a better one he would have done so previously, a similar one is pointless and god allowing disorder is not to be considered. If it does not perish, then it did not come into being either, another view that derives from Aristotle.[35] Sallustius asks the key question for teleological cosmogony:

> Where does the order of the *kosmos* come from if nothing sets it in order?[36]

This is rather a rhetorical question, as he does not propose any answer. Instead, he gives the standard providence arguments about the good design of animals, that teeth are designed for cutting, chewing etc. He is more concerned to oppose the Epicurean view that their gods are tranquil and do not bother themselves with the affairs of men than with opposing Epicurean ideas on cosmogony.[37] Sallustius takes a more original line:

> Everything which is generated is generated by skill, by nature or according to power.[38]

Those who generate by skill or by nature must precede what they create. If the gods generate the *kosmos* by skill, they can make its character but not its existence. If the gods generate the *kosmos* by nature, then the generator gives something of itself to the generated. As the gods are incorporeal, and the *kosmos* corporeal, this cannot be so. If such a *kosmos* were to perish, the gods would perish with it. However, anything generated according to power is generated with the possessor of that power, and cannot perish unless the possessor loses that power. The examples Sallustius uses are sun and sunlight, and body and shadow. Just as the sun has the power to produce sunlight and body has the power to produce shadow, so god has the power to produce the *kosmos*. So the *kosmos* exists as long as god exists. Conversely, if the *kosmos* ceases to exist, this will mean the gods have ceased to exist if the power of producing the *kosmos* is inherent to them.

Sallustius is concerned to oppose the idea that the *kosmos* could come to an end.[39] Most of his arguments are of the usual type, setting up dilemmas on how the *kosmos* might be destroyed (the *kosmos* destroys itself, or something else does, the cause of its destruction is either corporeal or incorporeal, etc.) and then refuting both arms of the dilemma. More interesting is his comment that if the *kosmos* destroys itself, then fire would have to burn itself and water dry itself. This may be a reference to Stoic ideas, the point being that even if the *kosmos* converted itself into fire or water, as it does in the Stoic cycle, for the *kosmos* to destroy itself from those stages fire would have to destroy themselves by burning themselves or drying themselves respectively. As this cannot happen, the *kosmos* cannot destroy itself.

Here I believe Sallustius makes an error which is quite common in later antiquity in accounts which seek to argue that the *kosmos* is indestructible. This is the conflation of the destruction of the *kosmos* with the destruction of matter, or the universe. If the *kosmos* is a good organisation of matter, clearly that organisation can be destroyed without matter, or the universe passing into non-existence. If the *kosmos* becomes fire or water, it is destroyed as a *kosmos*, unless one takes the Stoic view that this is a stage in a cycle of processes which could reasonably be called a *kosmos*. Sallustius does not take that view though, and showing that a universe comprised entirely of fire or water cannot destroy itself is irrelevant to whether a *kosmos* can maintain its order.

Sallustius believes he avoids the problem of there being an infinity of souls in the infinity of time by having them return to new bodies.[40] So there are a finite number of souls which keep returning. Unfortunately they would have to return an infinite number of times, a problem not helped by the fact that Sallustius hypothesises that if a soul lives virtuously it can escape reincarnation.[41]

Sallustius is neither very original, very perceptive nor very convincing in his arguments for a *kosmos* without beginning or end, but he does give a reasonable example of some later Platonist thought on cosmogony.

IV. Philo of Alexandria

We have seen something of Philo of Alexandria's criticisms of Aristotle and the Stoics in earlier chapters. He opposed Aristotle's eternal *kosmos* and the cyclical *kosmos* of the Stoics. Early in his work *On the Indestructibility of the Kosmos* he describes Plato's view in the *Timaeus* as being that the *kosmos* is generated and indestructible, rejecting the metaphorical interpretation.[42] He then goes on to say that this was the view of Hesiod before Plato,[43] and that:

> A long time previously Moses, the lawgiver of the Jews, said that the *kosmos* was generated and indestructible in the holy bible. There are five books, the first of which is called Genesis.[44]

Earlier in this chapter we saw Philo's position on cosmogony in the *Timaeus*. His view is that time, matter and the *kosmos* all come into being together, such that there is no disorderly matter, motion or time prior to the *kosmos*. As for what Philo himself believed about cosmogony, I agree with Winston and with Sorabji,[45] that Philo believed in an origin for matter, space, time and the *kosmos*.[46] These were all generated by god at the same instant, and there was no time before this act of creation. Philo says:

> At no time did god begin the generation of this *kosmos* as he was always attending to the generation of this very beautiful *kosmos*. They say it is not proper for god to ever be without something to do, as that would make god idle, and so god generated everything without a beginning.[47]

Philo's answer to the problem of what god was doing prior to the creation of the *kosmos* is then straightforward. There was no time for god to be idle in, and no matter he could have been working on. It is doubtful that Philo is entirely consistent on presenting this view throughout his many works. I would treat this as his core view though, on the grounds that this is what emerges from his most extended treatment of the topic and that this view agrees with the view he ascribes to Plato's *Timaeus*. The latter point is important as Philo claims that Moses, who he believed had reached the pinnacle of philosophy and was divinely instructed,[48] anticipated the view that Plato takes.[49]

One reason why Philo rejects Aristotle's anti-cosmogony is that he believes it would ascribe a huge inactivity to god.[50] Philo also does something which can be found in several discussions on cosmogony in later antiquity, which is to take some of Aristotle's own arguments and then attempt to disprove the idea that the *kosmos* had no beginning. Wolfson has argued that Philo believes Aristotle's idea of god as the prime mover is inconsistent with the idea that the *kosmos* has no beginning.[51] Wolfson reconstructs Philo's argument like this.[52] Aristotle hypothesises a god who is the reason for motion, as there cannot be an infinite regress of reasons for motion. As god is the reason for all movement, god must be more active than any other reason for motion in the *kosmos*. However, there are reasons for motion which generate things from non-existence, i.e. those reasons which produce change in substance. If god does not generate things, then, he is less of a reason for motion than some other reasons in the *kosmos*.

V. Philoponus, Proclus and Aristotle

John Philoponus of Alexandria (*c.* 490 – *c.* 570 AD) wrote *Against Proclus on the Eternity of the Kosmos*, a work which set out to refute Proclus' eighteen arguments in favour of eternity of the world.[53] This is a massive work, the translation alone running to four volumes in the Ancient Commentators on Aristotle series, so I can only hope to look at some of the more important points here. Philoponus also wrote *Against Aristotle on the Eternity of the Kosmos*, the complete text of which has been lost, but Simplicius, an implacable and at times vitriolic opponent of Philoponus, has preserved a good number of fragments.

Proclus (411-485 AD), in line with the mainstream of Neoplatonist thought, adopts a metaphorical view of Plato's views on creation in the *Timaeus*. The *kosmos* is eternal, having no beginning in time and no end, on the basis that, as Plato says in the *Timaeus*, god having generated something good would not wish for its downfall. Philoponus naturally adopts a literal position, though this does not sit easily with his rejection of the idea that the *kosmos*, once generated, will not undergo destruction. Attempts to argue that Plato was wrong on the fate of the *kosmos* will cut against any appeal to Plato's authority in attempting to establish that the *kosmos* had a beginning.[54] There is considerable disagreement between Proclus and Philoponus on the nature of god and the nature of forms. At root, this is a question concerning whether a beginning for the *kosmos* entails supposing a change in entities that might otherwise be thought changeless. In *Against Aristotle*, Philoponus has extensive arguments against many of Aristotle's ideas in cosmology, and against Aristotle's position on generation from opposites. Of greater interest are Philoponus' arguments on the nature of the infinities involved in a belief in an eternal *kosmos*, and his views on motion and time. Philoponus is keen to argue that Aristotle had a literal interpretation of creation in Plato's *Timaeus*,[55] and in arguing there can be no reconciliation between Plato and Aristotle on this issue, runs contrary to many commentators who seek a form of agreement between Plato and Aristotle.[56]

Aristotle does not believe there can be any actual or traversed infinities. Nor does he believe that there can be multiple infinities, or that infinities can increase. These views were widely accepted in antiquity. If there has been an infinity of past time, Philoponus argues, then there will have been an infinite number of individuals.[57] Thus there will be an actual infinity, and it is not possible for such an infinity to exist, whether we take the individuals to exist consecutively or simultaneously. The succession of individuals is impossible as this would amount to traversing an infinity, and by definition an infinity cannot be traversed. Further, the number of individuals up to Socrates will be infinite.[58] However, there will be more individuals who have come into existence between Socrates and now, so that there will be a greater number of individuals than infinity, and that is impossible. We can argue similarly with horses and dogs, and so infinity would be doubled or tripled.[59] Suppose the motions of the heavens have no beginning. Saturn must already have completed an infinity of cycles of the heavens, but Jupiter will have completed three times as many and the sun thirty times as many.[60] Simplicius in reply argues that time is only potentially, not actually, infinite, and the same goes for men. An infinite number of years, and infinite generations of men do not all exist at the same time. On this he is following Aristotle, who took the view that only one year can exist at once.[61] Nor can there be a question of the supposed infinities increasing, because they are not actual infinities in the first place.[62] So according to Philoponus, the *kosmos* must have a beginning, or such infinities occur.[63] Philoponus' goal is to set up a creation *ex nihilo* view. He says:

> If something were generated out of utter non-existence, they say that non-existence would then exist. If it is argued that things generated are generated from non-existence in the same manner as a ship from wood, which would mean that non-existence itself underlies what is generated and changes into it, then it will indeed follow that non-existence exists. No one though is so foolish as to understand generation from non-existence in this manner. Anything generated is brought into being only in so far as it is generated without having existed beforehand.[64]

In *Against Aristotle on the Eternity of the Kosmos* book VI Philoponus debates Aristotle's position in *Physics* VIII/1 on the eternity of motion and time, arguing that both motion and time can have a beginning.[65]

VI. The question of infinite past time

There is a modern philosophical debate about the possibility of infinite past time.[66] There are also strong modern physical arguments against infinite past time. The simplest consideration is the second law of thermodynamics. If entropy does increase over time, then an infinite past time would indicate a current state of maximum entropy, which we clearly do not have.[67] A related consideration is that if the *kosmos* has always been expanding it would now be infinitely large, which it is not. The evidence for a universe which is now expanding and has been expanding for all of its life is now extremely strong, so much so that it is one of the basic facts that any theory of cosmogony/ cosmology needs to be able to explain. One of the basic constants in cosmology is the Hubble constant, which gives a measure of the rate of expansion of the universe. The evidence of a very hot, dense phase for the universe is now virtually irrefutable as a result of the COBE (cosmic background explorer) experiments measuring cosmic background radiation. The evidence for a big bang itself is slightly less certain, as it involves extrapolating back beyond this hot, dense phase, but is nevertheless extremely strong, several methods all indicating very much the same time for the origin of the universe.

There does remain the possibility that our universe undergoes a cycle of expansion and contraction. If so, then some of the processes which we currently take to be irreversible would in fact be reversible, and so there would at least be the physical possibility of infinite past time. As we live in an expanding universe, it is hard to say quite what a contracting universe would behave like.[68] It is interesting that both ancients and moderns have speculated on the possibility of a reverse cosmic cycle. Arguably we find this in Empedocles, where love and strife undo each other's work, associating or dissociating to states of total love or total strife. Certainly we can find a reverse *kosmos* in Plato's myth from the *Politicus*. He has the Athenian Stranger say:

> that the motion of the universe at one time is the circular movement which it is now, and at another the opposite.[69]

He then says:

> First of all, all living things, whatever their stage of life, as far as they were mortal ceased to appear to be growing older, and changing back, on the contrary, growing so as to be younger and gentler. The white hair of the aged became black, and further the cheeks of the bearded became smooth, each returning to the prime of life, the bodies of those in their prime becoming smoother and smaller in body throughout both night and day, each returning to the nature of a new born child in respect of their souls and the man like parts of their bodies, before they wasted away entirely and then wholly disappeared. The bodies of those who had died by force at this time suffered the same misfortunes, and were destroyed with unheard of speed in a very small time.[70]

One interesting aspect of Plato's myth is that he conceives of time as being reversed. If our universe were to contract, there is a question of what would become of the second law of thermodynamics. This is particularly interesting as many believe that physical time is given its 'arrow', its directionality by the second law of thermodynamics. If this law were to reverse with the contraction of the universe, so might physical time, though what effect that would have on time as perceived by humans would be a quite different question.[71]

Whether the universe expands or contracts depends on a balance between the expansive energy of the universe, a measure of which is given by the Hubble constant, and the tendency to contract under gravity, a measure of which is given by the amount of matter/ energy in the universe.[72] At the moment, the recent determination of supernovae recession velocities (a good measure of the Hubble constant) indicates that our universe is expanding slightly more rapidly than the critical rate and so will continue to expand for its entire life.

VII. The fate of Aristotle

One might think Aristotle's position on cosmogony was entirely clear, and thinkers like Philo and Philoponus seem to accept that Aristotle believed in an eternal *kosmos* when they use some of Aristotle's own arguments to undermine that view. Proclus too accepted that Aristotle believed in the eternity of the *kosmos*, but argued that on Aristotle's own principles Aristotle should have considered his god to be a creator as well.[73] Iamblichus held that Aristotle did not disagree with Plato on forms. This is significant as Plato was construed by some as making forms efficient causes.[74] Dexippus is keen to emphasise that Aristotle brought all substances together into a single system and related them to a single originating principle,[75] and that indeed all things are in some way dependent upon that originating principle.[76] Syrianus argued that if Aristotle accepts that the heavens derive their motion from outside themselves, he should also accept that they derive their existence outside themselves. Without motion, the heavens would cease to be the heavens, as circular motion is part of what it is to be the heavens.

Ammonius goes further and argues that not only is a productive god implied by some things that Aristotle said, but Aristotle believed in a productive god himself. Ammonius' view was highly influential, with many following his line on Aristotle in later antiquity.[77] Ammonius quotes Aristotle's *Physics*:

> Something is the primary origin of change of stability, as the advisor is the reason, or the father of the child, or more generally the maker of that which is made.[78]

Ammonius takes it as entirely clear from this that the primary origin of motion, i.e. Aristotle's god, is also the maker of the physical world.[79] He also quotes from *On the Heavens*:

> Eternity took its name from always being, both immortal and godly. On this things depend, some more precisely, others more obscurely, for their being and life.[80]

So according to Ammonius, all things get their goodness due to the final reason, and they get their being and life due to the 'demiurgic' reason.[81] Ammonius has a battery of such arguments recorded by Simplicius, who himself took only a selection from Ammonius' book on Aristotle's god as a making cause.[82]

Ammonius also argues that, according to Aristotle, a finite body can possess only finite capabilities. So the eternal motion of bodies must come from the unmoved mover. Similarly, as these bodies have eternal existence, they must receive the ability to exist eternally from something non-material.[83] Actually, Ammonius says that the heavenly bodies receive eternal motion and being.[84] This argument, or variations on it, is quite commonly cited in relation to Aristotle. Ammonius believes the conclusion, that Aristotle's god is productive of the *kosmos*, is one Aristotle intended. Syrianus (d. 437 AD) is less committal, arguing that Aristotle's forms are productive causes, whether Aristotle and his followers accept this or not.[85] Syrianus cites Plato *Phaedrus* 245e where the world soul causes motion and the heavens would disintegrate without its influence, such that the world soul is the cause of the continuing existence of the heavens.[86] Proclus has an interesting variation on this argument, asking where the *kosmos* attains its unlimited power of existence, if it is not spontaneous in the manner of Epicurus?[87] This leaves Aristotle's followers with the awkward choice of agreeing with the Epicureans that matter is inherently or accidentally eternal, or being forced to give some reason for the eternal nature of matter, which would push them towards some explanation in terms of god.

Proclus also holds the view that the *kosmos* cannot receive the ability to exist perpetually all at once, or a finite body would have an infinite power.[88] Thus god must perpetually supply the ability to exist to the *kosmos*. He says:

> Something else will give it the ability to exist, but this will not be given all at the same time, as it would not be able to receive it all at the same time. It will be given in amounts that can be received, in a stream which flows and flows perpetually to it. So the *kosmos* is always coming to be and never has being.[89]

The *kosmos* must derive its existence from something according to Proclus, and it is likely that it derives its existence from the same source as that from which it derives its motion.[90]

Philoponus agrees with Aristotle that a finite body cannot house an infinite capability.[91] His argument for a beginning to the *kosmos* in relation to this view proceeds in four stages:[92]

(1) If the *kosmos* is finite, and the capabilities of all finite bodies are finite, as Aristotle shows in *Physics* VIII/10, then the *kosmos* must have been generated and have had a beginning in time.
(2) If someone replies that while it is impossible for bodies to have infinite capabilities, what has and what will sustain the *kosmos* is god, then he can be answered as follows.
(3) Is the *kosmos* indestructible of its own nature, or does something else, i.e. god, give it this indestructibility?
(4) If the *kosmos* is destructible of its own nature, then it is generable, and has an origin in time.

Philoponus used the third stage of this argument against Proclus. Proclus took the view that since Plato considered the *kosmos* indestructible, he considered it to be ungenerated as well. What Philoponus brings out is that Plato considered the *kosmos* destructible but not destroyed due to the will of god, and so the *kosmos* must be generable for Plato as well. A question then arises as to whether god can override the destructible nature of the *kosmos* and hold it in existence for ever. Alexander took the view that god cannot,[93] while Philoponus, a believer in an omnipotent Christian god, held that he could.[94]

Simplicius, attacking the view of Philoponus, took the more complex view that the *kosmos* is, in its own nature, fitted to accept an eternal existence from god, such that god in keeping the *kosmos* in eternal existence does not override any natural property of the *kosmos*. Simplicius says:

> As the *kosmos* is generated and finite, so too it is simultaneously fitted for perpetual existence and motion. It is then false to say, without qualification, that it is destructible of its own nature.[95]

So the *kosmos* is destructible should it be left to its own devices, but has the natural ability to last forever if supported by god.

A second line of attack came from a concession made by Aristotle on creation *ex nihilo*. Aristotle does not allow matter to be created, but he does allow that form can come into existence where a form did not exist before.[96] As Sorabji points out, this would make the Christian position of the generation of matter *ex nihilo* less radically different from Greek views.[97] Philoponus in particular is insistent that Aristotle allows creation *ex nihilo* for forms, and that if this is possible, then creation *ex nihilo* ought also to be possible for matter. Philoponus says:

> If it is necessary that god is immeasurably superior to nature, and nature generates by taking what exists in substrate and matter, and fashions only the forms into existence, then if god generates something out of nothing, then he must generate not only the forms of what is fashioned, but also the substrate which did not previously exist, or he would not be greater than nature. It is not necessary then, for the universe to be generated from something that exists, if it is generated by god.[98]

> If prime matter were to come into existence, it would not need any further matter or substrate in order to come into existence.[99]

VIII. Plutarch

Plutarch (46-127 AD) has an interesting discussion of the number of *kosmoi* in his work *The Obsolescence of Oracles*.[100] The discussion begins by distinguishing those who believe in one *kosmos*, those who believe in an unlimited number, and Plato who is said to have contemplated the possibility of five *kosmoi* but in the end settled on one. Those opting for one *kosmos* are said to have done so for fear of being overwhelmed by an embarrassing infinity if they had supposed more than one.[101] We then hear of another opinion, that of Petron that the number of *kosmoi* is:

> Not unlimited, nor one, nor five, but one hundred and eighty-three.[102]

These *kosmoi* are arranged in a triangle, with sixty on each side and one at each vertex. This is said to be offered by way of a myth without demonstration or argument. Homer's view that there are five *kosmoi* (heavens, earth, water, air, Olympus) is then discussed before there is a more rigorous assertion that Plato believed in a unique *kosmos*.[103] Not defending the idea of one *kosmos* leads to the universe being unlimited, and supposing there to be no more and no less than five is against reason and lacking all plausibility.[104]

The counter-argument to this is that the arguments which preclude there being an unlimited number of *kosmoi* do not preclude there being more than one *kosmos*. God and providence can exist in more than one *kosmos*, and the incidence of chance can be minimised. It is not likely that this *kosmos* is alone and neighbourless in an infinite space, when we see in nature that individual things come in classes, like seeds in a pod.[105] We should not worry about the destruction of the *kosmos* by anything external. The gods look after the *kosmoi*, and if there is any interaction it is likely to be harmless and peaceable. There may be physical objections to many *kosmoi*, such as those of Aristotle, but if we assume that each *kosmos* has its own centre for the purposes of natural motion, then there will be no overwhelming objection.

It is interesting that Plutarch describes such a radical disjunction between those believing in an unlimited number of *kosmoi* and those believing in one, with only Homer in a loose sense believing in five *kosmoi* and the bizarre theory of Petron as counter-examples. It is also significant that the worry of being overwhelmed by an infinity is given as the reason that many thinkers opt for a unique *kosmos*. The view that god could provide providence for a limited number of *kosmoi* reveals the usual later Greek attitudes about infinity and god. God could cope with a finite number of *kosmoi*, but not an unlimited number. Aristotle's physical arguments are clearly influential as well. Plutarch suggests that the benevolence of god would lead him to produce more than one *kosmos*. His god faces some sufficient reason questions though. If he is going to produce five *kosmoi*, why not more and why not less? Precedent of Homer and Plato will not help here, nor will the supposition that there are five elements. The distribution of those *kosmoi* is also problematic. In a universe with a centre, it is evident where a unique *kosmos* should go. But how is Plutarch's god going to distribute his five *kosmoi* within his universe? Nor does Plutarch seem to realise the regressive nature of his amity and class arguments. Would god leave the universe alone and neighbourless? And if there is nothing unique and everything comes in classes, why is the universe unique? Plutarch effects a separation of *kosmos* and universe which Plato would object to, as in the *Timaeus* universe and *kosmos* are co-extensive. That allows Plato to argue strongly for a unique *kosmos*, and for one that is entirely self-sufficient, in no need of anything external. So the *kosmos* for Plato simply is not something that requires friends of its own kind. Plutarch does not address the question of how good his *kosmoi* are, and how they differ from one another. If god creates the best to start with, why would he create more? They cannot be better; surely god would not create something worse; why would god produce more of the same *kosmos*? Plutarch provides us with some interesting material, but ultimately his arguments in relation to the number of *kosmoi* are rather superficial.

IX. Theophilus

Theophilus (fl. 177 AD), bishop of Antioch, has some interesting points to make in opposition to Platonism. In his second *Letter to Autolycus*, he is critical of all the Greeks, both philosophers and poets. He disapproves of the Stoics, who he believes say there is no god, or if there is a god then god cares only for himself. He disapproves of those who say the *kosmos* comes together by chance, or that the *kosmos* is ungenerated and matter eternal, or those who deny divine providence. His most detailed attack though is on the Platonists. Theophilus interprets Platonic cosmogony literally, there being pre-existing matter which god forms into a *kosmos*. Theophilus objects that god is ungenerated and immutable, and if matter is ungenerated it would also be immutable, so equal to god. This is a common Christian objection to pre-existing matter, but in my view a poor one. Following Aristotle's analyses of ungenerated/ indestructible things, it was common to assume that the ungenerated was unchangeable, but to the modern eye the notion of something which exists for all time and changes appears to be a perfectly reasonable possibility.

Theophilus also questions what wonder would it have been if god had created the *kosmos* out of pre-existing matter? Human craftsmen can produce things out of matter themselves. However, man, while he can make idols, cannot give speech, breath or understanding to his works as god can. Nor can man make something out of nothing, or generate whatever he chooses or wills. It is important for Theophilus that god's creative powers are radically different from man's and that there are no limits on these powers.

It is also significant that Theophilus' god can create whatever he wills. This is not quite so for the Platonic demiurge, who has to work with pre-existing matter which is in some way recalcitrant, and has to produce the best compromise possible between reason and necessity. While there is debate on what Plato actually meant by this, the Christian tradition takes this as a limitation of god's power that he cannot produce exactly what he wishes. If god produces his own matter, however, then he will be able to shape it to his own ends and produce whatever he wills, a theme we will see again in Origen. That god chooses and wills is also typical of Christian theology, as we saw in relation to Plotinus.

Theophilus is critical of Greek poetry too, attacking Homer's view that the gods and *kosmos* are descended from Okeanos and Tethys on the grounds that if god made everything, he made the oceans and the waters too. Hesiod is attacked for giving an account of the origins of the gods and the *kosmos* without giving an account of who these things were made by.

Hesiod is also criticised for supposing a pre-existing chaos without something to turn that chaos into the *kosmos*. 'Did matter form and beautify itself?', Theophilus asks.[106]

X. Tertullian and Hermogenes

According to Tertullian (160-225 AD), Hermogenes' view is that God made everything out of pre-existent matter. Hermogenes argues for this by presenting three alternatives:

(1) God generated everything from himself.
(2) God generated everything from nothing.
(3) God generated everything from pre-existent matter.[107]

Hermogenes then argues that the first two are impossible, so the third must be true. If god generated the *kosmos* from himself, it would be made from parts of god but god has no parts, being indivisible. God is immutable as well, and so cannot make *kosmos* from self.[108] God cannot have generated the *kosmos* from nothing, because there is evil in the *kosmos* and god could not possibly have generated anything evil. God, being immutable, cannot have begun the work of creation.[109] God has always been lord. So there must always have been something for him to have been lord of.[110] The pre-existent matter was unorganised and lacking in any characteristics. It was possible for an immaterial god to influence matter in the way that a magnet does.[111] God is always at work in the *kosmos*, as matter needs to be kept in order.[112] Hermogenes also leans on Genesis 1:2, 'the earth *was* without form and void', this being taken to imply pre-existent matter, as was Genesis 1:1, 'In the beginning', where 'in' is taken to denote pre-existent matter.[113]

Tertullian's reply is to defend the notion of creation *ex nihilo*. The Bible does not refer to god as lord until after creation *ex nihilo*. This is not a problem in terms of immutability, as god does not become god the father until after the birth of Christ.[114] Genesis 1:1 should be read as *from* the beginning rather than *in* the beginning.[115] In Genesis 1:2, even if matter existed prior to its organisation, god still created that matter.[116]

XI. Origen and successive *kosmoi*

Origen (185 – c. 254 AD) gives us a Christian version of the successive *kosmoi* view, though I find his position rather ill-thought-out and confused. Jerome tells us:

> Origen says there are innumerable *kosmoi*, though not many similar ones existing at the same time, as with Epicurus, but after the end of one *kosmos* comes the beginning of another.[117]

Theophilus tells us:

> No man dies over and over again, as Origen writes, in wishing to establish the impious view of the Stoics.[118]

There is also a record of Clement of Alexandria believing in successive *kosmoi*, as Photius tells us that he 'spoke of the transmigration of souls and of many *kosmoi* existing before Adam'.[119] However, it would seem that in later works he reverted to the idea of god creating everything, including matter.[120]

Why does Origen opt for successive *kosmoi*? Origen does seem to envisage some form of deterioration from the initial state of unity and harmony in the *kosmos*, which would be restored with the creation of a new *kosmos*.[121] As with the Stoics, there is an emphasis on things eventually turning into fire, though he does not envisage a conflagration.[122] He also believes that god comprehends all things, any other view being impious. That being so, all things must have a beginning and an end as things lacking beginning or end are incomprehensible. Hence the *kosmos* has a beginning and an end.[123] He is then concerned that if the *kosmos* had a beginning in time, god might have been idle prior to his creation of the *kosmos*. This is not so, as ours is not the first *kosmos*.[124] As Sorabji points out, if Origen supposes there to have been a finite number of *kosmoi*, there must have been a first *kosmos*. However far Origen pushes that first *kosmos* back, there will still remain the question of what god was doing beforehand.[125] If Origen then opts for an infinite number of *kosmoi*, he is faced with the problem that, on his account, god will be unable to comprehend them. One might also ask what god will do after the end of the last *kosmos*, assuming that the final *kosmos* in the finite sequence is not an indestructible one.

Origen has matter and the *kosmos* generated by god so certainly there is a first *kosmos* in the succession.[126] Ours is the only cycle with the birth, crucifixion and resurrection of Christ.[127] Origen speaks of the *kosmos* reaching a 'perfect end' and when things are no longer 'in an age' (i.e. not part of a cosmic cycle) but 'God is all and in all'.[128]

Origen does not believe, as with some of the Stoics, that each successive *kosmos* will be exactly the same as the last. The reason Origen gives is a strange one. He argues that if a bushel of corn was to be poured onto the floor twice, we would not expect to get exactly the same positions of all the grains of corn each time. Nor would we get exactly the same pattern if the bushels were poured incessantly for countless ages.[129] The Stoic reply here, I assume, would simply be to say that the generation of the *kosmos* is a matter of the providence of god and should not be compared to chance events such as the pattern arising from pouring a bushel of corn on the floor. Indeed, it seems very odd to me that a Christian should make this point, when he might be accused of siding with Empedocles or any of the atomists against the idea that the *kosmos* is generated by a benevolent god. Doubtless Origen wishes to avoid the problems of free will associated with the Stoic identical recurrence position, as well as defending the unique nature of the crucifixion and resurrection,[130] but his use of the bushel of corn analogy is clumsy and potentially highly misleading. That Origen does not consider god to be omnipotent is perhaps significant here. An omnipotent god could arrange identical falls of a bushel of corn and identical successive *kosmoi*. Origen does quote some biblical passages in support of successive *kosmoi*. From Isaiah:

> For as the new heavens and the new earth which I shall make shall remain before me, says the Lord.[131]

From Ecclesiastes:

> What has been is what will be, and what has been done is what will be done: and there is nothing new under the sun. Is there a thing of which it is said, 'See, this is new'? It has been already, in the ages before us.[132]

XII. Origen and matter

Origen's most interesting and perceptive argument is against the pre-existence of matter. He argues that God creates only enough matter to make a *kosmos* he could control and give providence to.[133] Origen does not make god omnipotent, as he holds that infinity is incomprehensible and an infinite god would not be able to comprehend himself. So god generates a finite *kosmos* which he can control. That god is not omnipotent is important for the following argument.

Origen firstly notes that matter is so wonderful as to be entirely sufficient for whatever god wishes to create out of it.[134] He then asks whether something so fitting for its purpose could exist by chance. Origen's answer to this is no, god produced matter *ex nihilo* that was conducive to his purpose.[135] This then gives Origen the opportunity to berate those who believe that god did not create matter. These people criticise others for arguing that the *kosmos* could come into existence by chance, without the guidance and providence of god, he says, but then they suppose that matter exists by nature and chance. Origen does seem to believe that there could be better and worse matter for god's purpose, as he questions whether god would have produced better or worse matter than he actually did, in the manner that Aristotle questions whether god would create a better or worse world.[136] God would of course create the best matter that he could. What Origen fails to point out though is in what way this primordial matter could be better or worse for god's purpose.

Origen's argument works against any conception of primordial matter where that matter has any character or quality. One can always then ask why primordial matter should have that character or quality as opposed to any other character or quality. Whether it works against an entirely characterless primordial matter is another question. One view might consider characterless matter as one state of matter on a par with all characterised states of matter, and so ask why out of all those states we get characterless matter. Another view would be that characterless matter co-exists with god and it is then god's task to impose character on that matter. Then one would have to decide whether characterless matter could actually exist on its own. A perfectly reasonable view would be that such matter does not actually exist, and is only the result of a thought experiment which strips all the characteristics away. In creating characteristics, god creates matter.

It is difficult to know if Origen's argument here is original, or if it is influenced by the sort of argument I attributed to Plato about what we are justified in assuming about matter prior to the organisation of the *kosmos*. Given the way that Origen frames his argument, it is a shame that he does not go one step further and ask whether it is not extremely fortuitous that something so supposedly wonderful as god exists without having been designed by a creator.

XIII. Augustine

We have looked at many of Augustine's views already, but I want to finish off this chapter by collecting them together as they were so influential. He displays a splendidly simple, sensible and trenchant manner in dealing with many of the issues of cosmogony. For Augustine, an omnipotent God creates everything in heaven and earth out of nothing:

> An omnipotent god has no aid from any material which he did not make himself to generate what he wished to make. If in generating the things he wished to make, he had any help from anything he had not generated, then he was not omnipotent, and that is sacrilege.[137]

Nothing can be coeternal with god, and god did not create matter out of himself, therefore he created it *ex nihilo*.[138] It is god who generates time and motion where there was none before.[139] In enacting this creation God is not pained, busy or industrious.[140] When he is not creating, God is not idle, slothful or sluggish. The creation is an exercise of God's free will, and is something that he wills, contra the Neoplatonists.[141] God creates all things simultaneously,[142] and the six days of creation are to be taken as a metaphor to aid human understanding.[143] God is the ultimate creator of all creatures, as he generates the 'seminal principles' which go on to generate all of the creatures.[144] Neither God nor his will are subject to change.[145] Willing a change is not the same as changing one's will. There is no new decision to generate the *kosmos*.[146] The 'why now' question is simply irrelevant if there is no time before creation. There can be no explanation of creation *ex nihilo*.[147] Can god know infinity? Augustine simply replies that the numbers are infinite, and surely god knows the numbers.[148]

God's omnipotence would allow him to create more, but this would run contrary to his sense of justice, so only a certain number of possibilities are actualised. There is only one *kosmos*. If there were successive *kosmoi*, the crucifixion and resurrection would become futile,[149] and improving our knowledge of god and his will pointless, as we would eventually only return to our previous miserable state of ignorance once again.[150]

XIV. Conclusion

Cosmogony is still very much a live subject in later antiquity. Platonists, Aristotelians, Stoics and Christians contribute to a lively debate involving the possible creation of matter, space and time, and god's role before, during and after the generation of the *kosmos*. While Plato had considered the idea of a start to orderly, measurable time, the Christians introduce the idea of an absolute origin for time, as well as for space and matter. These latter ideas are radical departures, and stem at least in part from Christian need to emphasise the omnipotent nature of their god and to separate themselves from both pagan religions and Greek philosophy. There is an interesting tension in Christian thought between those who wish to emphasise god's omnipotence and the unique, miraculous nature of his creation of the *kosmos*, matter, space and time out of absolutely nothing and those who wish to stress the orderly and invariant nature of god, having god generate the *kosmos* from pre-existing matter or having god create the *kosmos* at all times, so denying any unique moment of creation. The Neoplatonists staunchly defend their eternalist view, producing many interesting arguments in its favour and provoking a philosophically very productive debate with Philoponus on the eternity of the *kosmos*, with further contributions from Simplicius.

Conclusion

I. Ancient cosmogony questions

Why did the ancient Greeks hold their views on cosmogony? Clearly, their philosophical views on a wide range of subjects are important here. Theories of explanation, the nature of matter, the nature of change, infinity etc. all have a role to play in explaining why specific thinkers adopted specific cosmogonies. One thing that I have sought to do in this book is to show that cosmogony is not a single, undifferentiated issue for the ancient Greeks. Not only do they address several issues in cosmogony, but the problems and answer-types they formulate have strong structural similarities to perennial problems and answer-types in cosmogony, many of which are still under discussion today. So we have found discussions such as:

- How many *kosmoi* should we suppose there are?
- Should we explain our *kosmos* as one of many, or as a unique entity?
- Are the processes of cosmogony part of the ongoing physics of the *kosmos*, or do they form a special case?
- If *kosmos*-forming processes will also destroy the *kosmos* if unlimited or unopposed, how might they be limited or opposed?
- If *kosmos*-forming processes will also destroy the *kosmos* if unlimited or unopposed, can we make sense of their being unlimited or unopposed, perhaps with successive *kosmoi*?
- Does creation *ex nihilo* make sense?
- Can we have a causal explanation of creation *ex nihilo*?
- Could there be either a limited or everlasting cycle of *kosmoi*?
- Does the universe have an appropriate density for *kosmoi* to form?
- Could some of the physical processes of the *kosmos* reverse, and if so what sort of *kosmos* would that produce?
- Do the processes of cosmogony obey proportionality as other processes in physics do?
- Do the elements have their properties by chance or by design?
- Can we give sufficient reasons for cosmogony beginning at this place and at this time?
- If we assume a homogeneous initial state, are there sufficient reasons for its later differentiation?
- If we assume a non-homogeneous initial state, are we able to explain why that state and not another is the initial state?

- If the result of cosmogony is human beings living on earth, what do we then need to assume about the process of cosmogony?
- Could there have been infinite past time, or did time begin in some manner?
- Are there irreversible processes which in an infinite amount of time would generate unacceptable infinities or unobserved conclusions to those processes?
- Was space generated where none existed before?

It is significant that the Greeks, and no culture before them, formulated and addressed these questions. Philosophical cosmogony begins with the ancient Greeks. If we take our criteria to be parsimony, invariance, consistency and a rejection of the supernatural, then prior to the Milesians we have mythological accounts of creation and with the Milesians we have the first philosophical cosmogonies. It is also significant that no one among the ancient Greeks assumed there to be a unique *kosmos* which came about without any design principles.

II. One *kosmos* or many?

One contrast I have pushed throughout this book is between those who advocate a unique *kosmos* generated through design and those who advocate multiple *kosmoi* generated through chance, with no middle ground and no confusion of the issues among the Greeks. As we have seen, the question of the number of *kosmoi* and whether they come together through chance or design is very clearly formulated by the Greeks. For those who believe in one *kosmos*, and for those who believe in multiple co-existent *kosmoi*, that works very well. For those who believe in successive *kosmoi*, the matter is a little more complex but still fairly clear. The Stoics, whose successive *kosmoi* are either identical or have insignificant changes, I would place in the 'one *kosmos* by design' camp. Where they differ from ordinary 'one *kosmos*' theorists is that their *kosmos* is subject to inevitable decline, and so has to be remade. Empedocles I place in the 'multiple *kosmoi* by chance' camp. Successive *kosmoi* for him may be radically different, and his zoogony is very much of the 'species occur through a multiplicity of accidents' type. There is a case for a form of teleology in Empedocles, but it is a fairly weak case for what would be a very odd sort of teleology. Origen, though he possibly allows more variation between successive *kosmoi* than the Stoics, I would also place in the teleologist camp, as his god, while not exerting full control, is responsible for everything positive that is generated. Origen explains in terms of design, rather than a multiplicity of accidents.

I have argued a radical but consistent line on the Milesians, and for the more minor figure, Diogenes of Apollonia. Each generates the *kosmos* from something primordial, be it water, the unlimited or air. Each believes in some form of steering principle, inherent in matter, which guides the proper formation of the *kosmos*. The Milesians do not have, and do not have any need for, multiple *kosmoi*, either co-existent or successive, nor do they have or need vortices. There is no trace of the types of explanation consonant with co-existent *kosmoi* that we find in either group of atomists. The Milesians have no boundary or individuation criteria for such *kosmoi*. They have no need of successive *kosmoi* as they do not subscribe to Empedocles' ideas on chance, nor is there anything in Milesian thought that requires successive *kosmoi*. The 'out of x and into x' formula does not commit the Milesians to an end to the *kosmos*, nor does the idea that the *kosmos* has a beginning, or, using a biological metaphor, a birth, commit them to the idea that it has an end or death. Hesiod and the immortal gods having births are good precursors, and the steering principle may guide the *kosmos* from origins through development into a state of equilibrium with the surrounding primordial material. The Milesians attempt to bring cosmogony within the ongoing physics of the *kosmos*. Cosmogony for them is not a unique action, either incomprehensible or with laws of its own. Rather, the processes of cosmogony are part of the ongoing physics of the *kosmos*, contra Hesiodic theogony and mythological accounts of creation. The Milesians are the first to contribute to this problem of the nature of cosmogony. While there is considerable later evidence of multiple *kosmoi* in the Milesians, I have argued that this evidence is extremely suspect due to misunderstandings stemming from Aristotle and a tendency for later commentators to read atomist views where none existed.

Heraclitus is a good example where virtually all modern commentators believe that the ancient commentary tradition got something entirely wrong. Heraclitus did not believe in *ekpurôsis*, but believed that the *kosmos* had no beginning and no end. He effects a change of focus from ordered states to ordered processes. While the Milesians attempted to make the processes of cosmogony part of the ongoing physics of the *kosmos*, those processes had to change so that they did not destroy the *kosmos* they had created, a problem we see in a more extreme form in Empedocles. Heraclitus is able to have an unchanging process for the *kosmos* as long as he jettisons the idea of cosmogony, so the *kosmos* is an ever-living fire without beginning or end. This is not problematic as long as we recognise that fire has a causal priority for Heraclitus, rather than an ontological or temporal priority.

III. Types of ancient cosmogony

Prior to the engagement of Christianity with Greek philosophy, I would identify four main views of cosmogony.

(1) A unique *kosmos* governed by design principles.
(2) Multiple co-existent *kosmoi* generated by chance.

(3) Cyclical *kosmoi*, of the Empedoclean or the Stoic types.
(4) Anti-cosmogony, of the Heraclitean, Parmenidean or Aristotelian types.

Heraclitus, I have argued, rejected cosmogony, or at least the cosmogony of the Milesians, in order to be able to have an entirely invariant physical process for the *kosmos*. Parmenides gives the first formulation of many sufficient reason problems for cosmogony. This is so not only for creation *ex nihilo*, but also for the origins problem of whether the initial state is homogeneous or not, and, I have argued, he also poses sufficient reason questions for how his predecessors chose pairs of opposites for the generation of the *kosmos*. Aristotle also produces an interesting battery of anti-cosmogony arguments. Some are tied to his physical theory and stand or fall with it, but others have a more enduring appeal. He produces interesting regress arguments about matter and motion. He poses awkward questions for any cyclical conception of cosmogony. Most interesting and enduring, though, is his argument that all comprehensible physical processes are governed by some form of proportionality. Creation *ex nihilo* is then incomprehensible as nothing cannot enter into relationships of proportionality.

A common assumption with all of these cosmogonies (including the anti-cosmogony views) is that matter exists eternally, as does space. Most also assume that time has no beginning or end, the qualified exception being Plato and later Platonists who believe that orderly, measurable time comes into existence with the *kosmos*. Aristotle, most likely in reaction to this, has a conception of time which denies that there can be anything prior to time. Christianity brings with it challenges to the earlier Greek views on matter, space and time. I have argued that the Bible itself is ambiguous on the issue of whether god created the *kosmos ex nihilo* or from pre-existent matter. In all likelihood, this was not an issue prior to the engagement of Christian thinkers with Greek philosophy in the first and second centuries AD. It is only then that creation *ex nihilo* settles as a Christian doctrine.

With the Christian thinkers we get the first supposition that god created space as well as matter and the *kosmos*. So too we get a firmer notion of the creation of time. God generates time when he generates matter, space and the *kosmos* such that there can no longer be any question of what god was doing in the time before the creation.

IV. Resources in ancient cosmogony

The range of explanatory resources in cosmogony available to the ancient Greeks was very limited. In particular, they lacked anything like a modern conception of gravity. I have argued elsewhere that a key factor in the shaping of Greek astronomy and cosmology was the lack of a conception of gravity.[1] The Greeks were forced to explain celestial motion by other means. This deeply affected Greek cosmogony as well, especially where a *kosmos* is supposed to be ordered from pre-existing chaos. In modern cosmogony, gravity is the key factor. Essentially, it is gravitational effects which allow the generation of galaxies, stars and planets. Again, the Greeks needed to explain critical gravitational phenomena by other means. So we find, for instance, Leucippus and Democritus using vortices and a like-to-like principle. The lack of any real conception of force acting at a distance is also a problem. One might separate two senses of force. There is force as we use the term in mechanics, as in equations such as $F = ma$. In some isolated technical examples, some Greeks came close to this sort of conception of force without ever quite developing a general theory. The other sense of force is that of a fundamental force of nature, such as gravity or electro-magnetism. So we can find odd principles used in cosmogony – the like-to-like principle, or Empedocles' love and strife – without there being any real conception of force. Those who sought to explain the origins of the *kosmos* without recourse to design or teleology struggled to put forward a persuasive case for how a *kosmos* could be generated from a prior chaos. So too they struggled to explain how living beings came into existence in the absence of any really coherent theory of evolution. We should not be overly critical here, as coherent theories of cosmic and animal evolution are the products of the nineteenth and twentieth centuries. If we are to understand why teleology and design principles were used so extensively in ancient Greek cosmogony and zoogony, though, we do need to appreciate the weaknesses of the case put forward by the evolutionists.

As I indicated in the introduction to this book, while modern science believes it has the correct types of answers for many questions, it still struggles with the issue of what cosmogony is like. The Greeks struggled with the question of what the fundamental forces of nature were like, as well as what cosmogony was like.

V. Cosmogony and modelling

The Greeks and their predecessors drew on many different resources in their attempts to say what cosmogony was like, and here we can see many social, religious, political and geographical factors at work. With the Babylonians, it is no great surprise that they model their creation tales on their flood plain agriculture, which each year sees the emergence of earth from marsh and a proliferation of insects apparently generated spontaneously from the receding mud. Such ideas naturally carry over into Greek thought to some extent. As there is a good deal of biological modelling in Greek cosmology, it is no surprise that biological modelling is important in cosmogony too. I do not see this as necessarily crude anthropomorphism or zoomorphism. The Greeks sought analogues for what the *kosmos* was like. As it moved in a regular rather than a chaotic manner, as the parts when in motion kept relations with one another, as it seemed to constitute a whole, it was reasonable to liken the *kosmos* to an animal in some respects. As the movements of the heavens were orderly and regular, and intelligence was associated with order and regularity, it was reasonable even to consider the *kosmos* to be like an intelligent animal. Prior to the seventeenth century, since when there has been a dominance of explanation in purely physical terms, the use of biological analogues was perfectly reasonable, and this applies to the use of such analogues for the birth of the *kosmos*. Biological birth was one of the few processes in the ancient world in which something intelligent and mobile came into existence.

Given that Greek cosmology employs political metaphors, it is slightly surprising to find that Greek cosmogony does not.[2] One might have expected to find some cosmogonies based on how cities founded themselves, either as geographical or political entities, but this is not so. The closest we find to this is possibly Anaximander's notion of cosmic justice and retribution.

Changes in religious belief are reflected in changing roles for god in cosmogony. The drive of the Presocratic is away from the capricious gods of myth to the invariant, intellectual and entirely good gods of Xenophanes, Plato and Aristotle, or to the sort of pantheism/ panpsychism we see with the Milesians. What an entirely changeless and entirely good god is capable of becomes an important theme in the cosmogony of later antiquity. The rise of Christianity and its engagement with Greek philosophy brings new challenges and arguments. Creation *ex nihilo* only became a Christian doctrine, rather than one of several possibilities, during the period of this debate with Greek philosophy. The Christians undoubtedly brought something new. The generation of matter from nothing was not an idea that was seriously entertained by any Greek philosopher, before or after Parmenides. Nor was the idea of the generation of space, and although Plato had supposed the generation of orderly, measurable time with the *kosmos*, he and his followers were alone in doing so. The Christians were more radical on time, having no time whatsoever prior to god's creation of matter, space, time and *kosmos*. That the Christian god was thought to be omnipotent is important here in giving some credence to the notion that matter, space and time can be generated from nothing. Why Christianity formed this sort of notion of god and why it felt the need to separate itself from the pagan religions are questions beyond this book, but the Christian notion of god and the Christian engagement with Greek philosophy did have a profound effect on the nature of Christian cosmogony.

VI. Cosmogony and related fields

Understanding ancient cosmogony does help us to understand ancient views in related areas. I have argued that the split between those who explain using a multiplicity of accidents and those who prefer single designed entities runs deeper than is generally recognised in ancient thought, and can help to explain the diversity of ancient views on the origins and development of life and the origins and nature of the elements. On the teleologist side, this is perhaps most clearly seen with Plato. Plato is clear and adamant on three related issues. There is one *kosmos* generated by a god who designs and crafts that *kosmos* to be the best possible. There are individual animal species, existing without change, again generated by a designer/ craftsman god. There are two and only two types of basic triangle, once again chosen by the designer/ craftsman god, which are the best two triangles for the purpose. In each case Plato rejects a multiplicity of accidents leading to the current state of the *kosmos*, and in cosmogony, zoogony and stoichogony supposes there to be one, or a very limited number of entities which exist because they are the best. Throughout, I have attempted to show the links between cosmogony, zoogony, stoichogony, conceptions of the end of the *kosmos* and more general philosophical considerations concerning explanation.

Some issues in cosmogony, of whether cosmogony is an invariant process and whether the physics of cosmogony are part of the ongoing physics of the *kosmos*, may be important drivers of philosophical change among the Presocratics. The Milesians attempt to establish invariance against the previous collection of myths and theology, but Heraclitus, Parmenides and Empedocles all have interesting views relative to a perception that the Milesians have not been thorough enough on this issue.

VII. The Greeks and us

One concern of this book has been to show that, in one sense at least, the Greeks were like us in their thinking about cosmogony. They addressed similar problems and produced similar answers, albeit at a different level of sophistication. Those likenesses are worth stating as they have not been fully investigated or recognised previously, and because they sharply differentiate Greek thought from that of its predecessors and contemporaries. I hope I have also done something to bring out the differences between ancient Greek and modern thought on cosmogony. One might argue that I have examined only a tiny fraction of Greek thought on cosmogony, that of the major philosophical writers, and have left the cosmogony of the populace untouched. That is true, and I readily acknowledge that the Greek populace largely believed in mythology and theogony. Conversely, I have compared ancient Greek cosmogony with only a tiny fraction of modern cosmogony, that of scientists and philosophers, and to a lesser extent theologians.

There are of course many issues in modern cosmogony on which the ancients have nothing to say. The notion of the expansion of space is a completely alien one to them, as are questions about the rate of that expansion (the current critical rate, a previous inflationary period, etc.).

The Stoic *kosmos* may have expanded and contracted, but the space it was situated in did not, and the notion of the generation of space was absent from Greek thought until the engagement with Christianity. The question of the curvature of space, critical to many modern theories of cosmogony and cosmology, was also absent, as indeed it was from all thought prior to Einstein. As mentioned above, the key resource the Greeks lacked in cosmogony, and indeed cosmology, was a conception of gravity. Gravity plays the key role in modern cosmogony in bringing together all the significant features of the *kosmos* (galaxies, stars, planets etc).

At certain points in this book I have argued some fairly radical theses, perhaps most notably about the Milesians (no co-existent or successive *kosmoi*), Empedocles (chance having a very significant role in cosmogony) and Leucippus and Democritus (rejecting sufficient reason in favour of *ou mallon* considerations). I believe those theses are justified as isolated interpretations of those thinkers. I also believe they are to some extent justified as fitting those thinkers into a coherent pattern of discussion and development in ancient Greek cosmogony, and the engagement of Greek cosmogony with certain perennial problems in cosmogony. If one of these positions were to prove untenable, would it damage any of the theses of this book? I hope to have done enough to establish some general patterns within ancient Greek cosmogony, in terms of its development, its concerns, its relation to more general philosophical issues and its relation to perennial problem types and solution types in cosmogony. If there should prove to be some exceptions to these general patterns, some conclusions might need to be more qualified, but would still be of interest as general trends in Greek cosmogony.

VIII. Finis

In conclusion, then, philosophical cosmogony began with the ancient Greeks. They formulated a series of perennial problems and prospective answers in cosmogony which are not to be seen in any culture prior to the Greeks. In more sophisticated forms, these problems and types of answers are still discussed today. There is a long and interesting history of Greek thought on cosmogony. This was not merely a matter of teleologists against anti-teleologists, creationists against evolutionists or closed world theorists against open universe theorists. A great number of important questions in cosmogony were discussed, and a great diversity of prospective answers were produced, many of which were highly influential in the Western tradition, certainly down to the scientific revolution of the seventeenth century and in some cases to the present day.

Notes

Introduction

1. See Kahn (1960) pp. 219ff. and Vlastos (1975) on the meaning and earliest usage of *kosmos*. LSJ give 'good order, good behaviour', 'of states, order, government', 'ornament, decoration, esp. of women', 'world order, universe' for *kosmos*.

2. I transliterate rather than translate *kosmos* and its cognates as the translations (world, universe) are prone to generate confusion in systems where there are multiple *kosmoi* in one universe, and because I wish to retain the Greek sense of a well ordered entity, in contrast to the modern connotations of 'cosmos'.

3. For those who do not believe that philosophy or science make progress, I offer the following paradox. If you hold and argue for this belief, presumably you believe it to be a better view than its contrary (that there is progress). So the history of science or history of philosophy make progress, from believing in scientific or philosophical progress to not doing so. So the history of science makes progress but science does not? And perhaps worse (as arguably the history of philosophy is philosophical but the history of science is not scientific) the history of philosophy makes progress but philosophy does not? If you believe there is no progress at all, I don't see why I should accept that theory as presumably it is no progress on previous theories.

4. Parmenides Fr. 8,6-10. Parmenides' problem here is quite general rather than specifically cosmogonical, but does have implications for cosmogony.

5. The life-cycle of a star depends almost entirely on its mass. Stars of greater mass than our sun may become e.g. neutron stars or explode as supernovae. Our sun will turn red giant in around four billion years.

6. There is a list of the perennial questions I believe the ancients address in the first section of the conclusion to this book.

7. See Furley (1989) pp. 1-8. The terms are borrowed from Koyré (1957) *From the Closed World to the Open Universe*.

8. Furley (1989) pp. 3-4.

9. The only possible exception here being Hesiod.

10. This is so not only for Aristotle on the Presocratics, but for Aristotle on Plato as well.

11. Using Hubble's law, v = Hr (recession velocity equals distance times Hubble's constant), it is possible to calculate an age for the universe.

12. Plato *Meno* 80d.

13. I appreciate that this does not by any means solve the many fascinating philosophical problems that Meno's paradox poses, but there is an interesting contrast to be brought out between areas of science which believe they can give the correct type of answer and those areas, such as cosmogony, where it is far from clear what the correct type of answer is or how we would recognise it should we come across it.

14. See Gregory (2000) chs 5 and 6.

15. Plato *Timaeus* 34bff.

16. See Gregory (2000) 167ff.

1. Mythological Accounts of Creation

1. 'There is no one definition of myth, no Platonic form of a myth against which all actual instances can be measured.' Kirk (1970) p. 7.

2. See here e.g. Frankfort (1946), Vernant (1965).

3. Wilson, in Frankfort (1946) p. 49.

4. We may need to specify a relatively broad window in time and area for the beginnings, but that does not invalidate the process. We may need to specify a broad window for the origins of *Homo sapiens* without doubting that *Homo sapiens* at some stage came into being.

5. Vernant (1965) p. 343.

6. In my view, Dodds equates magic too easily with irrationality. Natural magic conforms to parsimony, invariance, consistency and a rejection of the supernatural, as opposed to supernatural magic. I would argue that some of Dodds' instances of magic are natural rather than supernatural magic.

7. Physical principles such as the like-to-like principle of *Timaeus* 62c-e.

8. Plato *Laws* 967b.

9. Plato *Timaeus* 30b.

10. Cf. Empedocles Fr. 31 (Hippolytus *Refutation of All Heresies* VII,29,13) 'Of the shape of the *kosmos*, what it was like when ordered by Love, he speaks like this: "No twin branches spring from its back, there are no feet nor nimble knees, no parts of generation, but it was a Sphere and in all directions equal to itself." ' See too Xenophanes Frr. 15 & 16 for early criticism of anthropomorphism in religion.

11. See e.g. Plutarch *On the Face in the Moon* 926dff.

12. Plato *Timaeus* 34a. The six motions are up, down, left, right, backwards and forwards.

13. See Gregory (2000) and (2003a).

14. Plato *Timaeus* 40ab.

15. Plato *Timaeus* 40a.

16. Cf. *Phaedrus* 245cff. where self-motion is the essence and definition of soul and *Timaeus* 37b where it is that which is self-moved.

17. Plato *Laws* 897b. Cf. *Laws* 895c and *Epinomis* 982b on intelligence and regularity.

18. Both Plato and Aristotle would hold that these entities are fully alive, intelligent, etc. but would want to distinguish them from terrestrial living, intelligent entities.

19. The demiurge, e.g., always wishes for the best and so the *kosmos* will continue in existence perpetually.

20. The *Laws* is also adamant that prayer or other offerings will be of no avail in trying to change the minds of the gods.

21. 'The sun will not overstep his measures. If he does, the Erinyes, the defenders of justice, will find him out.' Heraclitus Fr. 94. In some other cultures there is more of an emphasis on collectively righting a wrong, rather than on punishment, should a civilly framed physical law be broken.

22. The use of a civil law conception of physical law was both plausible and widespread prior to the seventeenth century. The standard Greek term for law, *nomos*, was used in political/ civil contexts and not in mathematical contexts.

23. The Creation of Atum, version 1, following J.A. Wilson's translation, see Wilson (1969).

24. Wilson (1969) pp. 1-2.

25. Heidel (1942) p. 5.

26. Tablet V is somewhat shorter at 22 lines.

27. This translation is that of Heidel (1942) p. 8.

28. Homer *Iliad* 14.200 and 14.301. Plato *Theaetetus* 152e, 'Homer, in saying "Okeanos generator of the gods and mother Tethys" declared all things to be generated from flux and motion.' Cf. *Cratylus* 402b, *Timaeus* 40d, Aristotle *Metaphysics* I/3, 983b27ff.

29. Homer *Iliad* 14.244.

30. Damascius *Principles* 124.

31. Orphic Rhapsodies 109. Aristotle *Metaphysics* XIII/6, 1071b27 talks of 'The theologians who generate from Night.'

32. See below on the translation of *chaos* as chasm rather than chaos. In this table I merely transliterate.

33. Athenagoras *Pro Christianis* 18.

34. Aristophanes *Birds* 693.

35. Euripides Fr. 484, cf. Diodorus 1,7,1, Apollonius of Rhodes I,496.

36. See KRS pp. 30-3 and their selection of Derveni material.

37. See e.g. Derveni papyrus Col. 8, 9, 14, cf. Betegh (2004) pp. 167ff.

38. See Gurney (1990) pp. 175ff., KRS pp. 45ff.

39. See West (1971) pp. 1-77, KRS pp. 50ff.

40. See Aristotle *Metaphysics* XIV/4, 1091b.

41. See Penwill (1974), West (1963) pp. 154ff., West (1967), Vernant (1970), West (1971) pp. 206-8, KRS pp. 47-9.

42. I follow KRS here in omitting line 118, which is in some medieval MSS but not in many ancient quotations. See KRS p. 35 n. 1.

43. Hesiod *Theogony* 116-33.

44. Cornford (1952) pp. 194ff.

45. See the discussion of KRS pp. 37ff. For the separation of earth and sky in later Greek literature, see KRS pp. 42-3, and in non-Greek sources see pp. 43-4.

46. Philo *On the Indestructibility of the Kosmos* IV.

47. Aristotle *Physics* IV/1, 208b30.

48. Philo *On the Indestructibility of the Kosmos* IV.

49. Hesiod *Works and Days* 303ff.

50. Anaximander Fr. 2.

51. Xenophanes Fr. 11.

52. Hesiod *Works and Days* 50ff.

53. Hesiod *Works and Days* 81ff.

54. It may be the case that Hesiod produces no new gods, that is gods who were not present in previous mythologies, and attempts to systematise these, but he still uses far more gods than a parsimonious account of cosmogony would require.

55. E.g. Earth bears Pontos parthenogenetically (132), Night bears doom, fate, death, sleep, dreams and misery (213ff.).

56. West (1971).

57. See Kuhn (1970). I by no means agree with all of Kuhn's work, but he has been instrumental in demonstrating that science occasionally makes radical leaps.

58. On this issue see Lloyd (1992).

59. West (1971) argues mainly for Iranian influence on Pherecydes, Anaximander and Heraclitus but suggests many other possibilities as well.

2. The Milesians

1. Aristotle *Metaphysics* I/3, 983b7ff.

2. Aristotle *Metaphysics* I/3, 983b21ff. Cf. Simplicius *Physics* 23,21ff. who gives Theophrastus' similar account of Thales.

3. Aristotle *On the Heavens* II/13, 294a28ff.

4. Heraclitus Homericus *Homeric Questions* 22.

5. Aristotle *On the Soul* I/2, 405a19.

6. Aristotle *On the Soul* I/2, 411a7, cf. Diogenes Laertius I,24. 'Aristotle and Hippias say that he gave a portion of soul to soulless entities, citing the Magnesian stone and amber as proofs.' Contra KRS p. 95, amber when rubbed produces static electricity, not a magnetic field. The earliest differentiation between electrostatic and magnetic effects though was not until the sixteenth century, with William Gilbert's *On the Magnet*.

7. Aetius I,7,11, cf. Stobaeus I,1,11.

8. Cicero *On the Nature of the Gods* I,10,25.

9. Something which Plato may allude to at *Laws* 899b 'Is there anyone who will hold this and maintain that all things are not full of gods?' Plato pushes the notion of invariance very strongly at this point in the *Laws*, cf. my Chapter 1/III.

10. KRS p. 97

11. Cf. O'Grady (2002) pp. 108ff.

12. Cf. West (1971) pp. 212-13, Frankfort (1946) p. 253.

13. Aristotle *On the Heavens* II/8, 294a29ff.

14. Cf. O'Grady (2002) pp. 71ff., KRS pp. 92ff.

15. Aristotle *Metaphysics* I/3, 983b22ff.

16. O'Grady (2002) pp. 64-5.

17. Pseudo-Plutarch *Lives* II,1.

18. Stobaeus II,1,2. As I believe the evidence of all the later commentators is questionable in relation to the Milesians, I do not rely on such lists very heavily.

19. Simplicius *Physics* 1120,28.

20. West (1963) pp. 172ff., cf. (1971) pp. 212-13.

21. West (1963) p. 173.

22. I do not believe that either Anaximander or Anaximenes employed vortices in their cosmogonies, see below.

23. West (1963) p. 174.

24. West (1963) p. 175. For examples of this in water-based creation tales, see Dahnhardt (1907) pp. 18ff., 43, 46, 54, 56, 66ff.

25. One might wonder though where the 'natural disturbances' in the water come from.

26. Simplicius *Physics* 24,13.

27. See KRS pp. 105ff., 117ff.

28. See Kahn (1960) pp. 231ff.

29. Simplicius *Physics* 24,21; cf. Aristotle *Physics* III/4, 187a20.

30. Pseudo-Plutarch *Stromateis* 2. 'The heavens are formed from a mixture of hot and cold.' Aetius II,11,5.

31. Hippolytus *Refutation of All Heresies* I,6.

32. Aristotle *Physics* III/4, 203b7ff.

33. Guthrie I, 88 n. 2; LSJ give 'steer', 'drive', 'guide, govern', 'act as pilot'.

34. Furley (1987) pp. 23-4.

35. Plato *Philebus* 28d.

36. Plato *Philebus* 30d.

37. Heraclitus Fr. 41, 64 though with a different verb.

38. Parmenides Fr. 12.

39. Hippocrates *On Regimen* I.

40. Diogenes of Apollonia Fr. 5.

41. Arguably it can also be found in the Derveni papyrus, again in relation to air, Col. 19 lines 3-4.

42. Zeller (1931).

43. Burnet (1920) pp. 50ff.

44. Cornford (1934).

45. Kirk (1955), KRS pp. 122ff., Kahn (1960).

46. McKirahan (2001), Finkelberg (1993).

47. Cornford (1934) p. 15

48. KRS p. 122.

49. Kirk (1955) p. 29.

50. By chance here for the Milesians I mean something like a random, uncaused or inexplicable event (as opposed to the mere absence of teleology). Later discussions of Leucippus and Democritus, Empedocles, Plato and Aristotle on the topic and its relation to cosmogony will throw the issue of chance and the Milesians into sharper relief.

51. See Ch. 7 for a fuller discussion of *ou mallon* arguments.

52. Aristotle *On the Heavens* II/13, 295b10ff.

53. Plato *Phaedo* 108e.

54. Furley (1987) pp. 23ff. has questioned this account of Anaximander and the earth's stability, citing Simplicius *On the Heavens* 532,13 ('Anaximander thought that the earth remains in place *because of the air that supports it* and because of its equilibrium and uniformity', Furley's translation and emphases). I side with Panchenko (1994) and Schofield (1999) against Furley here in giving greater weight to the position Aristotle describes, especially as it is supported by Hippolytus *Refutation of All Heresies* I,6,3, who says that for Anaximander 'the earth is aloft, not held in the power of anything but remaining where it is because of its like distance from all things'.

55. Aetius II,1,8.

56. McKirahan (2001) p. 60.

57. McKirahan (2001) p. 60.

58. McKirahan (2001) p. 61.

59. Cf. Guthrie I, p. 85.

60. One might be concerned that a cosmic intelligence could produce many *kosmoi*. I would turn that around, and simply ask why would it?

61. Finkelberg (1994) p. 498.

62. McKirahan (2001) p. 60.

63. I don't want to read a thermodynamic law in here – this would work equally well with wet and dry or any similar pairs – just a notion that separating off some of one part of a pair entails separating of an equal amount of the other part of the pair.

64. We would though have to treat 'the assessment of time' as poetic metaphor, as the separation of heat would simultaneously separate out cold as well (and the same for other pairs of contraries), although it is possible Anaximander believed some short time lag was involved in separation, when the assessment of time would be less metaphorical, but operate over a short time-span.

65. Engemann (1991) p. 20.

66. See Aristotle *Physics* I/4, 187a20ff., Pseudo-Plutarch *Stromateis* 2, Kahn (1960) pp. 39-42.

67. See below on Anaximander's zoogony, see KRS 91 n.1 on Thales' choice of water being due to the observation that corpses dry out.

68. Aristotle *Meteorology* II/1, 353b6ff.

69. Aristotle's own view is that any recession of the sea is only local, and is compensated for by an advance of the sea in other places. See *Meteorology* I/14 352aff.

70. Alexander *Meteorology* 67,11.

71. Cornford (1952) pp. 183-5, Kahn (1960) p. 185, KRS p. 139.

72. Hippolytus *Refutation of All Heresies* I,6,7. Anaximander makes great use of wind in explanation, cf. Aetius III,3,1 and Seneca *Natural Questions* II,18.

73. Simplicius *Physics* 1121,5.

74. Aetius I,3,3, see also Cicero *On the Nature of the Gods* I,10,25 and Augustine *City of God* VIII,2.

75. See e.g. Vlastos (1947) p. 76, Freudenthal (1986) pp. 197, 218.

76. Simplicius *Physics* 24,16, 1121,5, *On the Heavens* 202,14, 615,15, Aetius I,3,3, I,7,12, II,1,3, II,1,8, Hippolytus *Refutation of All Heresies* I,6,1, Pseudo-Plutarch *Stromateis* 2, Cicero *On the Nature of the Gods* I,10,25 and Augustine *City of God* VIII,2.

77. Aetius II,1,3

78. Pseudo-Plutarch *Lives* II,1.

79. Aristotle, *On the Heavens* III/5, 303b10-13.

80. Ross (1936) p. 482.

81. Aristotle *On the Heavens* I/9, 278b8ff.

82. Simplicius *On the Heavens* 523,13. This seems a rather better attempt to harmonise the central, stable earth and multiple *kosmoi* view than that of Aetius (that all the *kosmoi* are evenly spaced).

83. Aristotle *On the Heavens* I/10, 279b13ff.

84. Aristotle *On the Heavens* III/1, 298b29.

85. This is somewhat controversial in the case of Empedocles, see my later chapter on this.

86. See Aristotle *On Philosophy* Frr. 19, 20.

87. Aristotle *Physics* III/4, 203b18-20.

88. Aristotle *Physics* III/4, 203b27-9.

89. Simplicius *On the Heavens* 615,15, cf. Aetius I,3,3 above.

90. Aristotle *Physics* III/8, 208a8ff.

91. Finkelberg (1994) pp. 496 and 504, cf. Engemann (1991) p. 9.

92. Aristotle *On the Heavens* II/13, 295a7ff.

93. See Aristotle *On the Heavens* II/13, 295b10ff.

94. Aristotle *On the Heavens* II/13, 294a28ff.

95. See Aristotle *On the Heavens* IV.

96. KRS p. 132.

97. West (1963) p. 173.

98. Aetius V,19,4.

99. E.g. Kahn (1960) pp. 109ff., Guthrie (1965) pp. 90-1, KRS pp. 141-2, Freudenthal (1986) pp. 216-17 esp. n. 67.

100. KRS p. 142.

101. KRS p. 142.

102. See Aristotle *Meteorology* II/1, 353b6ff., Alexander *Meteorology* 67,11 and Aetius III,16,1.

103. Burnet (1920) p. 71, Kahn (1960) p. 113.

104. So Barnes (1979) p. 22.

105. Hippolytus *Refutation of All Heresies* I,6,6.

106. Pseudo-Plutarch *Stromateis* 2.

107. Simplicius *Physics* 24,26. Cf. Aristotle *Metaphysics* I/3, 984a, Hippolytus *Refutation of All Heresies* I,7,1.

108. Pseudo-Plutarch *Stromateis* 3. The last line here probably attributes a view Anaximander did not hold: see KRS p. 152.

109. Hippolytus *Refutation of All Heresies* I,7,5.

110. Aetius I,3,4.

111. *Kosmos* need not have the precise technical sense it had later, and while *pneuma* may look like a typical Stoic accretion, the term was in use at this stage.

112. KRS pp. 159ff.

113. KRS p. 159.

114. An interesting idea here might be that just as air/ soul holds us together over time and individuates humans, so air/ soul holds the *kosmos* together over time and individuates it from its surrounds.

115. Aristotle *Physics* III/4, 203b7ff.

116. Simplicius *Physics* 24,26, reporting Theophrastus; cf. Hippolytus *Refutation of All Heresies* I,7,5.

117. Aetius I,7,13. This is not essential; see my reservation above about whether Aristotle may call the unlimited divine and whether Anaximander does. There is also evidence, in my view suspect, that Anaximenes believed that the gods were generated from air: 'Anaximenes held that unlimited air was the *archê*, from which the things which are being generated, which have been generated and which will be, along with the gods and things divine are all generated' (Hippolytus *Refutation of All Heresies* I,7,1); 'Anaximenes attributed all the cause of things to infinite air, but did not deny there were gods, nor was he silent about them. He did not believe that air was made by them, but that they arose from air' (Augustine *City of God* VIII,2). Cicero *On the Nature of the Gods* I,10,25ff. seems completely confused (asserting both that air is a god and that it comes into being!).

118. Diogenes Laertius IX,57. Diogenes of Apollonia's dates would indicate he was a follower rather than a direct pupil of Anaximenes, as Laertius has it.

119. Diogenes of Apollonia Fr. 5.

120. It is interesting that the Derveni papyrus, Col. 19 line 3 also has a *kratein* cognate in relation to the dominance of air, in this case *epikratein*.

121. KRS pp. 159-61.

122. Perhaps also, in a less developed way, in Thales as well.

123. Simplicius *On the Heavens* 202,1, *Physics* 1120,28.

124. Simplicius *Physics* 1121,12.

125. Aristotle *On the Heavens* II/10, 279b13ff.

126. KRS p. 126 also suggest that Simplicius may have classed Anaximenes with Heraclitus in that they both believed in 'a specific primary substance', and then (mistakenly) supposed that they had the same view on the fate of the *kosmos*.

127. Aristotle *On the Heavens* II/13, 294b13.

128. See Hippolytus *Refutation of All Heresies* I,7,6 where the heavenly bodies are said to move like a cap being turned on a head in Anaximenes.

129. Hippolytus *Refutation of All Heresies* I,7,4.

130. See Aetius II,14,3 and II,22,1.

131. Aetius II,23,1.

132. Aristotle *Physics* III/4, 203b7ff.

133. Aristotle *Metaphysics* I/3, 984a21ff.
134. Aristotle *Metaphysics* I/3, 984b11ff.
135. Cornford (1952) p. 198.
136. Cf. Furley (1987) p. 18.
137. Guthrie I pp. 62ff., Barnes (1979) pp. 6ff., O'Grady (2002) pp. 108ff.
138. This table largely follows Stokes (1962) p. 14, cf. Cornford (1952) pp. 199ff.
139. Cornford (1952) p. 159.
140. Furley (1987) pp. 28ff.
141. Furley (1987) p. 29.
142. West (1971) p. 77.
143. West (1971) p. 77.
144. West (1971) p. 77. I would also disagree that Anaximander's system is visionary rather than logical in its foundations, and even if Pythagoras heard Anaximander, I do not see that this is evidence of theological thinking in Anaximander.
145. West (1971) pp. 87-93, cf. p. 78.
146. West (1971) p. 95.
147. There are increasingly difficult issues as we extrapolate back closer to the supposed origin. We have to extrapolate beyond what we know about high temperature and high density physics. The universe will eventually become smaller than its Compton wavelength, so that quantum effects predominate and a theory of quantum gravity is required. Related to the smallest packet of energy is the smallest length of time (10^{-43} s) beyond which it is physically impossible to probe. Finally, there are the issues of infinity related to the ultimate origin.
148. On SS cosmogony, see Hoyle (1951), (1955), (1966), (2005), and also Narliker (1983). For an account of the debate between BB and SS see Kragh (1992) and J. Gregory (2005).
149. Key issues here were the proper calculation of timescales relative to the Hubble constant, mechanisms of the generation of elements heavier than helium, the distribution of radio galaxies and the question of whether relics of a very hot dense phase of the universe (predicted by BB but denied by SS) still remained. The latter formed the most spectacular evidence in favour of BB, with the discovery of the microwave background radiation in perfect accordance with BB.

3. Heraclitus

1. If anything, Parmenides seems to react to Heraclitus. See Guthrie I, p. 408.
2. Taking *ton auton hapantôn* as genuine with Guthrie, Vlastos and Kahn but against Kirk and KRS.
3. Heraclitus Fr. 30.
4. Guthrie I p. 455, so too Barnes (1982) pp. 61-2.
5. See e.g. Arius in Eusebius *Evangelical Preparation* XV,15. Aristotle also takes *kosmos* in this sense, as an order that can persist through cycles, see Aristotle *Physics* VIII/1.
6. Heraclitus Fr. 54, cf. Fr. 123 'Nature loves to hide herself'.
7. Kirk (1962) p. 311. There are similar polar expressions prior to Heraclitus, Homer *Iliad* 8.27, 14.342, Xenophanes Fr. 23.
8. Kahn (1979) p. 134.
9. Kahn (1979) pp. 152-3.
10. Aristotle *On the Heavens* I/10, 279b12ff.
11. Aristotle *Physics* III/5, 205a3-4, cf. *Metaphysics* XI/10, 1067a4ff.

12. Diogenes Laertius IX,8.

13. Plutarch *The Obsolescence of Oracles*, 12,415f-416a.

14. Marcus Aurelius *Meditations* X,7.

15. Heraclitus Fr. 90.

16. Kirk (1962) pp. 349-61 considers this to be a gloss by Hippolytus and so does not consider it to be a fragment. Although I am suspicious of Hippolytus, I do consider this a fragment.

17. Heraclitus Fr. 66.

18. Plato *Sophist*, 242d-243a.

19. Heraclitus Fr. 10.

20. *On the Kosmos* 5, 396b20ff. *On the Kosmos* is usually taken to be by an unknown author in imitation of Aristotle.

21. Censorinus *The Day of Birth* 18.11 attributes this to Heraclitus, with a cycle of 1,800 years. Censorinus' source was almost certainly Stoic (Kahn 1979, p. 157), so we must be suspicious of this attribution.

22. See Plato *Timaeus* 39d.

23. Heraclitus Fr. 100.

24. Harmless too are Fr. 103 'The beginning and end of the circumference of a circle are common, according to Heraclitus', and 'A generation is thirty years according to Heraclitus, during which time the generator has generated a new generator out of himself. The cycle of life is in this space, when nature returns from human seed to seed', Plutarch *The Obsolescence of Oracles* 415e.

25. Kahn (1979) p. 136.

26. Kahn (1960) p. 225 n. 2.

27. See Frr. 40, 42, 81, 129.

28. That will be so whatever the outcome of the debate concerning the nature and number of the phases of the cosmic cycle in Empedocles.

29. Cf. Reeve (1982) p. 304 who says that 'In the worlds of Thales, Anaximander and Anaximenes change depends on stability; in Heraclitus' world stability depends on change' and Graham (1997) pp. 35, 44.

30. Kahn (1979) pp. 152-3 complains that if fire cannot have a cosmogonical role, its role will then be 'symbolical and perhaps metaphysical'. But Kahn, as Inwood (1984) p. 229 points out, cannot have things both ways – he cannot object to a symbolical or metaphorical priority for fire when he takes a strongly symbolical and metaphysical approach to reading the fragments in general.

31. Burnet (1920) p. 145.

32. Vlastos, Furley and Allen I, pp. 422-4.

33. See Fr. 64 and discussion below.

34. Barnes (1979) 64ff.

35. Reeve (1982) pp. 302-3.

36. Heraclitus Fr. 31.

37. Which might seem possible if one translates *tropai* as 'turnings'.

38. Heraclitus Fr. 76.

39. Heraclitus Fr. 36.

40. Reeve (1982) p. 302.

41. Heraclitus Fr. 60.

42. Reeve (1982) p. 300.

43. Reeve (1982) p. 303.

44. Reeve (1982) p. 304.

45. Reeve (1982) p. 303.

46. Graham (1997) pp. 31ff.

47. Graham (1997) p. 37.
48. Graham (1997) p. 45.
49. Graham (1997) pp. 45-6.
50. Graham (1997) p. 34.
51. Heraclitus Fr. 94.
52. Heraclitus Fr. 80. 'War is the father of all, and the king of all, some he shows to be gods, some to be men, some he makes slaves, others free men,' Heraclitus Fr. 53. 'Father of all' may be an allusion to Homer, where Zeus is frequently referred to in this way.
53. Heraclitus Fr. 125.
54. Barley meal, grated cheese and wine, Homer *Iliad* 9.638ff., with honey as well, Homer *Odyssey* 10.234.
55. Heraclitus Fr. 64.
56. Heraclitus Fr. 41.
57. Heraclitus Fr. 118, cf. Frr. 36, 117.
58. Cf. Vlastos, Furley and Allen I, p. 424.
59. Kahn (1979) p. 194.
60. Homer *Iliad* 18.107, part of a mourning speech by Achilles.
61. Aristotle *Eudemian Ethics* 1235a25.
62. Simplicius *Categories* 412.
63. Frr. 31, 36, 76, 90.
64. Diogenes Laertius IX,9.
65. Aristotle *Physics* I/4, 187a12ff.
66. Aristotle *Metaphysics* I/3, 984a1ff.
67. Aristotle *On the Heavens* III/1, 298b29.
68. Heraclitus Fr. 124.
69. Kahn (1979) p. 85; Kahn sees the cosmos as 'a random assortment of refuse' as a 'striking illustration of the paradoxical connection of opposites' (p. 285). The Greek here is *sarma eikê kechumnôn ho kallistos phêsin Herakleitos ho kosmos.*
70. One might also make a case, or at least a parallel, for no use of chance in Heraclitus from the Derveni papyrus Col. 4. Lines 3-4 speak of not being got by chance, line 4 of how this is how the *kosmos* attained order, and line 5 begins 'In the same way Heraclitus', and the remaining lines may allude to Heraclitus Fr. 89 ('For those who are awake the *kosmos* is one and common, but those who sleep turn away each into a private world. We should not speak and act like sleeping me.') and they give at least part of Fr. 94.
71. Barnes (1979) p. 61.

4. Parmenides

1. Parmenides Fr. 8,1-4. Cf. Melissus Fr. 1, 'It is always was what it was and it always will be. If it was generated, of necessity, before it was generated it was nothing. If indeed it was nothing, there is no way in which it could be generated from nothing.'
2. Parmenides Fr. 8,6-10.
3. One problem for modern physics here is it expresses a great number of its laws in the form of x y (x is proportional to y), where x and y may stand for more complex mathematical functions. If we begin from absolutely nothing, then we must set one side of these proportionality relations at zero, with the result that the other side can only be zero or infinity, and not a determinate quantity.
4. Parmenides Fr. 8,11-15.

5. Leibniz *Monadology* 32, 'The Principle of Sufficient Reason, in virtue of which we believe that no fact can be real or existent, and no statement true, unless it has a sufficient reason why it should be thus and not otherwise.'

6. I disagree with Mourelatos (1981) p. 653 that this sort of argument works for time but not place. More precisely, this argument works against some conceptions of space and place held by the ancients but not all.

7. Parmenides Fr. 8,22-5.

8. Parmenides Fr. 8,44-9.

9. Depending on whether one advocates 'top down' or 'bottom up' theories of galaxy and star formation. So either large clouds of matter accrete, and out of them condense the stars in a galaxy ('top down'), or individual stars form which then accrete together to form a galaxy ('bottom up').

10. For Zeno's role, see Plato *Parmenides* 127dff., though some would question how accurately Plato portrays Zeno here. Melissus, from pseudo-Aristotle *Melissus, Xenophanes, Gorgias* 974a12-14: 'As it is one, it is alike in all ways. If it were unlike, being multiple it would not be one but many.'

11. Parmenides Fr. 8,50-2. Cf. Fr. 1,29-32.

12. Parmenides Fr. 8,53-61.

13. Parmenides, Fr. 9.

14. Plutarch *Against Colotes* 1114b, cf. Simplicius *On the Heavens* 559,26.

15. See Parmenides Frr. 10, 12, Aetius II,7,1.

16. Aetius II,7,1.

17. Parmenides Fr. 12.

18. Zeller (1931), Burnet (1927) p. 185.

19. Long (1963) p. 90.

20. Beginning with Aristotle *Metaphysics* I/5, 986b18.

21. Owen (1975) p. 48.

22. It is not entirely clear in this context what the *morphai* are (see Long 1963, pp. 98ff.). Forms is the usual translation, types a possibility; I am not convinced given the strictures of the way of truth that shapes is possible here. *Morphos* can be used in the abstract sense of a form which has many instances.Three alternatives here are that the *morphai* are light and night (Mourelatos 1970), or being and non-being (Long, 1963), or light/ being and dark/ non-being (Vlastos 1946). See Granger (2002) p. 103.

23. Plutarch *Against Colotes* 1114b, cf. Simplicius *On the Heavens* 559,26.

24. Aristotle gives us a list of pairs of opposites at *Metaphysics* I/5, 986a22ff.

25. Parmenides, Fr. 8,2.

5. Empedocles

1. Those who have doubted the existence of the cycle have been conclusively refuted by O'Brien (1969).

2. So I would disagree with Burnet (1920) p. 227 who finds it hard to find a trace of Heraclitean ideas in Empedocles.

3. Parmenides Fr. 8,50-2.

4. Empedocles Fr. 17,26, cf. Fr. 35,1-2 'I shall turn back once more to the path of song recounted previously, as I draw off account upon account, in that way', which may be an allusion to Parmenides Fr. 6 where mortals wander on a backward turning path, see also Guthrie II pp. 138-9. The opening of Fr. 17 may also recall Parmenides – see here KRS p. 287.

5. Heraclitus Fr. 40.

6. Empedocles Fr. 17,14.

7. KRS p. 285, the relevant fragments being Empedocles Frr. 2 and 3, Heraclitus 1, 55 and 107.

8. Trepanier (2004), esp. pp. 178-9.

9. KRS p. 285 only suggest this, but in general I see Empedocles engaging more with Heraclitus than they do.

10. Heraclitus Fr. 125.

11. Empedocles Fr. 34.

12. Empedocles Fr. 8.

13. Empedocles Fr. 12.

14. Empedocles Fr. 13.

15. Empedocles Fr. 14.

16. Empedocles Fr. 17.

17. See Plato *Sophist* 242dff.

18. Empedocles, Fr. 35.

19. O'Brien (1969) pp. 138-40.

20. Agreeing here with Guthrie II pp. 180-1 and Osborne (1987a) and (2000).

21. See Osborne (2000) p. 337.

22. Panzerbieter (1844).

23. But this is contested by Holscher (1965) pp. 11ff., Solmsen (1965) pp. 130-2.

24. Aristotle *Generation and Corruption* II/6, 334a6-8.

25. O'Brien (1969) pp. 93-8, 177-9.

26. See *Timaeus* 44c, Aristophanes' speech in the *Symposium*, *Politicus* 270ff., 271cff., 273eff. For further details of the parallels with Empedocles see O'Brien (1969) pp. 94-8, 177.

27. Zeller (1920) pp. 969ff., Bignone (1916) esp. appendix II.

28. Tannery (1887) pp. 285ff., von Arnim (1902).

29. Bollack (1965), Solmsen (1965), Holscher (1965).

30. O'Brien (1969).

31. Long (1974). I agree with Graham (1988) that Long has an interesting but not compelling view of Fr. 17.

32. Guthrie II pp. 167ff., Barnes (1982) pp. 308ff., KRS pp. 288-9, n. 1.

33. Furley (1987) p. 79 praises Empedocles for 'The postulation of two physical forces, qualitatively different from each other.'

34. Cf. Guthrie II pp. 156-7.

35. This is how I read the admittedly slightly unclear evidence of Aristotle (*Physics* VII/5, 250b29, VIII/1, 252a9ff., *Generation and Corruption* I/1, 315a3ff., *On the Heavens* III/2, 301a11ff.) and others on this – see O'Brien (1995) pp. 406-13 on the ancient evidence.

36. Aristotle *Metaphysics* III/4, 1000a26ff.

37. O'Brien (1969) pp. 127-45.

38. Tannery (1887) pp. 319ff.

39. Plutarch *On the Face in the Moon* 926f.

40. See 53b. The link between Empedocles and Plato here is made explicitly by Plutarch *On the Face in the Moon* 926eff.

41. Empedocles Fr. 29, rubric Hippolytus *Refutation of All Heresies* VII,29.

42. Plato *Timaeus* 33bff.

43. Mansfeld (1972) reads *doiê* here as meaning 'ambiguous'.

44. Taking the orthodox reading (initiated by Panzerbieter) of *threphtheisa diepte*. On this see O'Brien (1969) pp. 164-7, Long (1974) pp. 409ff., Wright (1981) pp. 96-7 and 166ff.

45. Empedocles Fr. 17.
46. Plutarch *Quaestiones Convivales* 683e
47. Aristotle *Poetics* 1, 1447b17-20.
48. Graham (1988) p. 305. See also Trepanier (2004) pp. 4ff.
49. Five times in Fr. 17, but also in Frr. 20, 21, 22, 26.
50. Graham (1998) pp. 304-6.
51. Graham (1988) p. 312.
52. I would agree with Graham that this tells against Long's view as well, even though Long claims symmetry for his view (Long 1974, p. 408).
53. For the most recent argument and overview, see Trepanier (2004) pp. 1-30.
54. Trepanier (2003b) pp. 14ff., (2004) pp. 4ff.
55. Empedocles Fr. 124.
56. Empedocles Fr. 62.
57. Trepanier (2003a) p. 388. I largely follow the textual suggestions of Trepanier in the translation of ensemble a that follows. There is a problem with the Strasbourg papyrus known as the 'three thetas' problem. Empedocles typically uses a two-line formula in which the first line has the universe coming together under love and the second has it dissociating under strife. In the first line, he uses the verb *sunerchomai*, 'to come together'. In the Strasbourg papyrus, at a,i,6, and at c3 the copyist has written *sunerchometh*, 'we come together' rather than *sunerchomen*, 'coming together', the participle form employed by Empedocles in similar lines. Someone else has 'corrected' the copyist by changing the theta to a nu at a,i,6 and c3. At a,ii,17 we have the last part of a compound verb ending *erchometh*, which could be restored with either *eis* or *sun*. The view I take is that we should read *sunerchomen* in all three instances. See Martin and Primaversi (1999), Inwood (2000), Mansfeld and Algra (2001), Laks (2002), Trepanier (2003a) p. 388ff., (2003b) pp. 1ff., (2004) pp. 250ff.
58. Empedocles Fr. 16, *aspetos aiôn*, 'countless ages'.
59. Empedocles Fr. 110.
60. Reading *anotato*, with Simplicius and most modern versions. This may be a little strange as a construction, but there is no reason to emend to Cornford's *anô ê katô* which makes no philosophical sense.
61. Aristotle *Physics* II/4, 196a20ff.
62. Aristotle *Generation and Corruption* II/6, 334aff.
63. Aristotle says something very similar at *Physics* VIII/1, 252a17 ('The natural is absolute, and not sometimes thus and sometimes otherwise. Fire is always carried upwards and not sometimes like this and sometimes in another manner. When nature is not absolute, there is proportion.'), so we have three passages in Aristotle asserting this.
64. See Guthrie II pp. 161ff.
65. Aristotle *Metaphysics* III/4, 1000b10ff.
66. Aristotle *Generation and Corruption* II/6, 333b9-15.
67. Aristotle *On the Heavens,* III/2, 300b27ff.
68. Aristotle *Physics* II/8, 198b29. There may be some Aristotelian gloss here, but the multiplication is still evident.
69. Philoponus *Commentary on Aristotle's Physics* 261,17ff.
70. Aetius II,7,6.
71. Achilles Tatius *Isagoge* 4,34.
72. Simplicius *Physics* 331,10.
73. Simplicius *Physics* 331,15.
74. Empedocles Fr. 98.

75. Empedocles Fr. 96, Simplicius *Physics* 300,21, KRS translation.
76. Empedocles Fr. 75.
77. Aetius V,19,5.
78. Empedocles Fr. 57.
79. Empedocles Fr. 61, Aelian *On the Nature of Animals* XVI, 29.
80. Empedocles Fr. 59.
81. Aristotle *Physics* II/8, 198b29ff.
82. Empedocles Fr. 103.
83. Empedocles Fr. 104.
84. The standard view, prior to the nineteenth century, was that the earth was around 6,000 years old, this estimate being based on an analysis of the genealogies in the Bible, working back to Adam. Some went as far as to give a precise date of creation, 23 October 4003 BC. This dating was challenged on several fronts in the nineteenth century. Geological discoveries and analysis of geological processes seemed to indicate a far greater age for the earth, as did thermodynamic considerations, and Darwin's work demanded a much longer timescale for the theory of evolution to work.
85. Aetius V,18,1.
86. Empedocles, Strasbourg Papyrus, Ensemble d,11.
87. Empedocles Fr. 85.
88. Empedocles, Strasbourg Papyrus, Ensemble a,ii,3-4, and a,ii,8.
89. Pseudo-Plutarch *Stromateis* in Eusebius *Evangelical Preparation* I,8,10.
90. Aetius II,6,3.
91. Pseudo-Plutarch *Stromateis*, in Eusebius *Evangelical Preparation* I,8,10.
92. Trepanier (2003) p. 418.
93. As with Aristotle *Metaphysics* I/4, cf. *Metaphysics* III/4, 1000a26ff.
94. Eudoxus is sometimes thought to be a target in the *Philebus*; my suggestion here is that Empedocles may be a target as well.
95. On the order of the fragments see Wright (1981) pp. 239ff.
96. Empedocles Fr. 86.
97. Empedocles Fr. 87.
98. Empedocles Fr. 84.
99. Empedocles Fr. 95.
100. Sedley (1998) pp. 18-21, cf. Campbell (2003) p. 3.
101. Wright (1981) p. 240.
102. See Descartes *Principles of Philosophy*.
103. On this see in particular the works of Hobbes, Boyle and Locke.
104. See Furley (1987) and Berryman (2003).
105. The Newtonians were bitterly criticised by the Cartesians for re-introducing occult qualities and scholastic qualities (occult in the sense that this force of gravity was active, invisible and inexplicable, scholastic in that gravity could now not be reduced to the dimensions and motion of particles of matter and their contact actions).
106. This was a significant problem for cosmologists from Newton to Einstein, until it was recognised that the universe is not static but is expanding.
107. Why strife begins to gain on love is also unexplained, as Aristotle points out, *Metaphysics* III/4, 1000b12ff.
108. See Vlastos (1947), Lloyd (1964).
109. KRS p. 347.
110. Aetius I,5,2.

111. I would suggest that Aetius is importing atomist ideas into his account of Empedocles here. Cf. O'Brien (1969) p. 145 n. 3, Furley (1987) p. 90.

112. Aristotle *Metaphysics* I/4, 985a29ff.

6. Anaxagoras

1. Although no great matters hang on this, I take Anaxagoras to be a later thinker than Empedocles, and also a later thinker than Zeno.

2. Aristotle *Metaphysics* I/3, 984b15.

3. Anaxagoras Fr. 17.

4. Anaxagoras Fr. 1, cf. Fr. 4.

5. Parmenides Fr. 8,5.

6. Anaxagoras Fr. 3.

7. Zeno Fr. 1. Cf. Anaxagoras Fr. 5, 'When these things have been separated, it is necessary to know that all things are neither more nor less. It is not possible that there should be more than all, but all things are equal' and Zeno Fr. 3, 'If there is a plurality, things are as many as they are, neither more nor less.'

8. Diogenes Laertius II,6. According to Diogenes, this was the opening of Anaxagoras' book, but this is likely to be an abridgement rather than a direct quotation; cf. Simplicius *Physics* 155,26 on Fr. 1 being the opening passage.

9. Anaxagoras Fr. 12.

10. The key issue here was the shape of the earth – flattened at the poles and bulging at the equator, according to Newton, vice versa according to Descartes. Expeditions to determine the curvature of the earth settled this issue in favour of the Newtonians in the middle of the eighteenth century.

11. Anaxagoras Fr. 9.

12. Anaxagoras Fr. 2.

13. Anaxagoras Fr. 15.

14. Anaxagoras Fr. 16.

15. Cf. Simplicius *Physics* 24, 6, Hippolytus *Refutation of All Heresies* I,7,1.

16. Though perhaps here Anaxagoras is only talking of the stones which constitute the heavenly bodies as burning stones rather than of stones in general.

17. Aetius II,8,1.

18. Hippolytus *Refutation of All Heresies* I,8,12.

19. Theophrastus *History of Plants* III,1,4.

20. Plato *Phaedo* 97b.

21. Plato *Phaedo* 98a.

22. Plato *Phaedo* 98b.

23. See Guthrie II p. 317, cf. McCabe (1993) p. 54. See also Diogenes Laertius II,6.

24. Anaxagoras Fr. 13.

25. Hippolytus *Refutation of All Heresies* I,8,6.

26. Plato *Laws* 967b. Cf. *Apology* 26dff.

27. Aristotle *Metaphysics* I/4, 985a18.

28. Aristotle *Metaphysics* I/7, 988b9.

29. Simplicius *Physics* 177,9.

30. See Anaxagoras Fr. 12, 14. One might compare and contrast Plato's demiurge here.

31. See here Alexander (1956) pp. xviff., and the Leibniz-Clarke correspondence generally.

32. This is a standard difficulty for all static *kosmoi* with attractive forces.

33. See astronomical appendix for an explanation of the problem here. In the eighteenth century Laplace demonstrated the solar system to be stable within certain limits, and on being asked where was the place for god in his system, is reputed to have replied 'Sir, I have no need of that hypothesis'.

34. See e.g. Plato *Apology* 26cd, and cf. Hippolytus *Refutation of All Heresies* I,8,3-10 and Diogenes Laertius II,8-9.

35. Anaxagoras Fr. 4, translated to favour a multiple *kosmoi* interpretation.

36. Aristotle *Physics* I/4, 187a23.

37. Aristotle *Physics* VIII/1, 250b18ff.

38. Simplicius *Physics* 1121,5ff.

39. Simplicius *Physics* 178,25.

40. Simplicius *Physics* 157,9ff.

41. Aetius II,4,6 says Anaxagoras does have multiple *kosmoi*, II,1,2 says he does not.

42. Anaxagoras Fr. 8.

43. See Guthrie II p. 313.

44. Anaxagoras Fr. 2.

45. Burnet (1920), Barnes (1982).

46. See Anaxagoras Fr. 2.

47. See Aristotle *Physics* IV/6, 213a22ff., *On the Heavens* IV/2, 309a19ff.

48. There is no indication of any role for chance in any of the fragments. Aristotle at *Physics* II/4, 196a20ff. contrasts Empedocles who does have a role for chance with Heraclitus and Anaxagoras who do not, and then contrasts the atomists who have a role for chance with those who believe nature is the key reason for cosmic order (the Milesians?) and Anaxagoras.

49. There is no mention of any role for chance or *ou mallon* in the evidence on Anaxagoras. Aristotle explicitly contrasts Anaxagoras with those who do have a role for chance, Aristotle *Metaphysics* I/3, 984b15.

50. Diogenes Laertius IX,1.

51. See also here the relation of Anaxagoras to the Derveni papyrus, discussed in Chapter 8.

52. Hippolytus *Refutation of All Heresies* I,13,2.

53. KRS p. 380. Nor do I think there is anything to be gained by following Simplicius *Physics* 1121,21, where he suggests that Anaxagoras assumed a cosmogony only for the sake of illustration. Simplicius, a Neoplatonist, may well be imposing a view that was a take on Plato's *Timaeus* on Anaxagoras here.

54. Cornford (1934) pp. 6-9, Guthrie II.

55. Plato *Phaedo* 109c.

56. Similar proposals are that there are further human civilisations on the moon (Zeller 1920, 1239), the moon and the planets (Johrens 1939). Diogenes Laertius II,8 reports Anaxagoras as believing that there are dwellings on the moon.

57. Vlastos (1959) p. 355.

58. Furley (1989) p. 56.

59. This is Schofield's gloss on Furley, rather than his own formulation. See Schofield (1996) pp. 13-14.

60. Leon (1927).

61. Strang (1963).

62. Mansfeld (1980).

63. Schofield (1996).

64. Sisko (2003).

65. Anaxagoras Fr. 6.

66. Anaxagoras Fr. 3.

67. Schofield (1996) pp. 13-14, cf. Schofield (1980) pp. 79-94.

68. See Anaxagoras Fr. 3.

69. See Sisko (2003) p. 113 n. 46.

70. Sisko (2003) calls this the 'quasi-Leibnizian' view, p. 112.

71. It is interesting that Leibniz, who also believed in infinite divisibility, believed in a multiplicity of worlds within worlds. See Leibniz *Monadology* 65ff.

72. Sisko (2003) p. 112.

73. See Diogenes Laertius IX,31.

74. The division of Fragment 4 is down to Fränkel (1955) pp. 284ff.

75. With some slight translation changes to facilitate a different interpretation.

76. Anaxagoras Fr. 4. this is Simplicius *Physics* 34,29ff. in the order given by Simplicius, cf. the order given by KRS, which reverses (1) and (2), splitting (1) into two sections, (1) = KRS 483 pp. 368-9 and KRS 498 pp. 378-9, (2) = KRS 468, pp. 357-8.

77. Carter, B. (1974) pp. 291-8.

78. I do not believe there is any insuperable problem with construing the final sentence, with its *an* + optative construction, as a potential optative, as giving what would happen if there were to be another *kosmos*. The preceding set of five infinitives governed by *dokein* have their usual sense. It has been objected by Owen (see Strang 1963, p. 379 n. 28) against Fränkel (1955) that these infinitives cannot be taken as optatives. As Schofield (1996) p. 10 points out though, for Fränkel the infinitives are part of the thought experiment and so do not need to have a potential sense. My account requires only the final sentence, with *an* + optative to be taken in a potential sense.

79. Schofield (1996) p. 13. Arguably there are differences between an intelligent, a teleological and a providential ordering of the *kosmos*, though these need not concern us here.

80. Aristotle Fr. 16 is explicit on the idea that if there is a better, then there is a best. If Anaxagoras believes the *kosmos* to be a better arrangement (than the original state, than other possible states) he may well believe that there is also a best arrangement for the *kosmos*.

81. McCabe (1982).

82. Sisko (2003) pp. 109-10.

83. Aristotle *Physics* VIII/1, 250b18ff.

7. Leucippus and Democritus

1. Cicero *Prior Academics* II,37,118: 'Leucippus supposed there to be the full and the void, Democritus believed something similar, though on other respects he was more productive.'

2. Bailey (1928) has made the greatest effort to separate the views of Leucippus and Democritus, but virtually all modern commentators believe he is overly optimistic in this and treat the views of Leucippus and Democritus as one.

3. Beginning with Aristotle *Generation and Corruption* I/8, 325aff.

4. Cf. Barnes (1982) pp. 553ff.

5. See KRS pp. 411, 415 n. 2.

6. Aristotle *Metaphysics* I/4, 985b8-10.

7. Aristotle *Metaphysics* IV/5, 1009b10-11. As KRS comment, it is then attractive to see Democritus' treatment of secondary qualities (Fr. 9, 'By convention sweet, by convention bitter, by convention hot, by convention cold, by convention cold, but in reality atoms and the void') in relation to this view of perception. See Barnes (1982) pp. 555ff. on the epistemological possibilities here.

8. Simplicius *Physics* 28,8. Cf. Aristotle *Physics* III/4, 203b25.

9. Parmenides Fr. 8.

10. 'On this Democritus and Epicurus differed, that Epicurus believed that all the atoms are very small, and Democritus believed that some atoms are very large.' Dionysius at Eusebius *Evangelical Preparation* XIV,23,3. See Epicurus *Letter to Herodotus* 56,1-4 for Epicurus' view.

11. See Aristotle *Physics* III/4, 203b27.

12. Reading *epekrusin* here with the MSS tradition (the alternative being *epekkrisin*, see KRS). I disagree with KRS's 'attraction' here on two counts – there are no attractive forces for L&D, only contact actions, and the vortex takes in whatever it touches (so there is no need to suppose attraction).

13. Diogenes Laertius IX,31.

14. KRS use 'abcission'.

15. There is no need to read this as a generation of atoms, and such a reading would be flatly contrary to what the atomists say elsewhere about the nature of atoms.

16. Epicurus *Letter to Herodotus* 73,10

17. Democritus Fr. 167 'A vortex was separated off from the universe (*dinon apo tou pantos apokrithênai*)', Anaxagoras Fr. 9 'separated off (*apokrinoumenon*) by force and speed'.

18. KRS p. 417.

19. Epicurus *Letter to Pythocles* 89,4-5.

20. Metrodorus of Chios, KRS translation.

21. Epicurus will insist that *kosmoi* grow from seeds.

22. Aristotle *Physics* II/4, 196a24ff.

23. Simplicius *Physics* 327,24. See also Simplicius *Physics* 327, 330,14, Themistius *Physics* 49,13, Eusebius *Evangelical Preparation* XIV,22,3, Cicero *On the Nature of the Gods* I,24,66.

24. Leucippus Fr. 2. On the nature of necessity, 'Democritus means by it the resistance and motion and blows of matter.' Aetius I,26,2.

25. So Taylor (1999) pp. 191-2, Guthrie (1965) p. 419, see also Furley (1987) pp. 148ff., Bailey (1928) pp. 139-43.

26. KRS (1983) pp. 419-20.

27. Aristotle *Physics* II/4, 196a20ff.

28. Aristotle *Physics* II/4, 196a25ff.

29. Diogenes Laertius IX,45.

30. Eusebius *Evangelical Preparation* XIV, 2,3. See also Furley (1989) pp. 77ff.

31. Eusebius *Evangelical Preparation* XIV,22,3.

32. Democritus Fr. 164, Sextus Empiricus *Against the Mathematicians* VII,116-18.

33. Theories of gravitation based on vortices (e.g. Descartes) were based on vortex action (the observed effect being that a whirlpool will drag objects into its centre), and such theories survived into the eighteenth century, until a determination of the shape of the earth confirmed Newtonian ideas (Newton's theory of gravity predicted an earth slightly bulged at the equator and flattened at the poles, the Cartesian theory vice versa).

34. Homer *Odyssey* 17.218.

35. Cf. Berryman (2003).

36. Hippolytus *Refutation of All Heresies* I,13,2. Cf. Aristotle *Physics* II/4, 196a24ff.

37. Simplicius *On the Heavens* 10,280

38. Diogenes Laertius IX,33.

39. If the values are randomly allocated, then we might expect duplicates we would not get with the one universe to each permutation supposition.

40. Lucretius *On the Nature of Things* V,186-94. Marcus Aurelius *Meditations* VI,10 states the issue as 'The universe is either confusion, an interweaving and scattering of atoms, or unity, order and providence.'

41. Koyré (1957), Furley (1987).

42. Aetius II,7,2.

43. This, in my view, is what generates the multiple *kosmoi* view. I see very little in common with this view and any supposed Indian precedents of multiple *kosmoi* (see West (1971) pp. 94-5).

8. Some Other Presocratics

1. Xenophanes, Frr. 11, 14, 15, 16.

2. Xenophanes Fr. 23.

3. Xenophanes Fr. 26. Cf. Fr. 24, god is 'All seeing, all thinking, all hearing.'

4. See e.g. Plato *Republic* 378e-383e.

5. Xenophanes Fr. 29.

6. Xenophanes Fr. 33.

7. Hippolytus *Refutation of All Heresies* I,14,3, cf. Aetius II,13,4 that the stars are generated from clouds, and are extinguished at setting and kindle at rising, and cf. Aetius II,25,4 that the moon is compressed cloud and II,29,5 that the moon is extinguished once a month.

8. Aetius II,24,9.

9. Hippolytus *Refutation of All Heresies* I,4,5.

10. Achilles Tatius *Isagoge* 4,34,11, cf. Aetius II,11,2.

11. Aetius II,4,11.

12. Aristotle *Metaphysics* I/8, 989b30.

13. Aristotle *Metaphysics* I/8, 989b34ff., cf. XIV/3, 1091a13ff.

14. Aristotle *Metaphysics* I/5, 986a18ff.

15. Aristotle *Metaphysics* XIV/3, 1091a13ff.

16. Aristotle *Metaphysics* XIII/6, 1080b16-22.

17. Stobaeus I,8,1, cf. Aristotle *Physics* IV/6, 213b24-9: 'The Pythagoreans asserted the existence of the void, and held that it enters the heavens from the unlimited breath, so breathing the void, which demarcates natural things, forming a void space or distinction between things which are in succession. It is this void which distinguishes their nature', Aetius II,9,1: 'The Pythagoreans say that outside of the *kosmos* there is a void, into and out of which the *kosmos* breathes', and Simplicius *Physics* 651,26.

18. Meno, DK44A27.

19. Philolaus Fr. 7.

20. Philolaus Fr. 1.

21. Philolaus Fr. 4.

22. Simplicius *Physics* 732,26.

23. Sorabji (1983) p. 183.

24. Simplicius *Physics* 27,23.

25. Hippolytus *Refutation of all Heresies* I,9,1.

26. See Anaxagoras Fr. 12.
27. See Anaxagoras Fr. 12.
28. See Theophrastus *History of Plants* III,1,4, cf. Diogenes Laertius II,17.
29. Guthrie II p. 363.
30. Simplicius *Physics* 25,1.
31. Diogenes of Apollonia Fr. 3.
32. Diogenes Laertius IX, 57.
33. Pseudo-Plutarch *Stromateis* 12.
34. Simplicius *Physics* 1121,15.
35. Guthrie II p. 371.
36. KRS p. 445 give 'Where the dense ran together centripetally to form the earth' while Guthrie II p. 370 gives 'Where the dense comes together it has formed a tight mass'. With KRS I read *tên gên* in brackets.
37. Simplicius *Physics* 151,24.
38. It is noticeable that while Aristotle attacks others for their use of chance in zoogony, he does not attack Diogenes on these grounds.
39. Aristotle *Metaphysics* I/3, 984a6.
40. Alexander *Meteorology* 67,11.
41. On the Derveni author and Anaxagoras, see Burkert (1970) and Betegh (2004) ch. 7, on Diogenes see Betegh (2004) ch. 8.
42. Derveni papyrus Col. 16 lines 6-10, Col. 21, lines 13ff., cf. Col. 17 on air always existing.
43. Derveni papyrus Col. 9, cf. Col. 17.
44. Derveni papyrus Col. 9 lines 7-10.
45. Derveni papyrus Col. 15 lines 3-4, cf. Col. 13, possibly Col.14 lines 1-2 as well.
46. Derveni papyrus Col. 25.
47. Derveni papyrus Col. 25, lines 1-3.
48. Derveni papyrus Col. 25 lines 7-9
49. Cf. Burkert (1997) pp. 167ff.
50. Derveni papyrus Col. 4, lines 3-4, accepting Betegh's text and following his translation.
51. Derveni papyrus Col. 4, lines 5-6.
52. Heraclitus Fr. 89. Col. 4 goes on to give us an independent source, Heraclitus Fr. 94.
53. The sun is 'separated and encircled' (Col. 15 lines 3-4).
54. Cf. Burkert (1997) pp. 173ff.
55. See Betegh (1997) pp. 308ff.
56. Derveni papyrus Col. 17 lines 1-3, air always has and always will exist.
57. See esp. Derveni papyrus Col. 19.
58. Betegh (2004) pp. 271-2.
59. There is another option of an infinite universe with the gravitating masses precisely equally spaced, so there is no preferred point of collapse, leading to a 'knife-edge' stability. Cf. here the Leibniz-Clarke correspondence on whether god should have any active, ongoing role in the running of the universe.
60. Betegh (2004) pp. 271-2.
61. Burkert (1997) pp. 172-3 has some interesting comments on the language of this separation being similar to the language of political coercion and control of the populace.
62. There was nothing in the nature of the equations which opposed this term being set at zero or being given a value that accentuated the gravitational problem.

The subsequent discovery that the universe is expanding removed the need to hypothesise a cosmological constant. Einstein called the cosmological constant his greatest blunder. The cosmological constant has however been revived. There are good reasons to believe that the universe has a certain density, but recent determinations of the Hubble constant suggest that the universe is expanding a little too quickly for this model. So the density assumption is preserved, with the extra expansion thought to be due to a cosmological constant.

63. Betegh (2004) p. 272.

64. See KRS pp. 177-8.

9. Plato

1. I use the term 'particle' here slightly loosely, in the same way that we might refer to quarks as 'ultimate particles', being the fundamental entities without necessarily having the qualities one would usually associate with a particle.

2. Plato *Timaeus* 30a.

3. Cf. Sextus Empiricus *Against the Mathematicians* VII,116-18.

4. Plato *Timaeus* 63c.

5. Plato *Timaeus* 63e, cf. *Timaeus* 81a.

6. Plato *Timaeus* 52d.

7. A *plokanon* is anything woven, a *liknon* is specifically a winnowing-basket. See here Cornford (1937) p. 201 for discussion and a (marvellous!) illustration.

8. Cf. Aristotle and the idea that the four elements would settle out into four concentric spheres if it were not for the action of the sun.

9. Attraction, *holkê*, *Timaeus* 80c3.

10. Descartes will seek to explain magnetism and gravity purely in terms of contact action as well, again denying any action at a distance.

11. Plato *Laws* 889b. Vlastos (1975) p. 29.

12. Homer *Odyssey* 17.218

13. Plato *Timaeus* 53a.

14. Plato *Timaeus* 69c.

15. One might add an extra level here based on the theory of geometrical atomism – the ultimate particles have to form up into either complex triangles with six members (water, air, fire) or squares with four members (earth), before they form up into the three-dimensional shapes of the elements proper.

16. Plato *Politicus* 273e, cf. *Philebus* 64e.

17. So Cornford (1937) p. 200.

18. Plato *Timaeus* 53a.

19. Especially if there is no limit to the amount of duration prior to the formation of the *kosmos*.

20. Cf. Harte (2002) pp. 212ff. on how different parts of the cosmogony interplay with each other.

21. One might express the intrinsic order of the elements as the extrinsic order of the two basic triangles. As noted earlier there may also be an intermediate stage of individual triangles forming up into complexes of triangles.

22. Origen *On First Principles* II,1,4.

23. Origen *On First Principles* II,1,4. Cf. Eusebius *Evangelical Preparation* VII,22,6.

24. Plato *Timaeus* 35a1-6.

25. Plato *Timaeus* 35a6-8.

26. Plato *Timaeus* 35b4-35c2.

27. Plato *Timaeus* 35c. Harmonic mean = (2ab/a+b), arithmetic mean = (a+b/2). These are the modern definitions of harmonic and arithmetic means for a pair of numbers. Plato has a less technical expression of this.

28. Not, as Bury (1929) in the Loeb *Timaeus* suggests, into a 'great cross', with the arms at right angles.

29. There is also a reading of the *Timaeus* according to which time and the *kosmos* come into being together, there having been nothing before that (see Sorabji 1983, p. 264). This was supported by Philoponus, and one can see how such a view would fit well with the later Christian view of creation *ex nihilo*.

30. It is possible that Speusippus, Plato's immediate successor, also took the metaphorical view. Plutarch specifically names Xenocrates and Crantor as early Platonists taking this view in *The Generation of Animals* 1013a; a scholium also mentions Speusippus, though how much weight we can place on this scholium is open to question.

31. In the modern debate Taylor (1928), Cornford (1937), Taran (1971) and Cherniss (1975) among others have defended the metaphorical position, while Vlastos (1939, 1965), Hackforth (1959), Sorabji (1983) and Vallejo (1997) among others have defended the literal position, see also Baltes (1996).

32. Taylor (1928) p. 69.

33. Plato *Phaedrus* 245c, *Laws* 896b.

34. Plato *Phaedrus* 245c.

35. Hackforth (1959) p. 20, Taylor (1928) p. 69.

36. Plato *Timaeus* 28aff.

37. Taran (1971) p. 391.

38. See e.g. Hackforth (1959) p. 22, Skemp (1942) p. 111.

39. See below on some ideas in relation to astronomy and cosmology.

40. Plato *Laws* 892c, cf. 894d and 896b.

41. Aristotle at *Metaphysics* XII, 1072a1ff. seems to sense some inconsistency in what Plato says about the sources of motion.

42. See Cornford (1937).

43. '*gignomena kai gennêta*' *Timaeus* 27c1-2. On this see Hackforth (1959) pp. 18-19.

44. Taylor (1928) pp. 67-8.

45. One might say the same of Xenocrates, but it is easier to find motivation as to why Xenocrates may have adopted a metaphorical reading.

46. Aristotle is consistent in his attribution of a cosmogony to Plato, and groups Plato with Leucippus in believing that there was motion without reason at some point (*Metaphysics* XII/6, 1071b33ff.).

47. Campbell p. 158.

48. Plato *Timaeus* 44c.

49. See Gregory (2000).

50. Cf. Berryman (2003).

51. Plato *Timaeus* 53d4-54a6.

52. See R. Lloyd (2006).

53. The main isotope of hydrogen has a nucleus of one proton, and has one orbiting electron. Deuterium has a nucleus of one proton and one neutron, tritium has a nucleus of one proton and two neutrons. When deuterium and tritium undergo nuclear fusion, they form helium (nucleus of two protons, two neutrons, with two orbiting electrons), emitting one neutron and a great deal of energy.

54. According to Diogenes Laertius IX,40, Aristoxenus wrote that Plato wished to burn all the books of Democritus. Diogenes believes (correctly) that Plato never mentions Democritus by name in his writings, but I believe he is incorrect in saying that Plato never addresses issues where he needs to contradict Democritus.

55. Simplicius *Physics* 28,8, see above.

56. Plato *Philebus* 17e5, cf. *Philebus* 64e and *Theaetetus* 183b.

57. Plato *Timaeus* 55c.

58. *emmelôs* at *Timaeus* 55c7 is also interesting as a contrast to the *plêmmelôs* (discordant) behaviour prior to the ordering of the *kosmos* in the critical passage on the ordering of the *kosmos* at *Timaeus* 30a.

59. See e.g. Cornford (1937) p. 220ff.

60. Cf. Plutarch's discussions here, at *Moralia* XIII,1, and *The Obsolescence of Oracles*, 422bff.

61. See e.g. *Timaeus* 53a7.

62. See Keyt (1971), and cf. Parry (1979), (1991) and Patterson (1981).

63. Plato *Timaeus* 31a3-4.

64. Brisson and Meyerstein (1995), originally published in French as *Inventer l'univers*.

65. Brisson and Meyerstein (1995) p. 1, their italics.

66. Brisson and Meyerstein (1995) p. 4.

67. Brisson and Meyerstein (1995) p. 4.

68. Brisson and Meyerstein (1995) p. 1.

69. As they put it, the 'perpetual renewing of the question'. Brisson and Meyerstein (1995) p. 1.

70. Brisson and Meyerstein (1995) p. 1.

71. See Gregory (1999), Gregory (2000) chs 2, 3, 5 and 6, Gregory (2003).

72. See Gregory (2003). This is so for Eudoxus as well. Simplicius tells us that 'These [unrolling spheres] of Eudoxus' school do not save the phenomena, not only those that were found later, but also those known before and recognised by them' (Simplicius *On the Heavens* 504.17 ff.). The three phenomena Simplicius cites are that (1) Venus and Mars appear at times much brighter than at others, (2) there is variation in the apparent size of the moon, (3) there are variations in the type of solar eclipses relating to the apparent size of the moon.

73. Brisson and Meyerstein (1995) p. 6.

74. Brisson and Meyerstein (1995) pp. 9-10.

75. Brisson and Meyerstein (1995) p. 136.

76. Plato, *Timaeus* 41a, cf. *Timaeus* 32c, 38b,

77. See Gregory (2000) chs 4 and 5.

78. *agêrôn kai anoson, Timaeus* 33a7 cf. *Politicus* 273e4.

79. Plato *Republic* 530ab. I agree with Vlastos (1981) p. 22 n. 17 against Cornford (1941) and Duhem (1906) that *autôn* must mean 'about them' rather than 'in them' which would require the dative *autois*, and against Shorey (1935, vol. 2) p. 186 n. a that the final clause is not a 'trailing anacoluthon'.

80. *paralattein* does occur at *Timaeus* 22d in Critias' account, which in turn is an older myth that has come down to him (*Timaeus* 21b). As Lloyd (1991) p. 342 has pointed out, there are some oddities about the nature and status of Critias' account; perhaps all we need here though is to note that Critias is entirely willing to defer to Timaeus in matters of astronomy, i.e. *Timaeus* 27a 'Timaeus is the best astronomer among us, and has made a special study of the nature of the universe'.

81. Cf. *Laws* 821a, 897b, 898ab, 967a.

82. See Gregory (2000).

83. Plato *Timaeus* 47aff.

84. Plato *Politicus* 273bff.

85. See e.g. *Timaeus* 34a on motion and cf. *Politicus* 269d-270b, and *Timaeus* 32c, 38b, and 41a on indissolubility.

86. See Plato *Politicus* 273bff.

87. Cf. *Politicus* 270a9 and *Timaeus* 33d9 and 34a. It is worth noting that prior to the *Timaeus* Plato always has the *kosmos* supported; a pivot in the *Politicus*, the spindle of necessity in the myth of Er (*Republic* 616bff.) about which the *kosmos* rotates. While the fixed stars may be an integral part of the spindle, the hemispheres of the sun, moon and planets which sit in the bowl of the great spindle cannot be, as they are said to rotate slowly in the opposite direction (617a6-8). The spindle anyway is said to 'turn upon the knees of necessity' (617b4), so each moving part of this model has a pivot.

88. See Gregory (2000) Introduction.

89. I would also argue that Plato has strong epistemological reasons for advocating teleology as well. See Gregory (2000).

10. Aristotle

1. Aristotle *On the Heavens* II/1, 283b26-9.

2. Aristotle *Physics* I/9, 192a29ff.

3. Aristotle *On the Heavens* III/2, 301b34ff.

4. Aristotle *Physics* VIII/1, 250b11-12.

5. Aristotle *Physics* VIII/1, 250b16.

6. Cf. Aristotle *On the Heavens* III/2 300b8.

7. Whether there are two periods of stasis in Empedocles is open to debate, but two periods is Aristotle's view here.

8. Cf. Aristotle *Physics* VIII/7, 260a6ff., *Generation and Corruption* II/10, 336a14ff.

9. Aristotle, *On the Heavens* III/2, 300b9-17, cf. I/7, 275b30ff., *Metaphysics* XII/6, 1071b37ff.

10. Aristotle *On the Heavens* III/2, 300b7ff., cf. *Metaphysics* XII/6, 1072a1ff.

11. Aristotle *Metaphysics* XII/6, 1071b7. Cf. *Physics* VIII/1, 251b10-13.

12. Aristotle *Physics* VIII/1, 251b16.

13. Aristotle *Physics* VIII/1, 251b21.

14. Aristotle believes that only Plato before him countenanced a beginning for time, see *Physics* VIII/1, 251b17.

15. Aristotle *Physics* VIII/1, 252a13.

16. Aristotle also makes a 'why now ?' point at *On the Heavens* I/12, 283a11: 'Why is this point in time preferred for something that has always existed to perish, or for something that has not existed for an unlimited time to be generated? If there is no preference, and the points are unlimited, it is clear that for an unlimited amount of time something was generable or perishable.'

17. One might also note that if c b a then c' b' a', and that Aristotle believes time and space to be continua.

18. Aristotle *Physics* IV/8, 216a8-11, cf. *On the Heavens* I/6 274a7.

19. Aristotle, *Physics* VIII/10, 266a26-8, cf. *Physics* VII/5, 250a28ff. and *On the Heavens* I/7, 275a32ff.

20. Aristotle *On the Heavens* I/7, 275a1-10. Aristotle is clear that all sorts of change fall under this rubric: 'Is this the same with qualitative change and growth? Yes, for there is an increaser and a thing increased, in a certain amount of time and to a certain extent the increaser affects the increased. So too there is an alterer and an altered, and a certain amount of change, according to the greater and the lesser, taking place in a certain amount of time.' Aristotle *Physics* VII/5, 250a28-b1.

21. Aristotle *Physics* VIII/1, 252a20.

22. Aristotle *Physics* VIII/1, 252a30.

23. Aristotle *On the Heavens*, I/10, 280a11-280a24.

24. Aristotle *On the Heavens*, I/8, 277b12ff.

25. Aristotle *On the Heavens* I/8, 276a31ff.

26. 'There is proof that motion is not unlimited in that earth moves more quickly the nearer it is to the centre, and fire more quickly the higher it is. If the movement were unlimited, so would be the speed, and if the speed, then the heaviness or lightness too.' Aristotle *On the Heavens* I/8, 277a27-30, cf. *Physics* 230b23-5, 265b12-16 and *On the Heavens* 277b5-9.The problem here too would be a failure of proportionality, and the implication that unlimited speed would entail unlimited weight, unlimited weight already having been shown to be impossible.

27. Aristotle *On the Heavens* I/9, 279a11.

28. Aristotle *Physics* II/4, 196a23.

29. Aristotle *Parts of Animals* I/1, 640a18. Cf. *Physics* II/7, 198b27ff., *On the Heavens* III/2, 300b30ff.

30. Aristotle *On the Heavens* III/7-8.

31. Aristotle *On the Heavens* I/12, 281b25: 'Everything which always exists is absolutely indestructible.' Aristotle then proceeds to demonstrate that any such thing must be ungenerable as well.

32. There is also the conclusion that: 'It is clear that if something is generable or destructible, it is not eternal.' Aristotle *On the Heavens* I/12, 282a23.

33. Aristotle *On the Heavens* I/12, 283a32.

34. A logical error say Williams (1965), Mignucci (1990), more a methodological problem, Sorabji (1983), Judson (1983).

35. Hintikka (1973), Waterlow (1982), Bogen and McGuire (1986).

36. Aristotle *On the Heavens* I/12, 283b20.

37. Aristotle Fr. 18 (Philo *On the Indestructibility of the Kosmos* 3,10-11).

38. Aristotle Fr. 19b (Philo *On the Indestructibility of the Kosmos* 6,28-7,34).

39. Aristotle Fr. 19c (Philo *On the Indestructibility of the Kosmos* 8,39-43).

40. Aristotle Fr. 19c (Philo *On the Indestructibility of the Kosmos* 8,39-43).

41. Later in antiquity this argument will be deployed against the Stoics and their view of an unending cycle of *kosmoi*.

42. Effe (1970) pp. 23-31. The passages are Aetius I,7,8-9, which may contain a fragment of Aristotle *On Philosophy*, Lucretius *On the Nature of Things* V,156ff., Cicero *On the Nature of the Gods* I,21-2.

11. Epicurus and His Followers

1. Lucretius *On the Nature of Things* I,159ff.

2. Epicurus *Letter to Herodotus* 38,10.

3. Epicurus *Letter to Herodotus* 41,6-42,5.

4. Epicurus *Letter to Herodotus* 39,2-6, Lucretius *On the Nature of Things* I,540.

5. Epicurus *Letter to Herodotus* 45,3-9. Cf. Lucretius *On the Nature of Things* I,232-3.

6. Cf. Epicurus *Letter to Herodotus* 42,1-2.

7. Lucretius *On the Nature of Things* II,1077ff.

8. Lucretius *On the Nature of Things* II,1086ff.

9. Epicurus *Letter to Pythocles* 88,4-9.

10. Cf. Epicurus *Letter to Herodotus* 74,1-2 'We must believe that the *kosmoi* do not of necessity have one form.'

11. Epicurus *Letter to Herodotus* 42,5-12.

12. Epicurus *Letter to Herodotus* 56,1-4.

13. Epicurus *Letter to Pythocles* 89,1-6.

14. Epicurus *Letter to Pythocles* 89,6-11.

15. See Lucretius *On the Nature of Things* I,170ff.

16. Lucretius *On the Nature of Things* II,1049ff. effects a transition from cosmic seeds to plant/ animal seeds without any recognition of any serious distinction between them. Cf. Furley (1989) p. 229ff.

17. Lucretius *On the Nature of Things* II,1105ff.

18. Lucretius *On the Nature of Things* V,812ff.

19. Lucretius *On the Nature of Things* V,449.

20. Epicurus *Letter to Pythocles*, 90,6-91,1.

21. Lucretius *On the Nature of Things* V,437.

22. Epicurus *Letter to Pythocles* 90,1-5.

23. Lucretius *On the Nature of Things* V,416.

24. See Epicurus *Letter to Menoeceus* 122, *Key Doctrines* 11-13, *Vatican Sayings* 45.

25. Lucretius *On the Nature of Things* II,1095, cf. Cicero *On the Nature of the Gods* I,20.

26. Lucretius *On the Nature of Things* II,156, cf. Cicero *On the Nature of the Gods* I,9.

27. Lucretius *On the Nature of Things* II,174.

28. Lucretius *On the Nature of Things* II,181, cf. Cicero *On the Nature of the Gods* I,7.

29. Lucretius *On the Nature of Things* V,186-94. Marcus Aurelius *Meditations* VI,10 states the issue as 'The universe is either confusion, an interweaving and scattering of atoms, or unity, order and providence.'

30. Lucretius *On the Nature of Things* V,200.

31. Lucretius *On the Nature of Things* II,180, V,195

32. Lucretius *On the Nature of Things* II,222.

33. Cicero *On the Nature of the Gods* XX.

34. Cicero *On the Nature of the Gods* IX.

35. Cicero *On the Nature of the Gods* XX.

36. See Cicero *On the Nature of the Gods* XIX for the tranquil life of the Epicurean gods.

37. Russell (1986) pp. 203-4.

38. Bierce (1911).

39. 'All Things Dull and Ugly', verse 1, from Monty Python's Contractual Obligations Album. This should be sung to the tune of 'All Things Bright and Beautiful'.

40. Lucretius *On the Nature of Things* IV,823ff.

41. Lucretius *On the Nature of Things* IV,834.

42. Epicurus *Letter to Herodotus* 74,2-6.

43. Sedley (1998) pp. 121-2, Campbell (2003) p. 1.

44. Lucretius *On the Nature of Things* V,784.

45. Lucretius *On the Nature of Things* V,821.

46. Lucretius *On the Nature of Things* V,837ff., cf. Simplicius *Physics* 371,33ff.

47. See Cicero *On the Nature of the Gods* II,120-53.

48. See e.g. Cicero *On Fate* 21,5.
49. Lucretius *On the Nature of Things* II,216-25.
50. Long and Sedley (1987) vol. I p. 52.
51. O'Keefe (1996) p. 308.
52. Lucretius *On the Nature of Things* V,188-9.
53. Epicurus *Letter to Herodotus* 43,1-44,7.
54. Agreeing here with O'Keefe (1996) p. 309.
55. There is no wave/ particle duality, no Heisenberg uncertainty principle, no quantum tunnelling and none of the epistemological issues raised by Schrödinger's cat.
56. A further similarity with QM here is that there is nothing arbitrary about the theory of the swerve and the swerve does not affect large scale physical entities. QM may suppose probabilistic action at a very small scale, but does so very precisely and its effects are strictly limited in terms of scale.
57. Epicurus *Letter to Herodotus* 39,2-6.
58. Epicurus Fr. 55.
59. Lucretius *On the Nature of Things* I,1071.
60. If a universe is static, and contains gravitating matter, then it will contract.
61. Epicurus *Letter to Herodotus* 41,6-42,5.
62. Epicurus *Letter to Herodotus* 73,7-74,1.
63. Lucretius *On the Nature of Things* II,1112-74, V,449-94.
64. Lucretius *On the Nature of Things* III,972.
65. Lucretius *On the Nature of Things* V,235.
66. Lucretius *On the Nature of Things* V,324.
67. Lucretius *On the Nature of Things* V,351.
68. Lucretius *On the Nature of Things* V,380.
69. Lucretius *On the Nature of Things* II,1143ff., cf. V, 812ff.
70. Bailey (1928) p. 359.

12. The Stoics

1. Philo *On the Indestructibility of the Kosmos* II.
2. See *SVF* II,605.
3. Chrysippus *Physics* (*SVF* II,579).
4. Seneca *Natural Questions* 3,13,1. Cf. *SVF* I,102, II,580, II,581, II,590, II,605.
5. Diogenes Laertius VII,135 (*SVF* II,589). Cf. *SVF* II,582. On the thinner part of air igniting see *SVF* I,102, I,104, I,105, II,579, II,581.
6. See here Hahm (1977) p. 83 n. 4.
7. Alexander of Aphrodisias (*SVF* II,1047).
8. Sextus Empiricus *Against the Professors* IX,332 (*SVF* II,524).
9. Diogenes Laertius VII,137 (*SVF* II,526).
10. 'Treading the path' is a literal translation – this could also be rendered as 'proceeding in an orderly fashion'.
11. Aetius I,7,33 (*SVF* II,1027).
12. Diogenes Laertius VII,135 (*SVF* I,102).
13. Aetius I,28,4 (*SVF* II,917).
14. Gellius 7,2,3 (*SVF* II,1000).
15. Hahm (1977) pp. 57ff.
16. Hahm (1977) pp. 68ff.
17. Philo, *On the Indestructibility of the Kosmos* 90 (*SVF* I,511).
18. Chrysippus *SVF* II,604, cf. *SVF* II,605.

19. 'The Stoics say that only bodies exist, as what is acts and is acted upon.' *SVF* II,525. Cf. *SVF* I,90 and II,140.

20. Cicero *On the Nature of the Gods* II,37.

21. Cicero *On the Nature of the Gods* II,37.

22. Cicero *On the Nature of the Gods* II,47.

23. Cicero *On the Nature of the Gods* II,48ff.

24. Aetius *SVF* II, 597.

25. *SVF* II,618, II,619.

26. This would certainly be suggested by the use of *sunizein*, 'to sit down together', to describe the transformation of fire into air, then to water. Cf. Philo, *On the Indestructibility of the Kosmos* XXI and Seneca *Natural Questions* 3,13.

27. Nemesius 309,5 (*SVF* II,625).

28. Simplicius *Physics* 886,12 (*SVF* II,627).

29. Origen *Against Celsius* 5,20 (*SVF* II,626).

30. Alexander *Commentary on Aristotle's Prior Analytics* 180,33 (*SVF* II,624).

31. Simplicius *Categories* 350,15 (*SVF* II,510).

32. Aristotle *Physics* II/11, 219b1.

33. Eudemus quoted in Simplicius *Physics* 732,26.

34. Lucretius *On the Nature of Things* III,847.

35. Mansfeld (1979) p. 137.

36. Chalcidius *SVF* II,933, tells us that Chrysippus held that everything in accordance with fate is providential, and everything providential is fated, while Cleanthes held that everything providential was fated, but recognised that some things come about by fate which are not providential. A related issue here is how the Stoics deal with apparently non-beneficial aspects of the *kosmos*. One strategy is to argue that these are in fact beneficial (so bed bugs are there to wake us, mice to make sure we are tidy, Plutarch *On Stoic Self-contradiction* 1044D (*SVF* II,1163), giving Chrysippus' view), another (which can run in tandem with this) is that there cannot be goods without evils, as the two are polar opposites (Gellius *SVF* II,1169, on Chrysippus).

37. Philo *On the Indestructibility of the Kosmos* III.

38. Seneca *On Anger* II,27.

39. Mansfeld (1979) p. 159ff.

40. Chrysippus, *SVF* II,604. Cf. Plutarch, *SVF* II,605: 'When the *kosmos* is entirely fiery, it is its own soul and controlling principle. When it has changed to water and the soul that goes with that, in some way it has altered into body and soul such that it is constituted from both, and has another structure.'

41. Plato *Politicus* 273bff.

42. Plato *Republic* 528ff., cf. *Republic* passim on the changelessness of god.

43. Aristotle *Meteorology* I/3, 340a1.

44. See e.g. Furley (2000) pp. 438ff., White (2003) pp. 133ff.

45. I would agree with Lapidge (1978) pp. 180-3 and Sandbach (1975) pp. 78-9 that we can list a variety of possibilities for a Stoic justification of eternal recurrence, and none of those factors on its own is convincing. It is the conjunction of an inevitably degenerating *kosmos* and a providential god that is key here.

46. Aristotle does, in places, recognise that a fire will burn until all its fuel is used up.

47. Alexander of Lycopolis *Against the Opinions of the Manicheans* 19,4.

48. Plutarch *On Common Conceptions* 1075d (*SVF* I,510).

49. Stobaeus I,213 (*SVF* I,120).
50. Cleanthes in Cicero *On the Nature of the Gods* II,15.
51. Cicero *On the Nature of the Gods* III,14.
52. 'Zeno appears to have proposed no solution to this glaring contradiction.' Lapidge (1973) p. 273.
53. See Mansfeld (1979) pp. 152ff., Lapidge (1973) pp. 370ff., Gilbert (1907) pp. 243ff., Boyancé (1936) pp. 67ff.
54. Mansfeld (1979) p. 156.
55. The relevant timescale here is around four billion years.
56. Nemesius 309,5 (*SVF* II,625).
57. Plato *Timaeus* 39d: 'Nevertheless, it is perfectly possible to grasp the amount of time elapsed at the completion of the perfect year, when the relative speeds of all the eight circuits have brought them to a head simultaneously, as measured by the motion of circle of the same and similar. In this manner and for these reasons all of those stars which have turnings on their passage through the heavens were created, in order that this world might be as like as possible to the perfect and intelligent living being in respect of it imitating its eternal nature.'
58. See Barnes (1978) pp. 7ff. and (1979) pp. 501ff.
59. Chrysippus *SVF* II,283.
60. Wallis (1981) pp. 225-6.
61. Nemesius *SVF* II,625, cf. Epictetus I,14.
62. Sorabji (1983) pp. 185-6.
63. Sorabji (1983) p. 185 n. 45.
64. Philo *On the Indestructibility of the Kosmos* XV.
65. Panaetius is said to have rejected *ekpurôsis* as he preferred the Aristotelian view of an eternal *kosmos*, see Frr. 64-9. In Fr. 74 Cicero also tells us that Panaetius was the only Stoic to reject the predictions of the astrologers.
66. Philo *On the Indestructibility of the Kosmos* XVI.
67. Philo *On the Indestructibility of the Kosmos* XVI.
68. Philo *On the Indestructibility of the Kosmos* XVI.
69. Philo *On the Indestructibility of the Kosmos* XVII.
70. Philo *On the Indestructibility of the Kosmos* XVII.
71. Philo *On the Indestructibility of the Kosmos* XX.
72. Philo *On the Indestructibility of the Kosmos* XXI.
73. Philo *On the Indestructibility of the Kosmos* XXIII. In all likelihood these arguments have been doctored to some extent by transmission by Philo (see Kidd (1996), Sedley (1998) pp. 166ff.). Whose arguments they are has also been a matter of debate – the Stoic Zeno of Citium according to Zeller (1876) but this is unlikely, see Sedley (1998) pp. 167ff.
74. Philo *On the Indestructibility of the Kosmos* XXIII.
75. Philo *On the Indestructibility of the Kosmos* XXIII.
76. Philo *On the Indestructibility of the Kosmos* XXIV.
77. Philo *On the Indestructibility of the Kosmos* XXIV.
78. Philo *On the Indestructibility of the Kosmos* XXV.
79. Philo *On the Indestructibility of the Kosmos* XXVI.
80. Philo *On the Indestructibility of the Kosmos* XXVII.
81. Philo *On the Indestructibility of the Kosmos* XXVII. Lucretius has a reply also, see *On the Nature of Things* V,324ff.
82. Sedley (1998) p. 176.
83. Lucretius *On the Nature of Things* V,235-50.

13. Early Christianity and Creation

1. Sorabji (1983) p. 246 n. 65.

2. See Chapter 1.

3. Epicharmus Fr. 1, Epicurus as reported in Diogenes Laertius X,2, and Sextus Empiricus *Against the Mathematicians* X,18-19 (= *Against the Physicists* II,18-19).

4. Aristotle does not mention Hesiod in *On the Heavens* I/10 where he discusses his predecessors' views on the origins of the *kosmos*, which he surely would have done if he was aware that Hesiod had taught creation *ex nihilo*. Aristotle does mention Hesiod at *Metaphysics* I/4 in the context of previous cosmogonies, but sees nothing untoward in respect of creation *ex nihilo* in what Hesiod says.

5. Unknown Aristotelian author, *Melissus, Xenophanes and Gorgias* 4.

6. See Chapter 10.

7. Apocalypse of Baruch, 21,4.

8. Apocalypse of Baruch, 48,8.

9. Augustine *On Genesis against the Manicheans* I,6.

10. Winston (1971-2) p. 191, cf. May (1994) pp. xi-xiii, Grant (1980) p. 137.

11. So I disagree with Chrysostom *Homilies on Genesis* 2.5, who says: 'We therefore listen to these words not as the words of Moses but as the words of the God of all things coming to us through the tongue of Moses'; cf. 7.10, 'Blessed Moses, inspired by the divine Spirit, teaches us with great precision.'

12. Genesis 1:1-1:2. Quotations are from the Catholic Edition RSV of the Bible unless otherwise stated.

13. Rogerson (1991) p. 56.

14. Genesis 1:3.

15. John 1:1-1:3.

16. Genesis 1:1-1:2 (the Anchor Bible)

17. Jerome *Homilies* 10.

18. Job 38:4.

19. Job 28:25.

20. Isaiah 45:18.

21. Grant (1952).

22. 2 Maccabees 7:28.

23. See Wolfson (1948) pp. 303-3, Winston (1971-2) pp. 186-7.

24. E.g. Xenophon *Memorabilia* II,2.3. See May (1994) p. 8 and n. 30.

25. Revelation 22:13.

26. Plato *Laws* 715e. The source of this is probably Orphic. 'Zeus the *archê*, Zeus the middle, from Zeus all things are created, Zeus is the basis of the earth and the starry heavens.'

27. Book of Wisdom 11:17.

28. Also of interest is Wisdom's treatment of the parting of the Red Sea (19:6ff.), where 'the whole of creation was formed over again in its original nature' to effect the passage, but there is no mention of creation *ex nihilo*.

29. Jubilees II.2.

30. Various dates from the beginning of the Christian era to the twelfth to fifteenth century have been proposed, but as it would appear that the early Christian fathers had no knowledge of this book, it cannot have been earlier than the end of the first century AD. See Apocrypha pp. 323-5.

31. Genesis 1:6.

32. See Schmidt (1967), and a reply from Steck (1981).

33. The classic account here is Gunkel (1895).

34. Heidel (1942) p. 129. Capitalisation of the Christian deity and non-capitalisation of pagan deities are Heidel's.

35. God vs. chaos is a theme in many ancient mythologies, see Anderson (1984) p. 2.

36. Van Seters (1988) finds influences from Greek theogony, Jacobsen (1981) finds influences from the Sumerian culture in 'The Eridu Genesis'.

37. Oden (1981) finds interesting parallels between the Babylonian Atrahasis and further parts of Genesis, Shea (1984) between Genesis, the Atrahasis and the Eridu Genesis.

38. Gunkel (1895) p. 31.

39. Hebrews 11:3.

40. Frank (1945) p. 59.

41. Tatian *Oration to the Greeks* V, cf. Theophilus *Letter to Autolycus* II,4: 'If god generated the *kosmos* from ungenerated matter, god would be no better than a human craftsman who fashions things out of materials he receives rather than generates.'

42. Tatian *Oration to the Greeks* V.

43. Chalcidius *Commentary on the Timaeus* 276.

44. Basil *Six Days of Creation* II,2.

45. Ambrose *Six Days of Creation* I,7.

46. Augustine *On the Literal Interpretation of Genesis* IV,13.

47. Augustine *On Genesis against the Manicheans* I,10.

48. Sorabji (1983) pp. 290ff.

49. Gregory of Nyssa *Six Days of Creation PG* 44,69c., cf. *The Soul and Resurrection PG* 46,124c, *The Making of Man PG* 44,212ff.

50. Gregory of Nyssa *Six Days of Creation PG* 44,69.

51. Gregory of Nyssa *The Soul and Resurrection* PG 46,124d.

52. Origen *On First Principles* I,3,3 and II,1,4.

53. 1 Corinthians 8:6.

54. 1 Corinthians 11:12.

55. Romans 11:36.

56. See Philo *On the Creation of the Kosmos According to Moses* 8-9 and 21-2. See my next chapter for more on Philo and *ex nihilo* creation.

57. See here Sorabji (1983) pp. 233-4.

58. Augustine *City of God* XI,6.

59. Augustine *City of God* XI,6, cf. *On Genesis Against the Manicheans* I,2. We can find Philo saying: 'There was no time prior to the *kosmos*' (Philo *Construction of the World* 7,26). We can also find Ambrose saying: 'At the beginning of time god generated heaven and earth. Time starts from this world, not before. The day is a division of time, not its beginning.' Ambrose *Six Days of Creation* I,6. For Plotinus, see the next chapter.

60. See Augustine *City of God* XI,5.

61. Basil *Exegetic Homilies* I,5.

62. Ambrose *Six Days of Creation* I,5.

63. Augustine *On Genesis against the Manicheans* I,2.

64. Origen *On First Principles* I,4,5.

65. Origen *On First Principles* I,4,4.

66. Origen *On First Principles* I,4,5.

67. Origen *On First Principles* I,2,2.

68. Origen *On First Principles* I,4,5. Elsewhere in *On First Principles* he takes the Trinity to be timeless.

69. Irenaeus *Against Haereses* II,10,2, cf. 28,3 and 28,7.
70. Augustine *Confessions* XI,12.
71. Sorabji (1983) p. 249.
72. By omnipotent I mean a physically rather than a logically omnipotent god.
73. Simplicius *Physics* 1175,35, cf. 1177,29.
74. Simplicius *On the Heavens* 138,5.
75. Simplicius *On the Heavens* 138,2.
76. Simplicius *On the Heavens* 138,13.
77. See Augustine *City of God* XI,4, XII,15, XII,18 and *Confessions* XI, 10, XI,30, Philoponus *Against Proclus on the Eternity of the World* 568,6ff. See also Sorabji (1983) pp. 240-2.
78. See Philoponus *Against Proclus on the Eternity of the World* 566,4 and 567,23. See also Sorabji (1983) pp. 242-4.
79. Sorabji (1983) pp. 244-5.
80. See Aristotle *Physics* VIII/1.
81. Just to be clear here, I do not see a great deal of originality in the Bible on issues of cosmogony. Later Christians do have original ideas in cosmogony when trying to interpret biblical passages.
82. Philo *Confusion of the Languages* 136.
83. Augustine *Confessions* XI,7.
84. Augustine *Confessions* XI,8.
85. Philo *On the Creation of the Kosmos According to Moses* II.
86. Zacharias *Ammonius or On the Construction of the World* 1072A, cf. 1128C.
87. Cf. Eichrodt (1999), The Anchor Bible: Genesis, p. 3 and pp. 11-13, Day (2002) 147-50.

14. Later Platonism and the Debate with Christianity

1. Aristotle *On the Heavens* I/10, 279b32. Theophrastus considers that this view is possibly true, but it is more likely that Plato supposed there to be a beginning to the *kosmos*, see Philoponus *On the Eternity of the Kosmos Against Proclus* 145,20.
2. See Plutarch *On the Birth of the Spirit* 1031b, who also ascribes this view to another Platonist, Crantor. Speusippus headed the academy from 347 to 339, Xenocrates from 339 to 314.
3. See Plutarch *On the Birth of the Spirit* 1013aff.
4. Taurus in Philoponus *Against Proclus on the Eternity of the Kosmos* 146,8.
5. Taurus in Philoponus *Against Proclus on the Eternity of the Kosmos* 146,13.
6. Taurus in Philoponus *Against Proclus on the Eternity of the Kosmos* 146,20.
7. Taurus in Philoponus *Against Proclus on the Eternity of the Kosmos* 147,5.
8. Dillon (1977) p. 244.
9. See Alcinous, 169,26ff.
10. Aristotle *On the Heavens* III/2, 300b16, *Metaphysics* XII/6, 1071b37.
11. For Eudemus see Simplicius *Physics* 702,24, for Diogenes Laertius *Lives* 3,73-7. Plutarch too disagrees with the notion of time prior to the *kosmos*.
12. Plutarch *Platonic Questions* 1007c, cf. Proclus *Timaeus* I,276,31ff. III,37,7ff. Atticus, see Aeneas Gazaeus *Theophrastus* 46,16. Cicero *On the Nature of the Gods* I,21 has Velleius distinguish two forms of time in Plato, but here he is attacking Plato's views.
13. Philo *On Providence* I,20 and I,22. Cf. Grant (1980) p. 141, Wolfson (1947) p. 300ff., Runia (2001) pp. 152-5, 172.

14. On Proclus, see Philoponus *Against Proclus on the Eternity of the Kosmos* 104,21ff. On Philoponus, see Philoponus *Against Proclus on the Eternity of the Kosmos* 116, Philoponus *Physics* 17ff., Simplicius *Physics* 1158,31.

15. See e.g. Atticus Fr. 3,799aff.

16. Plotinus *Enneads* V,4,1.

17. Plotinus *Enneads* III,8,10. Porphyry also talks of god generating from himself, see Proclus *Timaeus Commentary* 1.300,2.

18. Plotinus *Enneads* V,7,1 and V,7,3.

19. Plotinus *Enneads* III,7,12.

20. See Wilberding (2006) pp. 48-57.

21. Plotinus *Enneads* III,8,4.

22. Proclus *Elements of Theology* Proposition 174, cf. 26, 167, 169.

23. Proclus *Parmenides Commentary* 791,14.

24. See e.g. Sallustius *On the Gods and the Kosmos* VII (body and shadow).

25. See e.g. Taurus in Philoponus *On the Eternity of the Kosmos Against Proclus* 14,24 and 147,8 (sun, moon and light).

26. Ambrose *Six Days of Creation* I,5.

27. Sorabji (1983) pp. 316ff.

28. Porphyry in Proclus *Timaeus Commentary* I,396,24.

29. Porphyry in Philoponus *Against Proclus on the Eternity of the Kosmos* 148,7.

30. Porphyry in Philoponus *Against Proclus on the Eternity of the Kosmos* 148,25.

31. Porphyry in Proclus *Timaeus Commentary* I, 96ff., cf. Plotinus *Enneads* III,2,2.

32. Augustine *City of God* XI,4.

33. Augustine *City of God* XI,5, XII,18

34. Sallustius *On the Gods and the Kosmos* VII.

35. Sallustius *On the Gods and the Kosmos* VII.

36. Sallustius *On the Gods and the Kosmos* IX.

37. Sallustius *On the Gods and the Kosmos* IX.

38. Sallustius *On the Gods and the Kosmos* XIII.

39. Sallustius *On the Gods and the Kosmos* XVII.

40. Sallustius, *On the Gods and the Kosmos* XX.

41. Sallustius, *On the Gods and the Kosmos* XXI.

42. Philo *On the Indestructibility of the Kosmos* IV.

43. Philo *On the Indestructibility of the Kosmos* V.

44. Philo *On the Indestructibility of the Kosmos* V.

45. Winston (1977) pp. 300ff., Sorabji (1983) pp. 203ff.

46. See also Runia (2001) pp. 152-5.

47. Philo *On Providence* I, 6.

48. Philo *On the Creation of the Kosmos According to Moses* II.

49. Philo *On the Indestructibility of the Kosmos* V.

50. Philo *On the Creation of the Kosmos According to Moses* II,7.

51. Wolfson (1947) p. 296.

52. Wolfson (1947) pp. 296-8.

53. Here I am interested only in Philoponus' later work, after his conversion to Christianity.

54. See Verrycken (1997) p. 317.

55. Philoponus cites several passages from Aristotle and discusses the views of other commentators as well. See Verrycken (1997) pp. 282ff., 317.

56. See Judson (1987) p. 179.

57. Presumably assuming that humankind has been in existence for all eternity as well. Philoponus *Against Proclus on the Eternity of the Kosmos* 9,14ff.
58. Philoponus *Against Proclus on the Eternity of the Kosmos* 11,2.
59. Philoponus *Against Proclus on the Eternity of the Kosmos* 11,6.
60. Philoponus in Simplicius *Physics* 1179,15ff.
61. Aristotle *Physics* III/6, 206a33ff., III/8, 208a20.
62. Simplicius *Physics* 506,3ff., 1180,29ff.
63. Philoponus *Against Proclus on the Eternity of the Kosmos* 11,14.
64. Philoponus in Simplicius *On the Heavens* 136,19 = Philoponus *Against Aristotle on the Eternity of the Kosmos* 73, cf. 74-6.
65. Philoponus *Against Aristotle on the Eternity of the Kosmos* Frr. 108-33.
66. See Sorabji (1983) pp. 217ff. and his references.
67. Stars convert hydrogen and helium into heavier elements by a process known as stellar nucleosynthesis. Unless there is a process by which there is a conversion of heavier elements to hydrogen and helium, then hydrogen and helium levels must gradually fall, a significant problem for an infinite past. A major problem for steady state cosmology postulating an infinite past was explaining the present abundance (on a cosmic scale) of hydrogen and helium. Big bang cosmology predicts that the initial stages of the cosmos will produce hydrogen, helium and very little else, and the current proportions and distributions of elements are as predicted by stellar nucleosynthesis working on an initial abundance of hydrogen and helium. There are several other physical processes, either irreversible or with an asymmetry between production and recycling, which also suggest the physical impossibility of an infinite past.
68. The closest analogues to this sort of idea in antiquity would be Empedocles, and perhaps Plato's myth of the reversed *kosmos* in the *Politicus*.
69. Plato *Politicus* 270b.
70. Plato *Politicus* 270d.
71. Time's arrow may be the result of the second law of thermodynamics or be due to causal or psychological factors.
72. There are other factors which come into play here, not least the question of the existence of any cosmological constant and the degree of the curvature of space.
73. Proclus *Timaeus Commentary* I,267,4 and IV,268,24.
74. Iamblichus in Elias *Categories Commentary* 123,1.
75. Dexippus *Categories Commentary* 41,15.
76. Dexippus *Categories Commentary* 41,1; the relevant Aristotle passage is *Metaphysics* XII/7, 1072b14. Cf. Aristotle *Metaphysics* XII/10, 1075a.
77. See here Sorabji (1988) pp. 254-5, (1990) pp. 182-3, Verrycken (1990) pp. 215-26, (1997) p. 271.
78. Aristotle *Physics* II/3, 194b29.
79. Simplicius *Physics* 1361,11ff.
80. Aristotle *On the Heavens* I/9, 279a28.
81. Simplicius *Physics* 1361,19ff.
82. Simplicius *Physics* 1363,8.
83. Simplicius *Physics* 1363,4.
84. Sorabji (1988) p. 253.
85. Syrianus *Metaphysics Commentary* 118,6, see 117,25ff. for argument.
86. Syrianus *Metaphysics Commentary* 117,25ff.
87. Proclus *Timaeus Commentary* I,267,12.
88. Proclus *Timaeus Commentary* I,268,1.
89. Proclus *Timaeus Commentary* I,268,1.

90. Proclus, in Philoponus *Against Proclus on the Eternity of the Kosmos*, 238, 3, 297,21.

91. Philoponus *Against Proclus on the Eternity of the Kosmos* 1,18ff., *Against Aristotle on the Eternity of the Kosmos* in Simplicius *On the Heavens* 142,22.

92. See Judson (1987), Sorabji (1988) pp. 254ff.

93. Alexander *Natural Questions* I.18,31,24ff., esp. 32,3.

94. See Pines *Arabic Summary of a Lost Work of Philoponus* pp. 324-5. As Sorabji (2004) p. 186 and others have pointed out, this creates a difficulty for Philoponus. If god can override the natural capabilities of the *kosmos* at the end of the *kosmos*, cannot he also override its capabilities at the beginning and have a beginningless creation? Philoponus at least needs an argument here as to why this should not be so, cf. Philoponus, *Against Proclus on the Eternity of the Kosmos* 600,14ff.

95. Simplicius *On the Heavens* 361,10.

96. See e.g. Aristotle *Metaphysics* VII/9, VII/15, VIII/5.

97. Sorabji (1983) p. 248.

98. Philoponus *Against Proclus on the Eternity of the Kosmos* 341,13.

99. Philoponus *Against Proclus on the Eternity of the Kosmos* 450,8, cf. 448,16, 449,13.

100. Plutarch *The Obsolescence of Oracles* 421fff.

101. Plutarch *The Obsolescence of Oracles* 422a.

102. Plutarch *The Obsolescence of Oracles* 422b.

103. See Homer *Iliad* 15.187.

104. Plutarch *The Obsolescence of Oracles* 423b.

105. Plutarch *The Obsolescence of Oracles* 423cff. Cf. Metrodorus of Chios on whether a plain produces one ear of corn or many, Lucretius *On the Nature of Things* II,1077ff. on objects coming in classes.

106. Theophilus *Letter to Autolycus* II,6.

107. Tertullian *Treatise against Hermogenes* II,1.

108. Tertullian *Treatise against Hermogenes* II,2.

109. Tertullian *Treatise against Hermogenes* II,4.

110. Tertullian *Treatise against Hermogenes* III,1.

111. Tertullian *Treatise against Hermogenes* IV,11.

112. Tertullian *Treatise against Hermogenes* V,9.

113. Tertullian *Treatise against Hermogenes* XXIII,1 and XIX,1.

114. Tertullian *Treatise against Hermogenes* III,3.

115. Tertullian *Treatise against Hermogenes* XIX,1.

116. Tertullian *Treatise against Hermogenes* XXIIIff.

117. Jerome *Letter to Avitus* V.

118. Theophilus from Jerome *Letter* 96.

119. Photius *Bibliotheca Codex* 109.

120. Clement *Tutor* 1,62.

121. Origen *On First Principles* II,1,1.

122. Origen *On First Principles* II,1,4.

123. Origen *On First Principles* III,5,2, cf. II,9,1.

124. Origen *On First Principles* III,5,3.

125. Sorabji (1983) p. 186.

126. Origen *On First Principles* II,9,1 and IV,4,8.

127. Origen *On First Principles* II,3,4.

128. Origen *On First Principles* II,3,5, God is all and in all is 1 Corinthians 15:28.

129. Origen *On First Principles* II,3,4.

130. Origen also needs to distance himself from the 'insignificant differences' position of the Stoics in order to avoid repeated crucifixions and resurrections of marginally different Christs.

131. Isaiah 66:22.

132. Ecclesiastes 1:9-10.

133. Origen *On First Principles* II,9,1.

134. Origen *On First Principles* II,1,4.

135. Origen *On First Principles* II,1,4. Cf. Eusebius *Evangelical Preparation* VII,22,6.

136. Origen *On First Principles* II,1,4. Aristotle *On Philosophy* Fr. 19c.

137. Augustine *On Genesis against the Manicheans* I,10, cf. I,6.

138. Augustine *Confessions* XII,7.

139. Augustine *City of God* XI,6, *The Literal Interpretation of Genesis* V,5,12.

140. Augustine *City of God* XII,17, *The Literal Interpretation of Genesis* IX,17,32.

141. Augustine *City of God* XI,21, XI,24.

142. Augustine *The Literal Interpretation of Genesis* VII,28.

143. Augustine *Confessions* XIII,29,44.

144. Augustine *The Literal Interpretation of Genesis* VI,10,17ff., VI,14,25ff., *City of God* XXII,24, *On the Trimity* III,9,16.

145. Augustine *City of God* XII,15,17 and 18, Confessions XI,10 and 30, XII,15 and 28.

146. Augustine *City of God* XI,4, XII,15 and 18, XXII,2, *Confessions* XI, 10, 12.

147. Augustine *Confessions* XII,38.

148. Augustine *City of God* XII,18.

149. Augustine *City of God* XII,14.

150. Augustine *City of God* XII,21.

Conclusion

1. Gregory (2000).

2. Philo comes close in *On the Creation of the Kosmos According to Moses* 17-20, but his talk of the founding of a city is conducted more in terms of architecture than of politics.

Bibliography

Adam, J. (1902) *The Republic of Plato* (2 vols). Cambridge: Cambridge University Press.

Adkins, A. (1985) 'Cosmogony and Order in Ancient Greece', in Lovin & Reynolds 1985, pp. 39-66.

Aiton, E.J. (1972) *The Vortex Theory of Planetary Motions.* London: Macdonald.

Alexander, H.G. (1956) *The Leibniz-Clarke Correspondence.* Manchester: Manchester University Press.

Algra, K. (1994) *Concepts of Space in Greek Thought* (*Philosophia Antiqua*, 65). Leiden: E.J. Brill.

Anderson, B.W. (1984) *Creation in the Old Testament.* London: SPCK.

Anton, J.P. (1978) *Science and the Sciences in Plato.* New York: Eidos.

Aravintinou, A. (1993) 'Anaxagoras' Many Worlds', *Diotima* 21, pp. 97-9.

Archer-Hind, R.D. (1888) *The* Timaeus *of Plato.* London: Macmillan.

Asmis, E. (1981) 'What is Anaximander's "Apeiron" ', *Journal of the History of Philosophy* 19, pp. 279-97.

Augustine, *Confessions,* Loeb Classical Library (1912).

——— *City of God,* Loeb Classical Library (1966).

——— *On Genesis: Two Books on Genesis Against the Manichees; On the Literal Interpretation of Genesis, an Unfinished Book*, ed. R.J. Teske (1991). Washington, DC: Catholic University of America Press.

Austin, S. (1986) *Parmenides, Being, Bounds and Logic.* New Haven, CT: Yale University Press.

Bailey, C. (1926) *Epicurus.* Oxford: Oxford University Press.

——— (1928) *The Greek Atomists and Epicurus.* New York: Russell & Russell.

Ballew, L. (1974) 'Straight and Circular in Parmenides and the "Timaeus" ', *Phronesis* 19, pp. 189-209.

Baltes, M. (1976) *Die Weltentstehung des Platonischen Timaios nach den antiken Interpreten* (2 vols). Leiden: E.J. Brill.

——— (1996) '*Gegonen* (Platon, *Tim.* 28B7) Ist die Welt real enstanden oder Nicht?', in K. Algra & D. Runia (eds), *Polyhistor: Studies in the History and Historiography of Ancient Philosophy.* Leiden: E.J. Brill, pp. 76-96.

Bargrave-Weaver, D. (1959) 'The Cosmogony of Anaxagoras', *Phronesis* 4, pp. 77-91.

Barnes, J. (1979, 1982^2) *The Presocratic Philosophers.* London and New York: Routledge & Kegan Paul.

——— (1987) *Early Greek Philosophy.* London: Penguin.

Barrow, J.D. (1990) *Theories of Everything.* London: Vintage.

——— & Tipler, F.J. (1986) *The Anthropic Cosmological Principle.* Oxford: Oxford University Press.

Berryman, S. (2002) 'Democritus on the Explanatory Power of the Void', in Caston 2002, pp. 183-91.

——— (2003) 'Ancient Automata and Mechanical Explanation', *Phronesis* 48, pp. 345-69.

Betegh, G. (2004) *The Derveni Papyrus.* Cambridge: Cambridge University Press.

Bicknell, P.J. (1966) 'Anaximenes' Simile', *Apeiron* 1, pp. 17-18.
Bierce, A. (1911) *The Devil's Dictionary*. New York: Neale.
Bignone, E. (1916) *Empedocle*. Torino: Bocca.
Blanquart, F. (1987) *La Création dans l'Orient Ancien*. Paris: Cerf.
Blumenthal, H.J. & Markus, R.A. (eds) (1981) *Neoplatonism and Early Christian Thought*. London: Variorum.
Bogen, J. & McGuire, J.E. (1986) 'Aristotle's Great Clock: Necessity, Possibility, and the Motion of the Cosmos in *De Caelo* I:12', *Philosophy Research Archives* 12, pp. 387-447.
Bollack, J. (1965-69) *Empédocle* (3 vols). Paris: Les Editions du Minuit.
——— (1980) 'La cosmogonie des anciens atomistes', in F. Romano (ed.) *Democrito e l'atomismo antico*, Atti del Convegno Internazionale (Catania, 18-21 April 1979), *Siculorum Gymnasium* NS XXXIII, 1, pp. 11-59.
Bowen, A.C. (1991) *Science and Philosophy in Classical Greece*. New York: Garland.
Bowra, C.M. (1937) 'The Proem of Parmenides', *Classical Philology* 32, pp. 97-112.
Boyancé, P. (1936) *Études sur le Songe de Scipion*. Paris: E. de Boccard.
Brandon, S.G.F. (1963) *Creation Legends of the Ancient Near East*. London: Hodder & Stoughton.
Brisson, L. & Meyerstein, F.W. (1995) *Inventing the Universe: Plato's Timaeus, the Big Bang, and the Problem of Scientific Knowledge*. Albany, NY: SUNY Press.
Broadie, S. (2001) 'Theodicy and Pseudo-history in the *Timaeus*', *Oxford Studies in Ancient Philosophy* 21, pp. 1-28.
Brown, G. (1984) 'The Cosmological Theory of Empedocles', *Apeiron* 18, pp. 97-101.
Brumbaugh, R.S. (1961) 'Plato and the History of Science', *Studium Generale* 14/9, pp. 520-7.
Bulmer-Thomas, I. (1984) 'Plato's Astronomy', *Classical Quarterly* 34, pp. 107-12.
Burkert, W. (1970) 'La Genèse des choses et des mots. Le papyrus de Derveni entre Anaxagore et Cratyle', *Études Philosophiques* 25, pp. 443-55.
——— (1972) *Lore and Science in Ancient Pythagoreanism*. Cambridge, MA: Harvard University Press.
——— (1997) 'Star Wars or One Stable World? A Problem of Presocratic Cosmogony', in Laks & Most 1997.
——— (1999) 'The Logic of Cosmogony', in Buxton 1999.
Burnet, J. (1920) *Early Greek Philosophy*. London: A&C Black.
Burnyeat, M.F. (1982) 'Idealism and Greek Philosophy: What Descartes Saw and Berkeley Missed', *Philosophical Review* 91, pp. 3-40.
Bury, R.G. (1929) Plato's *Timaeus*, Loeb Classical Library.
Buxton, R. (1999) *From Myth to Reason*. Oxford: Oxford University Press.
Callahan, J.F. (1948) *Four Views of Time in Ancient Philosophy*. Cambridge, MA: Harvard University Press.
Calvo, C. & Brisson, L. (1997) *Interpreting the Timaeus-Critias*. Proceedings of the IV Symposium Platonicum. Sankt Augustin: Academia Verlag.
Campbell, G.L. (1999) Review of Sedley (1998), *Bryn Mawr Classical Review* 1999.10.29.
——— (2000) 'Zoogony and Evolution in Plato's *Timaeus*: The Presocratics, Lucretius and Darwin', in Wright 2000, pp. 146-78.
——— (2003) *Lucretius on Creation and Evolution*. Oxford: Oxford University Press.
Carter, B. (1974) 'Large Number Coincidences and the Anthropic Principle in Cosmology', in M.S. Longair (ed.), *IAU Sympoisum 63: Confrontation of Cosmological Theories with Observational Data*. Dordrecht: Reidel, pp. 291-8.

Caston, V. (2002) *Essays in Honour of Alexander Mourelatos*. Aldershot: Ashgate.
Cavagnaro, E. (1997) 'The *Timaeus* of Plato and the Erratic Motion of the Planets', in Calvo & Brisson 1997, pp. 351-62.
Chalmers, W.R. (1960) 'Parmenides and the Beliefs of Mortals', *Phronesis* 5, pp. 5-22.
Cherniss, H.F. (1944) *Aristotle's Criticism of Plato and the Academy*. Baltimore: Johns Hopkins University Press.
——— (1975) 'The Sources of Evil According to Plato', *American Philosophical Quarterly* 12, pp. 244-58.
Chroust, A.H. (1975) 'Some Comments on Philo of Alexandria "De Aeternitate Mundi" ', *Laval Theologique et Philosophique* 31, pp. 135-45.
——— (1978) 'Aristotle's Doctrine of the Uncreatedness of the Universe', *New Scholasticism* 52, pp. 268-79.
Cicero, *De Natura Deorum*. Loeb Classical Library (1933).
——— *On Stoic Good and Evil*, ed. M.R. Wright (1991). Warminster: Aris & Phillips.
——— *De Finibus*. Loeb Classical Library (1961).
Claghorn, G.S. (1954) *Aristotle's Criticism of Plato's Timaeus*. The Hague: Nijhoff.
Classen, C.J. (1977) 'Anaximander and Anaximenes: The Earliest Greek Theories of Change?', *Phronesis* 22, pp. 89-102.
Clay, D. (1983) *Lucretius and Epicurus*. Ithaca, NY: Cornell University Press.
Clegg, J.S. (1976) 'Plato's Vision of Chaos', *Classical Quarterly* NS 26, pp. 52-61.
Clement of Alexandria. Loeb Classical Library (1919).
Coggins, R.L. & Houlden, J.L. (eds) (1990) *A Dictionary of Biblical Interpretation*. London: SCM Press.
Copan, P. (1996) 'Is *Creatio ex nihilo* a Post-biblical Invention? An Examination of Gerhard May's Proposal', *Trinity Journal* 17, pp. 77-93.
Cornford, F.M. (1932) 'Mathematics and Dialectic in the Republic VI-VII', *Mind* 41, (I) pp. 37-52, (II) pp. 73-190.
——— (1933) 'Innumerable Worlds in Presocratic Philosophy', *Classical Quarterly* 28, pp. 1-16.
——— (1935) *Plato's Theory of Knowledge*. London: Routledge & Kegan Paul.
——— (1937) *Plato's Cosmology*. London: Routledge & Kegan Paul.
——— (1939) *Plato and Parmenides*. London & New York: Routledge & Kegan Paul.
——— (1941) *The Republic of Plato*. Oxford: Clarendon Press.
——— (1950) *Before and After Socrates*. Cambridge: Cambridge University Press.
——— (1952) *Principium Sapientiae*. Cambridge: Cambridge University Press.
——— (1957) *From Religion to Philosophy*. New York: Harper & Row.
Couprie, D.L. (2001) 'Anaximander's Discovery of Space', in Preus 2001.
——— (2002) 'Imagining the Universe', *Apeiron* 35, pp. 47-59.
Coxon, A.H. (1986) 'The Fragments of Parmenides', *Phronesis* Suppl. 3.
Curd, P. (1991) 'Deception and Belief in Parmenides' Doxa', *Apeiron* 25, pp. 109-33.
——— (1998) *The Legacy of Parmenides*. New Haven, CT: Yale University Press.
——— (2001) 'A New Empedocles? Implications of the Strasburg Fragments for Presocratic Philosophy', *Proceedings of the Boston Area Colloquium in Ancient Philosophy* 17, pp. 27-49.
Dahnhadt, O. (1907) *Natursagen*. Berlin: Teubner.
Dales, R.C. (1982) 'Maimonides and Boethius of Dacia on the Eternity of the World', *New Scholasticism* 56, pp. 306-19.
Dancy, R.M. (1989) 'Thales, Anaximander and Infinity', *Apeiron* 22, pp. 149-90.

Davies, P.C.W. (1982) *The Accidental Universe*. Cambridge: Cambridge University Press.
——— (1992) *The Mind of God*. Harmondsworth: Penguin.
Day, J. (2002) 'Creation Narratives', in Coggins & Houlden (eds), *The Dictionary of Biblical Interpretation*. London: Morehouse, pp. 147-50.
De Blic, J. (1940) 'Platonisme et Christianisme dans la conception Augustinienne du Dieu Créateur', *Recherches de Science Religieuse*. Paris, pp. 172-90.
——— (1945) 'Les arguments de Saint Augustin contre l'éternité du Monde', *Mélanges de Science Religieuse*. Lille, pp. 33-44.
De Santillana, G. (1961) *The Origins of Scientific Thought*, vol. I, *From Anaximander to Proclus – 600 BC to 300 AD*. London: Weidenfeld & Nicolson.
De Witt, N. (1967) *Epicurus and His Philosophy*. Cleveland: World Publishing Co.
Descartes, R. (1985) *The Principles of Philosophy*, in J. Cottingham, R. Stoothoff and D. Murdoch (tr.), *The Philosophical Writings of Descartes*. Cambridge: Cambridge University Press.
Dexippus, *On Aristotle Categories*, tr. J. Dillon. London: Duckworth.
Dick, S.J. (1982) *The Plurality of Worlds*. Cambridge: Cambridge University Press.
Dicks, D.R. (1959) 'Thales', *Classical Quarterly* NS 9, pp. 294-309.
Diels, H. (1922) *Die Fragmente der Vorsokratiker*. Berlin: Weidmann.
Dillon, J. (1977) *The Middle Platonists*. London: Duckworth.
——— (1995) 'The Neoplatonic Exegesis of the Statesman Myth', in Rowe 1995, pp. 364-74.
Diogenes Laertius, *Lives*. Loeb Classical Library (1925).
Dodds, E.R. (1951) *The Greeks and the Irrational*. Berkeley: University of California Press.
Doria, C. (1976) *Origins: Creation Texts from the Ancient Mediterranean*. New York: AMS Press.
Duhem, P. (1954) *Le Système du Monde: Histoire des Doctrines Cosmologiques de Platon à Copernic*, vol. 1. Paris: Herman.
——— (1954) *The Aim and Structure of Physical Theory* [originally *La Théorie Physique, son Objet et sa Structure*, 1906], tr. by P. Wiener from the 2nd edn (Paris: Rivière, 1914). Princeton, NJ: Princeton University Press.
——— (1969) *To Save the Phenomena* [originally 'Sozein ta phainomena: essai sur la notion de théorie physique de Platon à Galilée', *Annales de Philosophie Chrétienne*, 1908], tr. by E. Doland & C. Maschler. Chicago: University of Chicago Press.
Eckschmidt, W. (1989) *Weltmodelle: Griechischen Weltbilder von Thales bis Ptolemäus*. Mainz: Philipp von Zabern.
Edmunds, L. (1972) 'Necessity, Chance and Freedom in the Early Atomists', *Phoenix* 26, pp. 342-57.
Effe, B. (1970) *Studien zur Kosmologie und Theologie der Aristotelischen Schrift Über die Philosophie*. Munich: Beck.
Ehrhardt, A. (1968) *The Beginning*. Manchester: Manchester University Press.
Eichrodt, W. (1999) *The Theology of the Old Testament*. Westminster: John Know Press.
Engemann, J. (1991) 'Cosmic Justice in Anaximander', *Phronesis* 36, pp. 1-25.
Farrington, B. (1936) *Science in Antiquity*. Oxford: Oxford University Press, 1936.
Ferguson, J. (1971) 'Dinos', *Phronesis* 16, pp. 97-115.
——— (1974) *Clement of Alexandria*. New York: Twayne.
Field, G.C. (1933) 'Plato and Natural Science', *Philosophy* 8, pp. 131-41.

Finkelberg, A. (1986) 'The Cosmology of Parmenides', *American Journal of Philology* 107, pp. 303-17.
——— (1993) 'Anaximander's Conception of the Apeiron', *Phronesis* 38, pp. 229-56.
——— (1994) 'Plural Worlds in Anaximander', *American Journal of Philology* 115, pp. 485-506.
——— (1997) 'Xenophanes' Physics, Parmenides' Doxa and Empedocles' Theory of Cosmological Mixture', *Hermes* 125, pp. 1-16.
Frank, E. (1945) *Philosophical Understanding and Religious Truth.* Washington: University Press of America.
Fränkel, H. (1955) *Wege und Formen frühgriechischen Denkens.* Munich: Beck.
——— (1975) 'Studies in Parmenides', in Furley & Allen 1975, vol. 2, pp. 1-48.
Frankfort, H & H.A. (1946) *Before Philosophy.* London: Penguin.
Freudenthal, G. (1986) 'The Theory of Opposites and an Ordered Universe: Physics and Metaphysics in Anaximander', *Phronesis* 31, pp. 197-228.
Friedlander, P. (1958) *Plato as Physicist. An Introduction to Plato,* London & New York: Routledge & Kegan Paul, ch. 14, pp. 246-60.
Furley, D.J. (1967) *Two Studies in the Greek Atomists.* Princeton, NJ: Princeton University Press.
——— (1970) 'Variations on Themes from Empedocles in Lucretius' Proem', *Bulletin of the Institute of Classical Studies* 17, pp. 55-64.
——— (1981) 'The Greek Theory of the Infinite Universe', *Journal of the History of Ideas* 42, pp. 571-86.
——— (1986) 'The Cosmological Crisis in Classical Antiquity', *Proceedings of the Boston Area Colloquium on Ancient Philosophy* 2, pp. 1-19.
——— (1987) *The Greek Cosmologists.* Cambridge: Cambridge University Press.
——— (1989) *Cosmic Problems.* Cambridge: Cambridge University Press.
——— & Allen, R.E. (eds) (1970) *Studies in Presocratic Philosophy I.* London: Routledge & Kegan Paul.
——— & Allen, R.E. (eds) (1975) *Studies in Presocratic Philosophy II.* London: Routledge & Kegan Paul.
——— (2000) 'Cosmology', in K. Algra (ed.), *The Cambridge History of Hellenistic Philosophy,* Cambridge: Cambridge University Press, pp. 412-51.
Gagarin, M. (2002) 'Greek Law and the Presocratics', in Caston 2002, pp. 19-24.
Gilbert, O. (1907) *Die meteorologischen Theorien des griechischen Altertums,* Leipzig: Teubner.
Gilbert, W. (1600) *De Magnete.* London: Petrus Short.
Godwin, J. (2004) *Lucretius.* London: Duckworth.
Graham, D.W. (1988) 'Symmetry in Empedocles' Cosmic Cycle', *Classical Quarterly* 38, pp. 297-312.
——— (1990) 'Socrates, the Craft Analogy and Science', *Apeiron* 23, pp. 1-24.
——— (1994) 'The Postulates of Anaxagoras', *Apeiron* 27, pp. 77-121.
——— (1997) 'Heraclitus' Criticism of Ionian Philosophy', *Oxford Studies in Ancient Philosophy* 15, pp. 1-50.
——— (1999) 'Empedocles and Anaxagoras: Responses to Parmenides', in Long 1999.
——— (2002) 'Heraclitus and Parmenides', in *Presocratic Philosophy: Essays in Honour of Alexander Mourelatos,* ed. V. Caston. Aldershot: Ashgate, pp. 27-44
——— (2003) 'A New Look at Anaximenes', *History of Philosophy Quarterly* 20, pp. 1-20.
Granger, H. (2002) 'The Cosmology of Mortals', in Caston 2002, pp. 101-16.

——— (2004) 'Argumentation and Heraclitus' Book', *Oxford Studies in Ancient Philosophy* 26, pp. 1-17.

Grant, M. (1980) *Miracle and Law in Graeco-Roman and Early Christian Thought*, Amsterdam: North Holland Publishing Co.

Gregory, A.D. (1996) 'Astronomy and Observation in Plato's Republic', *Studies in History and Philosophy of Science* 27, pp. 451-71.

——— (1999) 'Plato and Aristotle on Eclipses', *Journal for the History of Astronomy* 31, 245-59.

——— (2000) *Plato's Philosophy of Science*. London: Duckworth.

——— (2001) *Eureka! The Birth of Science*. London: Icon.

——— (2001a) 'Aristotle, Dynamics and Proportionality', *Early Science and Medicine* 6, 1-21.

——— (2003a) 'Eudoxus, Callippus and the Astronomy of the Timaeus', *Bulletin of the Institute of Classical Studies* 78, pp. 5-28.

——— (2003b) 'Aristotle and Some Commentators on the Timaeus' Receptacle', *Bulletin of the Institute of Classical Studies* 78, pp. 29-47.

Gregory, J. (2005) *Fred Hoyle's Universe*. Oxford: Oxford University Press.

Gribbin, J. (1986) *In Search of the Big Bang*. London: Heinemann.

——— (1993) *In the Beginning*. Harmondsworth: Penguin.

——— & Rees, M. (1990) *Cosmic Coincidences*. London: Heinemann.

Grunebaum, A. (1952) 'Some Highlights of Modern Cosmology and Cosmogony', *Review of Metaphysics* 5, pp. 481-98.

——— (1989) 'The Pseudo-Problem of Creation in Physical Cosmogony', *Philosophy of Science* 56, pp. 373-94.

Gunkel, H. (1895) *Schöpfung und Chaos in Urzeit and Endzeit*. Gottingen: Vandenhoeck &. Ruprecht.

Gurney, O.R. (1990) *The Hittites*. Harmondsworth: Penguin.

Guthrie, W.K.C. (1950) *The Greeks and their Gods*. Boston: Beacon Press.

——— (1957) *In the Beginning*. London: Routledge & Kegan Paul.

——— (1962-81) *A History of Greek Philosophy* (6 vols). Cambridge: Cambridge University Press.

——— (1992) 'Flux and Logos in Heraclitus', in A.P.D. Mourelatos (ed.), *The Presocratics: A Collection of Critical Issues*. Princeton, NJ: Princeton University Press, pp. 197-213.

Hackforth, R. (1959) 'Plato's Cosmogony', *Classical Quarterly* NS 9, pp. 17-22.

Hahm, D.E. (1977) *The Origins of Stoic Cosmology*. Ohio: Ohio State University Press.

Hall, T.S. (1969) *Ideas of Life and Matter: Studies in the History of General Physiology, 600 BC – 1900 AD* (2 vols). Chicago: Chicago University Press.

Harte, V.A. (2002) *Plato on Parts and Wholes*. Oxford: Oxford University Press.

Hatab, L.J. (1990) *Myth and Philosophy*. London: Open Court.

Hawking, S. (1988) *A Brief History of Time*. London: Guild.

——— (1993) *Black Holes and Baby Universes and Other Essays*. London: Bantam.

Heidel, A. (1942) *The Babylonian Genesis*. Chicago: Chicago University Press.

——— (1957) *In the Beginning*. London: Methuen.

Heidel, W.A. (1906) 'The *Dinê* in Anaximenes and Anaximander', *Classical Philology* 1, pp. 279-82.

Helleman, W. (1994) *Hellenization Revisited*. Lanham: University Press of America.

Hershbell, J.P. (1974) 'Empedoclean Influences on the Timaeus', *Phoenix* 28, pp. 145-66.

Hintikka, J. (1973) *Time and Necessity*. Oxford: Oxford University Press.
Holscher, U. (1965) 'Weltzeiten und Lebenskyklus; eine Nachprüfung der Empedokles-Doxographie', *Hermes* 93, pp. 7-33.
——— (1975) 'Anaximander and the Beginnings of Greek Philosophy', in Furley & Allen 1975, vol. 1, pp. 281-322.
Hoyle, F. (1951) *The Nature of the Universe*. Oxford: Blackwell.
——— (1955) *Frontiers of Astronomy*. London: Heinemann.
——— (1966) *Galaxies, Nuclei and Quasars*. London: Heinemann.
——— (2005) *A Different Approach to Cosmology*. Cambridge: Cambridge University Press.
Huby, P. (1978) 'Epicurus' Attitude to Democritus', *Phronesis* 23, pp. 80-6.
Hussey, E. (1999) 'Heraclitus', in Long 1999.
Inwood, B. (1984) Review of Kahn 1979 in *Ancient Philosophy* 4, pp. 227-34.
——— (1986) 'Anaxagoras and Infinite Divisibility', *Illinois Classical Studies* 11, pp. 17-36.
——— (1992) *The Poem of Empedocles*. Toronto: Toronto University Press.
——— (1994) *The Epicurus Reader*. Indianapolis: Hackett.
——— (2000) Review of Martin and Primaversi 1999 in *Classical Review* 50, pp. 5-7.
——— (2003) *The Cambridge Companion to the Stoics*. Cambridge: Cambridge University Press.
Jacobsen, T. (1981) The Eridu Genesis. *Journal of Biblical Literature* 100, pp. 513-29.
Johansen, T.K. (2004) *Plato's Natural Philosophy*. Cambridge: Cambridge University Press.
Johrens, O. (1939) *Die Fragment des Anaxagoras*. Bochum-Langendreer: Druck H. Pöppinghaus.
Jones, H. (1989) *The Epicurean Tradition*. London: Routledge.
Judson, L. (1983) 'Eternity and Necessity in *De Caelo* I/12', *Oxford Studies in Ancient Philosophy* 1, pp. 217-55.
——— (1987) 'God or Nature? Philoponus on Generability and Perishability', in Sorabji 1987.
Kahn, C.H. (1960) *Anaximander and the Origins of Greek Cosmology*. New York: New York University Press.
——— (1970) 'On Early Greek Astronomy', *Journal of Hellenic Studies* 90, pp. 99-116.
——— (1974) 'Pythagorean Philosophy Before Plato', in A.P.D. Mourelatos (ed.), *The Pre-Socratics*. Garden City: Anchor Books.
——— (1979) *The Art and Thought of Heraclitus*, Cambridge: Cambridge University Press.
——— (1991) 'Some Remarks on the Origins of Greek Science and Philosophy', in Bowen 1991, pp. 1-10.
——— (1999) 'Review of Greek Philosophy from the Beginning to Plato', *Oxford Studies in Ancient Philosophy* 17, pp. 325-41.
Keyt, D. (1961) 'Aristotle on Plato's Receptacle', *American Journal of Philology* 82, pp. 291-300.
——— (1963) 'The Fallacies of the *Phaedo*', *Phronesis* 8, pp. 167-71.
——— (1969) 'Plato's Paradox that the Immutable is Unknowable', *Philosophical Quarterly* 19, pp. 1-14.
——— (1971) 'The Mad Craftsman of the *Timaeus*', *Philosophical Review* 80, pp. 230-5.

Kingsley, P. (1994) 'Empedocles and his Interpreters: The Four Element Doxography', *Phronesis* 39, pp. 235-54.
——— (1995) *Ancient Philosophy, Mystery and Magic: Empedocles and the Pythagorean Tradition*. Oxford: Oxford University Press.
——— (1996) 'Empedocles' Two Poems', *Hermes* 124, pp. 108-11.
Kirk, G.S. (1955) 'Some Problems in Anaximander', *Classical Quarterly* NS 5, pp. 21-38.
——— (1959) ' "Ecpyrosis" in Heraclitus: Some Comments', *Phronesis* 4, pp. 73-6.
——— (1962) *Heraclitus, The Cosmic Fragments*, Cambridge: Cambridge University Press.
——— (1970) *Myth*. Cambridge: Cambridge University Press.
——— (1974) *The Nature of Greek Myths*. London: Penguin.
——— (1992) 'Natural Change in Heraclitus', in A.P.D. Mourelatos (ed.), *The Presocratics: A Collection of Critical Issues*. Princeton, NJ: Princeton University Press, pp. 189-96.
———, Raven, J.E. & Schofield, M. (1983) *The Presocratic Philosophers*, 2nd edn. Cambridge: Cambridge University Press.
Knuuttila, S. (2001) 'Time and Creation in Augustine', in Stump & Kretzel 2001, pp. 103-15.
Koyré, A. (1957) *From the Closed World to the Infinite Universe*. New York: Harper.
Kragh, H. (1992) *Cosmology and Controversy*. Princeton: Princeton University Press.
Kramer, S.N. (1961) *Mythologies of the Ancient World*. Garden City, NY: Doubleday.
Kuhn, T.S. (1970) *The Structure of Scientific Revolutions*. Chicago: Chicago University Press.
——— (1979) *The Essential Tension*. Chicago: Chicago University Press.
Laks, A. (1983) *Diogène d'Apollonie. La Dernière Cosmogonie Présocratique*. Lille: Lille University Press.
——— (2002) 'Reading the Readings: On the First Person Plurals in the Strasbourg Empedocles', in Caston 2002.
——— & Most, G.W. (1997) *Studies on the Derveni Papyrus*. Oxford: Oxford University Press.
Lapidge, M. (1973) 'A Problem in Stoic Cosmology', *Phronesis* 18, pp. 240-78.
Lebedev, A. (1985) 'The Cosmos as a Stadium: Agonistic Metaphors in Heraclitus' Cosmology', *Phronesis* 30, pp. 131-50.
Leher, J.H. (1995) 'Xenophanes of Colophon: Fragments', *Phoenix* 49, pp. 182-4.
Leibniz, G.W.F. (1686) *Discourse on Metaphysics*.
——— (1687) *Letter Explaining the Laws of Nature*.
——— (1714) *Monadology*.
Lennox, J.G. (1985) 'Plato's Unnatural Teleology', in D.J. O'Meara (ed.) *Platonic Investigations*, Washington, DC: Washington University Press, pp. 195-218.
Leon, P. (1927) 'The Homoeomeries of Anaxagoras', *Classical Quarterly* 21, pp. 133-41.
Lerner, P. (1997) *Le Monde des Sphères* (2 vols). Paris: Belles Lettres.
Lesher, J.H. (1995) 'Mind's Knowledge and Powers of Control in Anaxagoras DKB12', *Phronesis* 40, pp. 125-42.
Leslie, J. (1989) *Universes*. London: Routledge.
Lewis, E. (2000) 'Anaxagoras and the Seeds of a Physical Theory', *Apeiron* 33, pp. 1-24.
Liddle, A.R. (1999) *An Introduction to Modern Cosmology*. Chichester: Wiley.

Lilla, S.R.C. (1971) *Clement of Alexandria*. Oxford: Oxford University Press.
Liritzis, I. (2003) 'Anaxagorean Nous and its Analogies in Orphic Cosmogony', *Philosophia* 33, pp. 257-65.
Lloyd, G.E.R. (1964) 'Hot and Cold, Wet and Dry in Early Greek Thought', in Furley & Allen 1970, 255-80.
——— (1966) *Polarity and Analogy*. Cambridge: Cambridge University Press.
——— (1968) 'Plato as a Natural Scientist', *Journal of Hellenic Studies* 28, pp. 78-92.
——— (1970) *Early Greek Science to Aristotle*. London: Chatto & Windus.
——— (1978) 'Saving the Appearances', *Classical Quarterly* 88, pp. 202-22. See also Lloyd 1991, pp. 248-53.
——— (1979) *Magic, Reason and Experience*. Cambridge: Cambridge University Press.
——— (1987) *The Revolutions of Wisdom*. Berkeley: California University Press.
——— (1990) 'Plato and Archytas in the Seventh Letter', *Phronesis* 35, pp. 159-74.
——— (1991) *Methods and Problems in Greek Science*, Cambridge: Cambridge University Press, pp. 333-51.
——— (1992) 'Methods and Problems in the History of Ancient Science: The Greek Case', *Isis* 83, pp. 564-77.
Lloyd, R. (2006) 'Symmetry and Asymmetry in the Construction of the "Elements" in the *Timaeus*', *Classical Quarterly* 56, pp.1-16.
Long, A.A. (1963) 'The Principles of Parmenides' Cosmogony', *Phronesis* 8, pp. 90-107.
——— (1974) *Hellenistic Philosophy: Stoics, Epicureans, Sceptics*. New York: Scribner's.
——— (1976) 'Heraclitus and Stoicism', *Philosophia* 6, pp. 133-56.
——— (1977) 'Chance and Law in Epicureanism', *Phronesis* 22, pp. 63-88.
——— (1993) 'Empedocles' Cosmic Cycle in the Sixties', in Mourelatos 1993.
——— (1999) *The Cambridge Companion to Early Greek Philosophy*. Cambridge: Cambridge University Press.
——— & Sedley, D. (1987) *The Hellenistic Philosophers* (2 vols). Cambridge: Cambridge University Press.
Longrigg, J. (1964) 'A Note on Anaximenes' Fragment 2', *Phronesis* 9, pp. 1-4.
Louth, A. (ed.) (2000) *Ancient Christian Commentary on Scripture*. Illinois: Inter-Varsity Press.
Lovin, R.W. & Reynolds, F.E. (1985) *Cosmogony and Ethical Order*. Chicago: Chicago University Press.
Lucretius, *De Rerum Natura*. Loeb Classical Library (1924).
MacKenzie/ McCabe, M.M. (1982) 'Parmenides' Dilemma', *Phronesis* 28, pp. 1-13.
——— (1993) 'Myth, Allegory and Argument', *Apeiron* 25, *The Language of the Cave*, pp. 47-67.
——— (1994) *Plato's Individuals*. Princeton: Princeton University Press.
——— (2000) *Plato and his Predecessors*. Cambridge: Cambridge University Press.
McGrath, A.E. (1994) *Christian Theology*. Oxford: Oxford University Press.
McKirahan, R.D. (1994) *Philosophy Before Socrates*. Cambridge: Cambridge University Press.
——— (2001) 'Anaximander's Infinite Worlds', in Preus 2001, pp. 49-65.
Malcolm, J. (1991) 'On Avoiding the Void', *Oxford Studies in Ancient Philosophy* 9, pp. 75-94.
Mansfeld, J. (1964) *Die Offenbarung des Parmenides und die menschliche Welt*. Assen: Van Gorcum.

——— (1972) 'Ambiguity in Empedocles B17, 3-5: A Suggestion', *Phronesis* 17, pp. 17-39.
——— (1979) 'Providence and the Destruction of the Universe in Early Stoic Thought', in M.J. Vermaseren (ed.), *Studies in Hellenistic Religions*, Leiden: E.J. Brill.
——— (1980) 'Anaxagoras' Other World', *Phronesis* 25, pp. 1-4.
——— (1992) '*Peri Kosmou*: A Note on the History of a Title', *Mnemosyne* 46, pp. 391-411.
——— (1995) 'Empedocles and his Interpreters', *Phronesis* 40, pp. 109-15.
——— (2002) 'Theophrastus and Simplicius on Anaximander', *Philosophia* 32, pp. 25-46.
——— & Algra, K. (2001) 'Three Thetas in the Empédocle de Strasbourg', *Mnemosyne* 54, pp. 78-84.
Marcovich, M. (1967) *Heraclitus, Editio Maior*, Venezuala: Merida.
——— (1982) Review of Kahn 1979 in *Gnomon* 11, pp. 417-36.
Martin, A. & Primaversi, O. (1999) *L'Empédocle de Strasbourg*. Bibliothèque Nationale et Universitaire de Strasbourg/ Berlin: de Gruyter.
May, G. (1994) *Creation Ex Nihilo*. Edinburgh: T&T Clark.
Miller, J. (1989) 'Thales on Water: The Egyptian Connection', *Southwest Philosophical Studies*, pp. 41-8.
Miller, M. (1980) 'The Implicit Logic of Hesiod's Cosmogony', *Independent Journal of Philosophy* 4, pp. 131-42.
——— (2001) ' "First of All": On the Semantics and Ethics of Hesiod's Cosmogony', *Ancient Philosophy* 21, pp. 251-76.
Minar, E. (1963) 'Cosmic Periods in the Philosophy of Empedocles', *Phronesis* 8, pp. 127-45.
Mohr, R.D. (1977) 'Plato, *Statesman* 284c-d: An Argument from the Sciences', *Phronesis* 22, pp. 232-4.
——— (1978a) 'The Formation of the Cosmos in the Statesman Myth', *Phoenix* 27, pp. 250-2.
——— (1980b) 'The Sources of Evil Problem and the *arche kineseos* Doctrine in Plato', *Apeiron* 14, pp. 41-56.
——— (1981) 'Disorderly Motion in Plato's *Statesman*', *Phoenix* 35, pp. 199-215.
——— (1986) 'Plato on Time and Eternity', *Ancient Philosophy* 6, pp. 39-46.
——— (1985) *The Platonic Cosmology*. Leiden: E.J. Brill.
——— (1985a) 'Plato's Theology Reconsidered: What the Demiurge Does', *History of Philosophy Quarterly* 2, pp. 131-44.
Moran, J. (1975) 'The Priority of Earth in the Cosmogony of Anaximenes', *Apeiron* 9, pp. 17-19.
Morriston, W. (2002) 'Creation *ex nihilo* and the Big Bang', *Philo* 5, pp. 34-61.
Morrow, G.R. (1950) 'Necessity and Persuasion in Plato's *Timaeus*', in R.E. Allen (ed.), *Studies in Plato's Metaphysics*.. London: Routledge & Kegan Paul, pp. 421-38.
Most, G.W. (1999) 'From Logos to Muthos', in Buxton 1999.
Mourelatos, A.P.D. (1970) *The Route of Parmenides*. New Haven, CT: Yale University Press.
——— (1981) 'Presocratic Origins of the Principle that there are No Origins from Nothing', *Journal of Philosophy* 78, pp. 649-65.
——— (1987) 'Quality, Structure and Emergence in Later Pre-Socratic Philosophy', in J. Cleary (ed.), *Proceedings of the Boston Area Colloquium in Ancient Philosophy* 2, 127-94.

——— (1991) 'Plato's Science – His View and Ours of His', in Bowen 1991, pp. 11-42.

——— (1993) *The Presocratics: A Collection of Critical Essays*. Princeton, NJ: Princeton University Press.

Naddaf, G. (1998) 'On the Origin of Anaximander's Cosmological Model', *Journal of the History of Ideas* 59, pp. 1-28.

——— (2005) *The Greek Concept of Nature*. New York: New York University Press.

Narliker, V.J. (1983) *Introduction to Cosmology*. Sudbury, MA: Jones & Bartlett.

Nehamas, A. (1980) 'The Eternal Recurrence', *Philosophical Review* 89, pp. 331-56.

——— (2002) 'Parmenidean Being, Heraclitean Fire', in Caston 2002, pp. 45-54.

Neugebauer, O. (1957) *The Exact Sciences in Antiquity*, 2nd edn. Providence, RI: Brown University Press.

——— (1975) *A History of Ancient Mathematical Astronomy*. Berlin: Springer-Verlag.

Nussbaum, M.C. (1972) '*Psuche* in Heraclitus I + II', *Phronesis* 17, pp. 1-16, 153-69.

O'Brien, D. (1968) 'The Relations of Anaxagoras and Empedocles', *Journal of Hellenic Studies* 88, pp. 93-113.

——— (1969) *Empedocles' Cosmic Cycle*. Cambridge: Cambridge University Press.

——— (1990) 'Heraclit et l'unité des opposes', *Révue du Metaphysics et Morale*, 2, pp. 147-67.

——— (1995) 'Empedocles Revisited', *Ancient Philosophy* 15, pp. 403-70.

——— (2000) 'Herman Diels on the Presocratic Empedocles' Double Destruction of the Cosmos', *Phronesis* 45, pp. 1-18.

Oden, R.A. (1981) 'Divine Aspirations in Atrahasis and Genesis', *Zeitschrift für die Alttestamentliche Wissenschaft* 93, pp. 197-216.

O'Grady, P. (2002) *Thales of Miletus: The Beginnings of Western Philosophy and Science*. Aldershot: Ashgate.

O'Keefe, T. (1996) 'Does Epicurus Need the Swerve as an Archê of Collisions?', *Phronesis* 41, pp. 305-17.

O'Neill (2002) 'How Early is the Doctrine of *Creatio Ex Nihilo*?', *Journal of Theological Studies* 53, pp. 49-465.

Osborne, C. (1987a) 'Empedocles Recycled', *Classical Quarterly* 37, pp. 24-50.

——— (1987b) *Rethinking Early Greek Philosophy: Hippolytus of Rome and the Presocratics*. Ithaca, NY/ London: Cornell University Press/ Duckworth.

——— (1988) 'Topography in the Timaeus: Plato and Augustine on Mankind's Place in the Natural World', *Proceedings of the Cambridge Philological Society* 214, pp. 104-14.

——— (1989) 'Philoponus on the Origin of the Universe and Other Issues', *Studies in History and Philosophy of Science* 20, pp. 389-95.

——— (1996) 'Space, Time, Shape and Direction: Creative Discourse in the *Timaeus*', in Gill & McCabe 1996, pp. 179-211.

——— (2000) 'Rummaging in the Recycling Bins of Upper Egypt', *Oxford Studies in Ancient Philosophy* 18, pp. 329-56.

Owen, G.E.L. (1966) 'Plato and Parmenides on the Timeless Present', *Monist* 50, pp. 317-40.

——— (1975) 'Eleatic Questions', in Furley & Allen 1975, pp. 48-81.

Palmer, J.A. (1998) 'Xenophanes' Ouranian God in the Fourth Century', *Oxford Studies in Ancient Philosophy* 16, pp. 1-34.

Panzerbieter, F. (1844) *Beiträge zur Kritik und Eklärung des Empedokles*. Meiningen: F.W. Gadow & Sohn.

Parry, R.D. (1979). 'The Unique World of the *Timaeus', Journal for the History of Philosophy* 17, pp. 1-10.
——— (1983) 'The Craft of Justice', *Canadian Journal of Philosophy* 9, pp. 19-38.
——— (1991) 'The Intelligible World-Animal in Plato's *Timaeus*', *Journal for the History of Philosophy* 29, pp. 13-32.
——— (1996) *Plato's Craft of Justice*. Albany, NY: SUNY Press.
Partenie, C. (1998) 'The "Productionist" Framework of the Timaeus', *Dionysius* 16, pp. 105-21.
Patterson, R. (1981) 'The Unique Worlds of the Timaeus', *Phoenix* 35, pp. 105-19.
——— (1985) 'On the Eternality of Platonic Forms', *Archiv für Geschichte der Philosophie* 67, pp. 27-45.
Pedersen, O. (1974) *Early Physics and Astronomy*. Cambridge: Cambridge University Press [2nd edn 1993].
Pegis, A.C. (1943) 'Cosmogony and Knowledge: St. Thomas and Plato', *Thought* 18, pp. 643-64.
——— (1944) 'Cosmogony and Knowledge I', *Thought* 19, pp. 269-90.
——— (1945) 'Cosmogony and Knowledge II', *Thought* 20, pp. 473-98.
Penwill, J.L. (1974) 'Alkman's Cosmogony', *Apeiron* 8, pp. 18-39.
Peters, E. (1984) 'What was God Doing Before He Created the Heavens and the Earth?', *Augustiniana* 34, pp. 53-74.
Philip, J.A. (1966) *Pythagoras and Early Pythagoreanism*. Toronto: Toronto University Press.
Phillips, J.F. (1997) 'Neoplatonic Exegesis of Plato's Cosmogony', *Journal of the History of Philosophy* 35, pp. 173-97.
Philo, *On the Eternity of the World*. Loeb Classical Library (1941).
Philoponus, *Against Aristotle on the Eternity of the World*, tr. C. Wildberg (1987). London: Duckworth.
———, *Against Proclus on the Eternity of the World*, ed. H. Rabe (1984). Leipzig: Teubner.
———, *Against Proclus on the Eternity of the World 1-5*, tr. M. Share (2005). Duckworth, London.
———, *Against Proclus on the Eternity of the World 6-8*, tr. M. Share (2005). Duckworth, London.
Pingree, D. (1992) 'Hellenophilia vs. the History of Science', *Isis* 83, pp. 554-63.
Plato, *Platonis Opera*, ed. J. Burnet (1900-7). Oxford: Oxford University Press.
Plotinus, *Plotini Opera,* ed. P. Henry & H-R. Schwyzer (1951-73). Paris: Desclée de Brouwer.
Popper, K.R. (1963) 'Kirk on Heraclitus, and on Fire as the Cause of Balance', *Mind* 72, pp. 386-92.
Potts, R. (1984) 'Anaxagoras' Cosmogony', *Apeiron* 18, pp. 90-6.
Preus, A. (2001) *Essays in Ancient Greek Philosophy VI: Before Plato*. Albany, NY: SUNY Press.
Pritchard, J.B. (1969) *Ancient Near East Texts Relating to the Old Testament*. Princeton, NJ: Princeton University Press.
Proclus, *Elements of Theology*, ed. & tr. E.R. Dodds (1933). Oxford: Oxford University Press.
———, *Commentaire sur le Timée* (5 vols), ed. A.J. Festugière (1966). Paris: Libraire Philosophique, J. Vrin.
Reale, G. (1997) 'Plato's Doctrine of the Origin of the World, with Special Reference to the *Timaeus*', in Calvo & Brisson 1997.

Reeve, C.D.C. (1981) 'Anaxagorean Panspermism', *Ancient Philosophy* 1, pp. 89-108.
——— (1982) '*Ekpurosis* and the Priority of Fire in Heraclitus', *Phronesis* 27, pp. 299-305.
Reinhardt, K. (1942) 'Heraklits Lehre vom Feuer', *Hermes* 77, pp. 1-27.
——— (1974) 'The Relation Between the Two Parts of Parmenides' Poem', in Mourelatos 1993, pp. 293-311.
Rist, J.M. (1967) *Plotinus: The Road to Reality*. Cambridge: Cambridge University Press.
——— (1969) *The Stoic Philosophy*. Cambridge: Cambridge University Press.
Robinson, J. (1961) 'The Framework of Greek Cosmology', *Review of Metaphysics* 14, pp. 676-84.
Robinson, T.M. (1993) 'Aristotle, the *Timaeus* and Contemporary Cosmology', *Philosophical Inquiry* 15, pp. 48-58.
Rogerson, J. (1990) 'Myth', in Coggins & Houlden 1990, pp. 479-82.
——— (1999) *Genesis 1-11*. Sheffield: Sheffield University Press.
Rosen, S. (1979) 'Plato's Myth of the Reversed Cosmos', *Review of Metaphysics* 33, pp. 60-85.
Ross, D. (1936) *Aristotle's Physics*. Oxford: Oxford University Press.
Rowan-Robinson, M. (1996) *Cosmology*, 3rd edn. Oxford: Oxford University Press.
Rowe, C.J. (1995) 'Reading the Statesman', *Proceedings of the III Symposium Platonicum*. Sankt Augustin: Academia Verlag.
Runia, D. (2001) *Philo of Alexandria. On the Creation of the Cosmos according to Moses*. Leiden: E.J. Brill.
Russell, B. (1986) *On God and Religion*, ed. A. Seckel. Buffalo, NY: Prometheus Books.
Sallust, *De Diis et Mundo*, ed. A.D. Nock (1924). Cambridge: Cambridge University Press.
Sambursky, S. (1956) *The Physical World of the Greeks*. London: Routledge & Kegan Paul.
——— (1959) *The Physics of the Stoics*. London: Routledge & Kegan Paul.
——— (1965) 'Plato, Proclus, and the Limits of Science', *Journal of the History of Philosophy* 3, pp. 1-11.
Sandars, N.K. (1960) *The Epic of Gilgamesh*. London: Penguin.
Sandbach, F.H. (1975) *The Stoics*. London: Chatto & Windus.
Schofield, M. (1980) *An Essay on Anaxagoras*. Cambridge: Cambridge University Press.
——— (1996) 'Anaxagoras' Other World Revisited', in K. Algra & D. Runia (eds), *Polyhistor: Studies in the History and Historiography of Ancient Philosophy*. Leiden: E.J. Brill.
——— (1997) 'The Ionians', in Taylor 1996.
——— (2002) 'Leucippus, Democritus and the *ou mallon* Principle', *Phronesis* 47, pp. 253-63.
Schmidt, W.H. (1967) *Die Schöpfungeschichte der Priesterschrift*. Neukirchen-Vluyn: Neukirchener Verlag.
Schulz, D.J. (1966) *Das Problem der Materie in Platons Timaios*. Bonn: Bouvier.
Sedley, D. (1982) 'Two Conceptions of Vacuum', *Phronesis* 27, pp. 175-93.
——— (1989) 'Teleology and Myth in the *Phaedo*', *Proceedings of the Boston Colloquium on Ancient Philosophy* 5, pp. 359-83.
——— (1991) 'Is Aristotle's Teleology Anthropocentric?', *Phronesis* 36, pp. 179-96.

——— (1997) 'Becoming Like God in the *Timaeus* and Aristotle', in Calvo & Brisson 1997, pp. 327-40.
——— (1998) *Lucretius and the Transformation of Greek Wisdom*. Cambridge: Cambridge University Press.
——— (2003) *The Cambridge Companion to Greek and Roman Philosophy*. Cambridge: Cambridge University Press.
Seligman, P. (1962) *The Apeiron of Anaximander*. London: Athlone Press.
——— (1978) 'The Soul and Cosmos in Presocratic Philosophy', *Dionysius* 2, pp. 5-17.
Seneca, *Naturales Quaestiones*. Loeb Classical Library (1972).
Sextus Empiricus, *Against the Mathematicians*. Loeb Classical Library (1931).
Shea, W.H. (1984) 'A Comparison of Narrative Elements in Ancient Mesopotamian Creation-Flood Stories with Genesis 1-9', *Origins* 11, pp. 9-29.
Shelley, C. (2000) 'The Influence of Folk Meteorology in the Anaximander Fragment', *Journal of the History of Ideas* 61, pp. 1-17.
Shipton, K.M.W. (1985) 'Heraclitus fr. 10: A Musical Interpretation', *Phronesis* 30, pp. 111-30.
Shorey, P. (1927) 'Platonism and the History of Science', *Proceedings of the American Philosophical Association* 46, pp. 159-82.
——— (1935) *The Republic* (2 vols). Loeb Classical Library (1935), commentary.
Shuttermayr, G. (1973) " 'Schöpfung aus dem Nichts' in 2 Makk 7,28?'. *Biblische Zeitschrift* 17, pp. 203-28.
Sider, D. (1981) *The Fragments of Anaxagoras*. Meisenheim am Glan: Verlag Anton Hein.
Silk, J. (1989) *The Big Bang*, 2nd edn. New York: W.H. Freeman.
Simplicius, *Simplicii in Aristotelis De Caelo Commentaria*, ed. J.L. Heiberg (1894). Berlin: Reimer.
——— *Aristotelis in Physicorum Libros Quattuor Priores Commentaria*, ed. H. Diels (1882). Berlin: Reimer.
Sisko, J.E. (2003) 'Anaxagoras' Parmenidean Cosmology: Worlds Within Worlds Within the One', *Apeiron* 36, pp. 87-114.
Skemp, J.B. (1942) *The Theory of Motion in Plato's Later Dialogues* [2nd edn. Amsterdam: Adolf M. Hakkert, 1967].
——— (1952) *Plato: The Statesman* [2nd edn. Bristol: Bristol Classical Press, 1987], commentary.
Smith, Q. (1998) 'Why Stephen Hawking's Cosmology Precludes a Creator', *Philo* 1, pp. 75-93.
Smoot, G. (1993) *Wrinkles in Time*. London: Little, Brown.
Snell, B. (1944) 'Die Nachrichten über die Lehren des Thales', *Philologus* 96, pp. 170-82.
Solmsen, F. (1940) 'Plato and the Unity of Science', *Philosophical Review* 44, pp. 566-71.
——— (1958) 'Aristotle and Presocratic Cosmogony', *Harvard Studies in Classical Philology* 63, pp. 265-82.
——— (1960) *Aristotle's System of the Physical World*. Ithaca, NY: Cornell University Press.
——— (1963) 'Nature as Craftsman in Greek Thought', *Journal of the History of Ideas* 24, pp. 473-96.
——— (1965) 'Love and Strife in Empedocles' Cosmology', *Phronesis* 10, pp. 109-48.

——— (1975) 'Eternal and Temporary Beings in Empedocles' Physical Poem', *Archiv für Geschichte der Philosophie* 57, pp. 123-45.
——— (1977) 'Epicurus on Void, Matter and Genesis', *Phronesis* 22, pp. 263-81.
Sorabji, R. (1980) *Necessity, Cause and Blame. Perspectives on Aristotle's Theory.* London: Duckworth.
——— (1983) *Time, Creation and the Continuum: Theories in Antiquity and the Early Middle Ages.* London: Duckworth.
——— (ed.) (1987) *Philoponus and the Rejection of Aristotelian Science.* London: Duckworth.
——— (1988) Matter, Space and Motion; Theories in Antiquity and their Sequel. Duckworth, London 1988.
——— (ed.) (1990) *Aristotle Transformed.* London: Duckworth.
——— (2004) *The Philosophy of the Commentators* (3 vols). London: Duckworth.
Sorel, R. (1994) *Les Cosmogonies Grecques.* Paris: Presses Universitaires de France.
Steck, O.H. (1975) *Der Schöpfungsbericht der Priesterschrift.* Gottingen: Vandenhoeck & Ruprecht.
Steiger, K. (1986) 'Die Kosmologie des Parmenides und Empedokles', *Oikoimene* 5, pp. 173-236.
Stenzel, J. (1959) *Zahl und Gestalt bei Plato und Aristoteles.* Darmstadt: Wissenschaftliche Buchgesellschaft.
Stevens, A. (1989) 'La Physique d'Empedocle selon Simplicius', *Revue Belge de Philologie* 67, pp. 65-74.
Stewart, J.A. (1905) *The Myths of Plato* [re-edited with new introduction by G.R. Levy (1960). Fontwell, Sussex: Centaur Press Ltd].
Stokes, M.C. (1962) 'Hesiodic and Milesian Cosmogonies I', *Phronesis* 7, pp. 1-37.
——— (1963) 'Hesiodic and Milesian Cosmogonies II', *Phronesis* 8, pp. 1-34.
——— (1965) 'On Anaxagoras, Part II: The Order of Cosmogony', *Archiv für Geschichte der Philosophie* 47, pp. 217-50.
——— (1971) *One and Many in Presocratic Philosophy.* Cambridge. MA: Harvard University Press.
Strang, C. (1963) 'The Physical Theory of Anaxagoras' , *Archiv für Geschichte der Philosophie* 45, pp. 101-18.
Stump, E. & Kretzman, N. (2001) *The Cambridge Companion to Augustine.* Cambridge: Cambridge University Press.
Tannery, P. (1887) 'La Cosmogonie d'Empedocle', *Revue Philosophique de la France et de l'Etranger* 24, pp. 285-300.
Taran, L. (1971) 'The Creation Myth in Plato's *Timaeus*', in J.P Anton & G.L. Kustas (eds), *Essays in Ancient Greek Philosophy.* Albany, NY: SUNY Press, pp. 372-406.
Tatian, *Oratio ad Graecos*, ed. M. Whittaker (1982). Oxford: Oxford University Press.
Taylor, A.E. (1926) *Plato: The Man and his Work.* Cleveland: Meridian Books.
——— (1928) *A Commentary on Plato's Timaeus.* Oxford: Oxford University Press.
Taylor, C.C.W. (1997) *From the Beginning to Plato.* Routledge History of Philosophy, vol. 1. London: Routledge.
——— (1999) *The Atomists: Leucippus and Democritus.* Toronto: Toronto University Press.
Timothy, T.B. (1973) *The Early Christian Apologists and Greek Philosophy.* Assen: Van Gorcum.
Toulmin, S. & Goodfield, J. (1961) *The Fabric of the Heavens.* London: Hutchinson.

Tracy, T. (1979) 'The Soul as the Boatman of the Body: Presocratics to Descartes', *Diotima* 7, pp. 195-9.
Trepanier, S. (2000) 'The Structure of Empedocles Fragment 17', *Essays in Philosophy* 1, pp. 1-16.
——— (2003a) 'Empedocles on the Ultimate Symmetry of the World', *Oxford Studies in Ancient Philosophy* 29, pp. 1-57.
——— (2003b) ' "We" and Empedocles' Cosmic Lottery', *Mnemosyne* 56, pp. 385-419.
——— (2004) *Empedocles: An Interpretation*. New York: Routledge.
Turbayne, C.M. (1976) 'Plato's Fantastic Appendix: The Procreation Model of the *Timaeus*', *Paideia* 50, pp. 125-40.
Vallejo, A. (1997) 'No, It's Not a Fiction', in Calvo & Brisson 1997.
Van der Ben, N. (1975) *The Proem of Empedocles' Peri Phuseos*. Amsterdam: Gruener.
——— (1978) 'Empedocles' Fragments 8, 9, 10 DK', *Phronesis* 23, pp. 197-215.
Van Seters, J. (1988) 'The Primeval Histories of Greece and Israel Compared', *Zeitschrift für die Alttestamentliche Wissenschaft* 100, pp. 1-22.
Van Steenbeghen, F. (1978) 'Le mythe d'un monde éternal', *Revue Philosophique de Louvain* 76, pp. 157-79.
Vernant, J-P. (1965) *Myth and Thought among the Greeks*. London: Routledge & Kegan Paul.
——— (1970) 'Thetis et la poème cosmogonique d'Alcman', *Collection Latomus* 114.
Verrycken, K. (1990) 'The Metaphysics of Ammonius Son of Hermeias', in Sorabji 1990, pp. 215-26.
——— (1997) 'Philoponus' Interpretation of Plato's Cosmogony', *Documenti e Studi Sulla Tradizione Filosofica Medievale* 8, pp. 269-318.
Vidal-Naquet, P. (1978) 'Plato's Myth of the Statesman, the Ambiguities of the Golden Age and History', *Journal of Hellenic Studies* 98, pp. 132-41.
Vlastos, G.L. (1939) 'The Disorderly Motion in the *Timaeus*', in R.E. Allen (ed.), *Studies in Plato's Metaphysics*. London: Routledge & Kegan Paul, pp. 379-420.
——— (1946) 'Parmenides' Theory of Knowledge', *Transactions and Proceedings of the American Philological Association* 77, pp. 66-77.
——— (1947) 'Equality and Justice in Early Greek Cosmologies', reprinted in Furley & Allen 1975, pp. 56-91.
——— (1950) 'The Physical Theory of Anaxagoras', *Philosophical Review* 59, pp. 31-57.
——— (1955) 'On Heraclitus', reprinted in Furley & Allen 1970, pp. 413-29.
——— (1959) 'One World or Many in Anaxagoras?', reprinted in Furley & Allen 1975, pp. 354-60.
——— (1965) 'Creation in the *Timaeus*: Is it a Fiction?', in R.E. Allen (ed.), *Studies in Plato's Metaphysics*. London: Routledge & Kegan Paul, pp. 401-19.
——— (1975) *Plato's Universe*. Oxford: Clarendon.
——— (1981) 'The Role of Observation in Plato's Conception of Astronomy', in P. Anton (ed.), *Science and the Sciences in Plato*. New York: Eidos, pp. 1-31.
Von Arnim, H. (1902) 'Der Weltperiod bei Empedocles', *Festschrift Theodor Gomperz*. Vienna, pp. 16-27.
Von Fritz, (1971) *Grundproblem der Geschichte der antiken Wissenschaft*. Berlin: de Gruyter.
Von Leyden, W. (1964) 'Time, Number and Eternity in Plato and Aristotle', *Philosophical Quarterly* 14, pp. 35-52.
Von Staden, H. (1992) 'Affinities and Elisions', *Isis* 83, pp. 578-95.

Wallis, R.T. (1981) *Neoplatonism*. London: Duckworth.
Wasserstein, A. (1962) 'Greek Scientific Thought', *Proceedings of the Cambridge Philological Society* NS 8, pp. 51-63.
Waterlow, S. (1982) *Passage and Possibility*. Oxford: Oxford University Press.
West, M.L. (1963) 'Three Presocratic Cosmologies', *Classical Quarterly* NS 13, pp. 154-76.
——— (1967) 'Alcman and Pythagoras', *Classical Quarterly* NS 17, pp. 1-15.
——— (1971) *Early Greek Philosophy and the Orient*. Oxford: Oxford University Press.
Weinberg, S. (1993) *The First Three Minutes*, 2nd edn. London: Fontana.
White, M.J. (2003) 'Stoic Natural Philosophy', in Inwood 2003.
Wiggins, D. (1982) 'Heraclitus' Conceptions of Flux, Fire and Material Persistence', in M. Schofield & M.C. Nussbaum (eds), *Language and Logos*. Cambridge: Cambridge University Press, pp. 1-32.
Wilberding, J. (2006) *Plotinus' Cosmology*. Oxford: Oxford University Press.
Wilcox, J. (2001) 'Whole-Natured Forms in Empedocles' Cosmic Cycle', in Preus 2001, pp. 109-22.
Wildberg, C. (2001) Commentary on Curd 2001, *Proceedings of the Boston Area Colloquium in Ancient Philosophy* 17, pp. 50-6.
Wilford, F. (1968) 'Embryological analogies in Empedocles' Cosmogony', *Phronesis* 13, pp. 108-18.
Williams, C.J.F. (1965) 'Aristotle and Corruptibility', *Religious Studies I*, pp. 95-107.
Wilson, J.A. (1969) 'Egyptian Myths and Tales', in Pritchard, J.B. *Ancient Near East Texts Relating to the Old Testament*. Princeton, NJ: Princeton University Press, 1-27.
Winston, D. (1971-2) 'The Cosmogony of the Book of Wisdom', *History of Religion* 11, pp. 185-202.
——— (1977) *The Wisdom of Solomon*. Garden City, NY: Doubleday.
Wolfson, H.A. (1947) *Philo*. (2 vols). Cambridge, MA: Harvard University Press.
——— (1948) 'The Meaning of *ex nihilo* in the Church Fathers, Arabic and Hebrew Philosophy and St. Thomas', in *Medieval Studies in Honour of J.D.M. Ford*, Cambridge, MA: Harvard University Press, pp. 355-70.
——— (1970) 'The Identification of *ex nihilo* with Emanation in Gregory of Nyssa', *Harvard Theological Review* 63, pp. 53-60.
Wolters, A.M. (1994) '*Creatio ex nihilo* in Philo', in Helleman 1994.
Wright, L. (1976) *Teleological Explanations*. Berkeley: University of California Press.
Wright, M.R. (1981) *Empedocles: the Extant Fragments*. New Haven, CT: Yale University Press.
——— (1995) *Cosmology in Antiquity*. London: Routledge.
——— (2000) *Reason and Necessity: Essays on Plato's Timaeus*. London: Duckworth/ Classical Press of Wales.
Xenophon: *The Fragments of Xenophon,* ed. J.-H. Lesher (1992). Toronto: Toronto University Press.
Zeller, E. (1920) *Die Philosophie der Griechen*, Leipzig: Reisland.
——— (1931) *Outlines of the History of Greek Philosophy*. London: Routledge & Kegan Paul.
Zeyl, D. (2000) *Timaeus*. Indianapolis: Hackett.
Zhmud, L. (1994) 'Die Beziehungen zwischen Philosophie und Wissenschaft in der Antike', *Sudhoffs Archiv* 78, pp. 1-13.

——— (1997) *Wissenschaft, Philosophie und Religion im frühen Pythagoreismus*. Berlin: Akademie Verlag.
——— (1999) 'Plato as "Architect of Science" ', *Phronesis* 43, pp. 211-44.

Index Locorum

Index of Names

General Index